J. Hromkovič   R. Klasing   A. Pelc
P. Ružička[†]   W. Unger

Dissemination of Information in Communication Networks

Broadcasting, Gossiping, Leader Election, and Fault-Tolerance

With 73 Figures

 Springer

Authors

Prof. Dr. Juraj Hromkovič
ETH Zentrum, RZ F2
Department of Computer Science
Swiss Federal Institute of Technology
8092 Zürich, Switzerland
juraj.hromkovic@inf.ethz.ch

Prof. Dr. Peter Ružička[†]

Dr. Ralf Klasing
CNRS - I3S & INRIA Sophia-Antipolis
2004 Route des Lucioles, BP 93
06902 Sophia Antipolis Cedex, France
Ralf.Klasing@sophia.inria.fr

Dr. Andrzej Pelc
Départment d'informatique
Université du Québec en Outaouais
C.P. 1250, Succursale Hull
Gatineau (Québec) J8X 3X7, Canada
andrzej.pelc@ugo.ca

Dr. Walter Unger
Lehrstuhl für Informatik I
Aachen University RWTH
Ahornstrasse 55
52074 Aachen, Germany
quax@i1.Informatik.RWTH-Aachen.de

Series Editors

Prof. Dr. Wilfried Brauer
Institut für Informatik der TUM
Boltzmannstrasse 3
85748 Garching, Germany
Brauer@informatik.tu-muenchen.de

Prof. Dr. Arto Salomaa
Turku Centre for Computer Science
Lemminkäisenkatu 14 A
20520 Turku, Finland
asalomaa@utu.fi

Prof. Dr. Grzegorz Rozenberg
Leiden Institute of Advanced Computer Science
University of Leiden
Niels Bohrweg 1
2333 CA Leiden, The Netherlands
rozenber@liacs.nl

Illustrations
Ingrid Zamečnikova

ACM Computing Classification (1998): B.4.3, C.2.0-1, F.2.2, G.2.1-2, G.3

ISBN 978-3-642-05648-2   e-ISBN 978-3-540-26663-1

Springer is a part of Springer Science+Business Media
springeronline.com

© Springer-Verlag Berlin Heidelberg 2010
Printed in Germany

Cover design: KünkelLopka, Heidelberg

To our dear friend
and coauthor Peter Ružička,
in memoriam.

It is necessary to live in such a way
that death might seem
to be the greatest injustice done to us.
What is really important
is to give life a real sense:
to identify oneself with the yearnings of people,
to be responsible for the way things are
and to be present in all the struggles
which shape our existence.

Emmanuel Robles

Preface

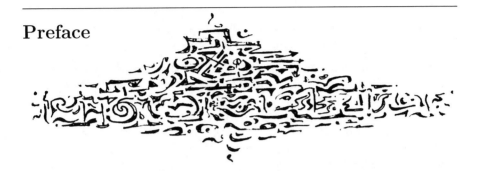

Due to the development of hardware technologies (such as VLSI) in the early 1980s, the interest in parallel and distributive computing has been rapidly growing and in the late 1980s the study of parallel algorithms and architectures became one of the main topics in computer science. To bring the topic to educators and students, several books on parallel computing were written. The involved textbook "Introduction to Parallel Algorithms and Architectures" by F. Thomson Leighton in 1992 was one of the milestones in the development of parallel architectures and parallel algorithms. But in the last decade or so the main interest in parallel and distributive computing moved from the design of parallel algorithms and expensive parallel computers to the new distributive reality – the world of interconnected computers that cooperate (often asynchronously) in order to solve different tasks. **Communication** became one of the most frequently used terms of computer science because of the following reasons:

(i) Considering the high performance of current computers, the communication is often more time consuming than the computing time of processors. As a result, the capacity of communication channels is the bottleneck in the execution of many distributive algorithms.

(ii) Many tasks in the Internet are pure communication tasks. We do not want to compute anything, we only want to execute some information exchange or to extract some information as soon as possible and as cheaply as possible. Also, we do not have a central database involving all basic knowledge. Instead, we have a distributed memory where the basic knowledge is distributed among the local memories of a large number of different computers.

The growing importance of solving pure communication tasks in the interconnected world is the main motivation for writing this book. The main goals of this material are:

(i) to provide a monograph that surveys the main methods, results and research problems related to the design and analysis of communication algorithms (strategies) under different technological constraints; and

(ii) to provide an introductory textbook in the field of information dissemination in interconnection networks with a special emphasis on broadcast, information collection, gossip, leader election, and related tasks.

Our work is divided into two parts. This first textbook is devoted to the classical, direct communication between connected pairs of nodes of a communication network and to the related communication tasks such as broadcasting, gossiping, and leader election. The forthcoming part focuses on the fast communication via fixed paths between senders and receivers, which is based on new technologies such as optical networks, ATM networks, and wireless networks (for instance, mobile phones and radio networks).

This book aims to be a textbook accessible for students as well as a monograph that surveys the research on communication, presents the border between the known and the unknown, and can so be of interest to researchers and professionals, too.

We would like to thank Manuel Wahle for the successful embedding of our manuscripts into a common style, and for his technical help concerning the typical problems that occur when several people try to write something together. We are indebted to Ingrid Zámečniková for her original illustrations. The (as always) excellent cooperation with Alfred Hofmann, Ingeborg Mayer, and Ronan Nugent from Springer is gratefully acknowledged.

This textbook is devoted to our dear friend and coauthor Peter Ružička, who died during the work on this project. Slovakian computer science lost in him one of its greatest personalities, one of those who wrote the computer science history in Czechoslovakia. Peter was an excellent researcher, and a beloved teacher, who was able to inspire his students for the study and the investigation of the topic of his interest in a fascinating way. But, first of all, he was a man, and to express what we mean and feel by this, it is impossible to find anything better than the following words of Antoine de Saint-Exupéry:

Only he who carries inside a greater personality
than himself deserves to be called a man.

All quotations forthcoming in this textbook remember Peter and present his views and his way of living as we were able to understand them.

Juraj Hromkovič, Ralf Klasing,
Andrzej Pelc, Walter Unger

December 2004

Contents

1

Introduction

By living always with enthusiasm,
being interested in everything
that seems inaccessible,
one becomes greater
by striving constantly upwards.
The sense of life is creativity,
and creativity itself is unlimited.

Maxim Gorky

1.1 Motivation and Aims

If one were required to characterize the recent development of the known human civilization using one word, then one would probably choose the term "communication". The interconnected world became reality. It does not matter whether one considers the Internet, telephone networks, mobile phones, e-mail, TV and radio broadcasting, parallel computer architecture or computer networks, or even something else, the dynamics of our current lives are essentially influenced by the communication facilities provided. Thus, it is not surprising that topics related to communication belong among the research areas of main interest in computer science as well as in electrical engineering. While engineering mainly focuses on the development of hardware technologies, computer science develops algorithms and communication strategies that efficiently use the structure (topology) of communication networks.

This work is devoted to the design of communication strategies (algorithms) in different frameworks and under distinct constraints determined by the communication technologies used. Presenting this topic we follow the following two aims making this book interesting for students and beginners as well as for researchers and practitioners:

1. to provide an introductory textbook on information dissemination in communication networks; and
2. to provide a monograph that surveys the main methods, results and research problems related to the design of communication algorithms (strategies, protocols) under different technological constraints.

To fulfill both these aims we start very slowly, assuming only elementary knowledge on algorithmics, and so have made this material self-contained.

Going to more complex matters we partition the track into a large number of steps that are as small as possible and provide enough exercises supporting learning by doing. We try to avoid any interruptions in thought. We spent a lot of time with the development of informal ideas, concepts and the creation of new terms that extend our formal language and become transparent instruments for solving problems posed. Presenting complex, technical arguments and proofs we will first explain the ideas in a simple and transparent way before providing technical details. In several cases we present hard proofs in two stages. When a transparent argument of a weaker result can bring across the idea succinctly then we first give it, and later provide the interested reader with a strong but technically demanding and probably confusing argument of the best known results. In this way we hope to be able to communicate also complex matters in a well readable way.

Our whole presentation does not only focus on a survey of known results, but on a transparent presentation of methods and techniques that can be useful for discovering new communication strategies or improving the best known results. This feature can make this book of special interest for Ph.D. students and researchers.

1.2 Organization of the Book

Our work consists of two books. This first one is devoted to frameworks when two parties communicate by a direct link between them, while the second one focuses on frameworks when one reserves a path (via several nodes of the network) between the two parties for the communication. Thus, the first book works mainly with the telephone and telegraph communication modes in either synchronous or asynchronous manner. The second book works with technologies such as switching networks, optical networks, ATM networks, and wireless networks such as radio networks and mobile networks.

This first textbook is organized as follows. It is divided into two parts. Part I consists of 5 chapters (Chapters 2 to 6) and Part II consists of 3 chapters (Chapters 7 to 9).

Part I deals with synchronous communication performed by the classical telegraph and telephone communication modes. These modes provide a simple communication framework, which is very suitable as a starting point for introducing communication tasks (problems) and techniques for designing and analyzing communication algorithms.

The aim of Chapter 2 is to provide all necessary preliminaries, to fix the notation, and to introduce the fundamental communication tasks. Here, we first present fundamentals of graph theory and combinatorics in Sections 2.2 and 2.3. Section 2.4 introduces broadcasting, accumulation and gossiping as fundamental communication tasks and explains what is considered to be a solution to these tasks. Here we consider the complexity of communication algorithms as the number of synchronous rounds. Section 2.5 presents some elementary general relations between the complexity of broadcasting, accumulation and gossiping in a way so that one gets the first touch with the matters studied in the chapters that follow.

Section 2.6 surveys the fundamental interconnection networks that were intensively investigated in the past, especially in relation to the design of parallel architectures with good communication facilities.

Chapter 3 is devoted to different aspects of broadcasting. After presenting some fundamental observations and techniques in Section 3.2, we introduce some design methods for broadcast algorithms in Section 3.3 and apply them for some common network structures. Section 3.4 is devoted to methods for proving lower bounds on the number of rounds necessary for broadcasting in concrete degree-bounded networks. An overview on results in broadcasting in fundamental networks is given in Section 3.5. Section 3.6 considers a completely different scenario. The task is to design an optimization algorithm that, for any given network, derives a good (possibly) optimal broadcast strategy. This problem is an NP-hard optimization task and we survey the current knowledge about this topic there.

Chapter 4 is devoted to gossiping. After presenting some basic facts about the complexity of gossiping and its relation to broadcasting in Section 4.2, the next sections deal with the design of gossip algorithms in concrete networks. In Section 4.3 one considers communication structures with a small bisection width. The main result here is optimal gossiping in cycle in both telegraph and telephone communication modes. The optimal gossip algorithms in cycle are used in Section 4.4 for designing efficient gossip algorithms in the hypercube-like networks. An optimal gossip algorithm in complete graphs is designed in Section 4.5. An important point is that in Sections 4.3 and 4.5 we develop nontrivial techniques for proving lower bounds on the complexity of gossiping and in this way the optimality of the designed algorithms. Finally, Section 4.6 provides an overview on gossiping in concrete networks.

So-called systolic communication as a very regular, distributed form of communication is introduced in Chapter 5. The idea is to allow the processors (nodes) of the network to repeat a short sequence of communication activities only and the main research question is how much communication complexity has to be paid for this regularity of algorithms. Chapter 5 is organized as follows. First the concept of systolic communication is introduced and some relations to the general communication modes are established. Then systolic gossip algorithms in different networks are designed. Surprisingly, for several networks the systolic algorithms can be as efficient as the general ones.

Chapter 6 is devoted to fault-tolerance, which is one of the central topics of current interest. The task is to design robust communication algorithms in the sense that they solve a given communication task even when some connections or nodes do not work correctly (or at all). Section 6.1 introduces the different kinds of faults that may appear in a communication network. Some basic facts about fault-tolerance are presented in Section 6.2. Then, Section 6.3 focuses on the results related to the bounded-fault model, and Section 6.4 is devoted to the probabilistic fault model.

In contrast to the simple, nice world of synchronous communication rounds in Part I, Part II is devoted to asynchronous networks that are a better model of the current reality. In this general network model there are inherently three sources of nondeterminism. Processes (performing within network nodes) are inherently asynchronous since there is no universal clock available for synchronizing their actions. The uncertainties in communication delays imply nondeterministic behavior in the sending and receiving of messages. And if FIFO requirements on links are not necessary, messages via the same link can overtake each other. These three forms of nondeterminism make the network model not only substantially complex, but the basic terms such as communication tasks and complexity change in meaning.

Part II is divided into three chapters. Chapter 7 is devoted to broadcasting in the asynchronous world. In Section 7.1 the general broadcast strategies in arbitrary networks are considered. Section 7.2 shows lower and upper bounds on broadcasting on tori, and Section 7.3 deals with broadcast tasks in hypercubes.

Chapter 8 is devoted to leader election in asynchronous distributed networks. The first two sections fix the communication model and the leader election problem as a communication task. Sections 8.1 and 8.2 present the algorithms for leader election in rings. The leader election algorithm on complete graphs and a lower bound on the complexity of this problem are given in Section 8.3. Leader election on hypercubes is studied in Section 8.4. Chapter 8 finishes with leader election in synchronous rings in Section 8.5.

Chapter 9 is, analogous to Chapter 6, devoted to fault-tolerance, but here in the general framework of distributed networks. Section 9.1 deals with the consensus problem with unsigned and signed messages. Broadcasting in synchronous networks with dynamic faults on hypercubes, tori and star graphs is studied in Section 9.2.

The exercises in this book are distributed in the text and not presented in separate sections. In this way the reader knows when it is the optimal time for dealing with them. The exercises marked by * are considered to be essentially nontrivial.

Part I

The Telegraph and Telephone Modes

2

Fundamentals

He who is constantly developing his own self contributes to the benefit of mankind.

Anton Srholec

2.1 Introduction

The goal of this chapter is to provide an elementary introduction to the theory of communication algorithms. To do this we first need to define some fundamental notions from graph theory and combinatorics, and to agree on the notation. Then, we explain what a communication algorithm is and specify the categories of communication tasks (problems) that will be covered in this book. Some elementary results and observations regarding the complexity of communication algorithms are presented to build a basis for the design of effective communication algorithms. Finally, some fundamental interconnection networks are introduced.

This chapter is organized as follows. Section 2.2 provides the basic definitions about graphs and their fundamental characteristics like radius, diameter, bisection width, separator, etc. Section 2.3 presents some elementary knowledge of combinatorics. Section 2.4 introduces the basic notions such as the dissemination problem, communication algorithm and communication modes. Moreover, two basic communication modes – telegraph (one-way) communication mode and telephone (two-way) communication mode – are presented and the notation of their complexity measures is stated. Section 2.5 presents some fundamental properties of telephone and telegraph communication modes focusing on the relation between the complexity of communication algorithms and the graph characteristics of the underlying communication networks. Some fundamental interconnection networks are described in Section 2.6.

2.2 Graphs

The central topic of this book is to study the communication in interconnection networks. As we have already noted an interconnection network may be viewed as a connected undirected graph whose vertices execute some communication actions via their incident edges. Here we define graphs and some of their characteristics and properties useful for the study of communication algorithms. We start with the definition of a graph. In what follows, $|A|$ denotes the cardinality of A for any set A, and $\mathbb{N} = \{0, 1, 2, \ldots\}$ denotes the set of natural numbers. For all sets A, B, $A \subset B$ denotes A is a proper subset of B $(A \subseteq B, B - A \neq \emptyset)$.

Definition 2.2.1. *An* **undirected graph**, *in short* **graph**, *is a pair* $G = (V, E)$, *where*

(i) V is a finite **set of vertices (nodes)**, *and*
(ii) $E \subseteq \{\{u, v\} \mid u, v \in V, u \neq v\}$ is a **set of edges**.

For every graph G, $\boldsymbol{V(G)}$ *denotes the set of vertices of G and* $\boldsymbol{E(G)}$ *denotes the set of edges of G.*

Note that we use graphs to model interconnection networks. The vertices of a graph correspond to computers (or processors of a parallel computer) and the edges represent the communication links between pairs of computers.

Later we shall also use (u, v) instead of $\{u, v\}$ to denote an edge of an undirected graph. Obviously it does not matter whether we write (u, v) or (v, u). We will use the terms vertex and node interchangeably here, with the preference for the term "vertex" when referring to graphs and the term "node" when dealing with communication networks.

Definition 2.2.2. *Let* $G = (V, E)$ *be a graph. Two vertices* $v, w \in V$ *are* **adjacent** *if* $\{v, w\} \in E$. *We say that the edge* $\{v, w\}$ *is* **incident** *upon the vertices v and w. A* **path** *is a sequence of vertices* $v_1, v_2, \ldots, v_n$ *from V such that* $\{v_i, v_{i+1}\} \in E$ *for* $1 \leq i < n$. *A path* $v_1, v_2, \ldots, v_n$ *for* $n \geq 2$ *is* **simple** *if all vertices on the path are distinct* $(|\{v_1, v_2, \ldots, v_n\}| = n)$, *with the exception that v_1 and v_n may be identical* $(v_1 = v_n$ *and* $|\{v_1, v_2, \ldots, v_n\}| = n - 1)$. *The* **length of the path** $v_1, v_2, \ldots, v_n$ *is* $n - 1$ *(the number of edges along the path). A* **(simple) cycle** *in G is a (simple) path* $v_1, \ldots, v_n$ *of length three or more with* $v_1 = v_n$.

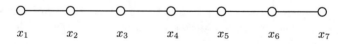

$$x_1 \qquad x_2 \qquad x_3 \qquad x_4 \qquad x_5 \qquad x_6 \qquad x_7$$

Fig. 2.1.

The graph

$$G = (\{x_1, x_2, x_3, x_4, x_5, x_6, x_7\}, \{\{x_1, x_2\}, \{x_2, x_3\}, \ldots, \{x_6, x_7\}\})$$

of seven vertices (see Figure 2.1) and the graph

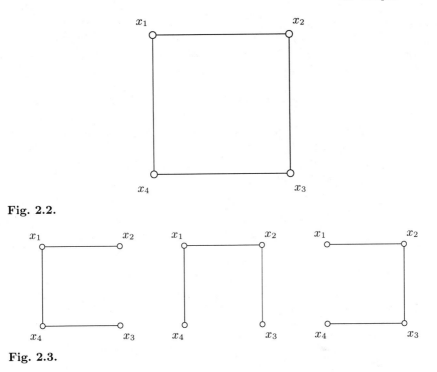

Fig. 2.2.

Fig. 2.3.

$$G' = (\{x_1, x_2, x_3, x_4\}, \{\{x_1, x_2\}, \{x_2, x_3\}, \{x_3, x_4\}, \{x_4, x_1\}\})$$

of four vertices (see Figure 2.2) are examples of graphs. G is called the path of seven vertices, and G' is called the cycle of four vertices. As we can see in Figures 2.1 and 2.2 one can consider a graphical representation for any graph. Graphs are drawn as follows. The vertices of the graph are depicted as points in the plane, and an edge $\{u, v\}$ can be depicted as a line connecting the points u and v. The sequence of vertices $x_2, x_3, x_4, x_5, x_4, x_3, x_4, x_5, x_6$ is a path in G, but it is not a simple path. The sequence x_1, x_2, x_3, x_2, x_1 is a cycle in G', but it is not a simple cycle.

Exercise 2.2.3. List all simple cycles of the graph in Figure 2.4.

Exercise 2.2.4. For any positive integer n, estimate exactly the number of graphs of n vertices $x_1, x_2, \ldots, x_n$. Note that the graphs depicted in Figure 2.3 are considered to be different because of the labelling of the vertices.

Exercise 2.2.5. For all positive integers n, and k, $k \leq n \cdot (n-1)/2$, estimate exactly the number of all graphs of n labelled vertices and k edges.

Definition 2.2.6. *Let $G = (V, E)$ be a graph. Two vertices u and v in V are* **connected** *if and only if $u = v$ or there exists a path between u and v in G. We say that G **is connected** if, for all $u, v \in V$, u and v are connected.*

G is cyclic *if it contains at least one simple cycle. If G does not contain any simple cycle, then* **G is acyclic.**
We say that $G = (V, E)$ *is* **bipartite** *if there exist* $V_1, V_2 \subseteq V$ *such that* $V_1 \cap V_2 = \emptyset$, *and* $E \subseteq \{\{u, v\} \mid u \in V_1 \text{ and } v \in V_2\}$.

The graphs in Figures 2.1, 2.2 and 2.3 are connected graphs. The graphs in Figures 2.1 and 2.3 are acyclic, the graph in Figure 2.2 is cyclic because x_1, x_2, x_3, x_4, x_1 is a simple cycle of this graph. All the graphs in Figures 2.1, 2.2 and 2.3 are bipartite. To see it consider $V_1 = \{x_1, x_3, x_5, x_7\}$ and $V_2 = \{x_2, x_4, x_6\}$ for the path in Figure 2.1 and $V_1 = \{x_1, x_3\}$ and $V_2 = \{x_2, x_4\}$ for all graphs in Figures 2.2 and 2.3.

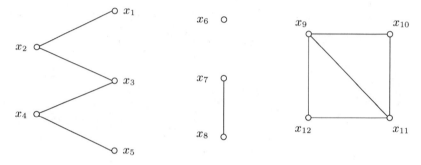

Fig. 2.4.

The graph in Figure 2.4 is not connected. For instance, there are no paths between x_1 and x_7. There are also no paths between x_6 and any other vertex. As such x_6 is referred to as an **isolated vertex**. The graph in Figure 2.4 is not bipartite because if x_9 were to be in V_1 (V_2), then both x_{11} and x_{12} would have to be in V_2 (V_1). However, this is impossible because the edge $\{x_{11}, x_{12}\}$ connects x_{11} and x_{12}.

Note that the terms connected, acyclic, bipartite and several other graph properties introduced later have nothing to do with the labelling of the vertices. Therefore we will often omit labelling of the vertices in the following figures.

Definition 2.2.7. *Let n be a positive integer. The graph*

$$K_n = (\{x_1, \ldots, x_n\}, \{\{x_i, x_j\} \mid i, j \in \{1, \ldots, n\}, i \neq j\})$$

is called the **complete graph** *of n vertices. Let k and m be positive integers. The graph*

$$K_{k,m} = (\{x_1, \ldots, x_k, v_1, v_2, \ldots, v_m\}, \{\{x_i, v_j\} \mid i \in \{1, \ldots, k\}, j \in \{1, \ldots, m\}\})$$

is called the **(k, m)-complete bipartite graph.**

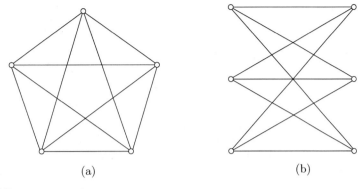

Fig. 2.5.

The graph in Figure 2.5(a) is the complete graph of five vertices and the graph in Figure 2.5(b) is $K_{3,3}$.

Exercise 2.2.8. Show that if a graph G contains a path between two vertices u and v, then G contains a simple path between u and v.

Exercise 2.2.9. Prove that every connected graph $G = (V, E)$ satisfies $|E| \geq |V| - 1$.

Exercise 2.2.10. Prove that every acyclic graph is bipartite.

Exercise 2.2.11. Let $G = (V, E)$ be a graph such that $|E| > |V| - 1$. Prove that G is cyclic.

Definition 2.2.12. *An acyclic graph is called a* **forest**. *A connected acyclic graph is called a* **tree**.

The graphs in Figures 2.1 and 2.3 are trees. The graph in Figure 2.4 is not a forest, but by removing the edges $\{x_{10}, x_{11}\}$ and $\{x_{11}, x_{12}\}$ one obtains a forest.

The class of trees is one of the most important graph classes, especially in the study of communication algorithms in networks. Hence we will next focus on several equivalent definitions of trees.

Lemma 2.2.13. *Let* $G = (V, E)$ *be a graph. The following statements are equivalent:*

 (i) *G is a tree.*
 (ii) *Any two vertices in G are connected by a unique simple path.*
 (iii) *G is connected, but if any edge is removed from E, the resulting graph is disconnected.*
 (iv) *G is connected, and $|E| = |V| - 1$.*
 (v) *G is acyclic, and $|E| = |V| - 1$.*

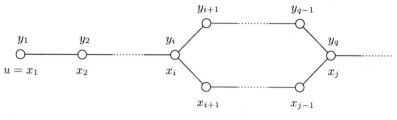

Fig. 2.6.

(vi) G is acyclic, but any graph $G' = (V, E')$ with $E \subset E'$ is cyclic.

Proof. To prove the equivalence of these six statements, we prove the following six implications $(i) \Rightarrow (ii)$, $(ii) \Rightarrow (iii)$, $(iii) \Rightarrow (iv)$, $(iv) \Rightarrow (v)$, $(v) \Rightarrow (vi)$, $(vi) \Rightarrow (i)$ separately.

(1) **$(i) \Rightarrow (ii)$** If G is a tree, then G is acyclic and connected (Definition 2.2.12). Therefore, any two vertices of $V(G)$ are connected. We have to show that every pair of vertices in G is connected by a unique simple path. Let us assume otherwise. Assume that two distinct vertices u and v exist that are connected by two distinct simple paths $P_1 = x_1, x_2, \ldots, x_k$ ($u = x_1, v = x_k$) and $P_2 = y_1, y_2, \ldots, y_m$ ($y_1 = u, y_m = v$). Let $i = \min\{l \in \{1, 2, \ldots, \min\{k, m\} - 2\} \mid x_{l+1} \neq y_{l+1}\}$, $j = \min\{p \in \{i+2, i+3, \ldots, k\} \mid x_p = y_r$ for some $r \in \{i+2, \ldots, m\}\}$, and $q = \min\{l \in \{i+2, i+3, \ldots, m\} \mid y_l = x_r$ for some $r \in \{i+2, i+3, \ldots, k\}\}$. Obviously, such i, j, and q exist, $x_{i+1} \neq y_{i+1}$ are the first vertices of P_1 and P_2 respectively in which these two paths diverge, and $x_j = y_q$ is the first vertex in which P_1 and P_2 again reconverge. Now, one can easily observe (Figure 2.6) that

$$x_i(= y_i), x_{i+1}, x_{i+2}, \ldots, x_j(= y_q), y_{q-1}, y_{q-2}, \ldots, y_i(= x_i)$$

is a simple cycle in G. Clearly, this contradicts the assumption that G is a tree.

(2) **$(ii) \Rightarrow (iii)$** That any two vertices of $V(G)$ are connected by a unique simple path implies that the graph G is connected. We have still to show that deleting any edge from $E(G)$ disconnects G. Let $\{x, y\}$ be an arbitrary edge from $E(G)$. This edge $\{x, y\}$ is the unique path from x to y. By removing $\{x, y\}$, there is no path between x and y in $G' = (V(G), E(G) - \{\{x, y\}\})$, and hence G' is not connected.

(3) **$(iii) \Rightarrow (iv)$** The assumption that G is connected is common for (iii) and (iv). It remains to show that $|E| = |V| - 1$. Exercise 2.2.9 claims $|E(H)| \geq |V(H)| - 1$ for every connected graph H. Thus, it suffices to show that $|E| \leq |V| - 1$. We prove this by contradiction. Let $|E| > |V| - 1$. Following the claim of Exercise 2.2.11, G is cyclic. Let $x_1, x_2, \ldots, x_k, x_1$, $k \geq 3$, be a cycle of G. Consider the graph $G' = (V, E - \{\{x_1, x_2\}\})$. We shall show that G' is connected, which contradicts (iii). Let u and v be

two arbitrary distinct nodes of V. Since G is connected, there is a simple path $P = u, y_1, y_2, \ldots, y_r, v$ in G. If P does not contain the edge $\{x_1, x_2\}$, then P is a simple path in G' too. If P contains $\{x_1, x_2\}$ (i.e., $y_j = x_1$ and $y_{j+1} = x_2$ for some j), then

$$P' = u, y_1, \ldots, y_j (= x_1), x_k, x_{k-1}, \ldots, x_2 (= y_{j+1}), y_{j+2}, \ldots, y_r, v$$

is a path connecting u and v in G'.

(4) $(iv) \Rightarrow (v)$ The assumption that $|E| = |V| - 1$ is common for both (iv) and (v). It remains to show that if G is connected, then G is acyclic. We give an indirect proof by showing that if G is cyclic, then G is not connected. Let $x_1, x_2, \ldots, x_k, x_1$ be a cycle in G. Consider the subgraph $G_k = (\{x_1, x_2, \ldots, x_k\}, \{(x_1, x_2), (x_2, x_3), \ldots, (x_{k-1}, x_k), (x_k, x_1)\})$ consisting of this cycle. Obviously, $|V(G_k)| = k = |E(G_k)|$. Now, we look for a vertex $u \in V - V(G_k)$ such that $(u, x_i) \in E$ for some $i \in \{1, \ldots, k\}$. If such a vertex does not exist, then G is not connected. If such a vertex exists, consider $G_{k+1} = (V(G_k) \cup \{u\}, E(G_k) \cup \{(u, x_i)\})$. Obviously, we have again $|V(G_{k+1})| = |E(G_{k+1})|$. One can easily observe that adding a new vertex of V to G_{k+1} in the way described above results in a graph G_{k+2} with $|V(G_{k+2})| = |E(G_{k+2})|$, etc. If, for some $i < |V| - k$, G_{k+i} cannot be extended to G_{k+i+1} by adding a vertex $v \in V - V(G_{k+i})$ adjacent to at least one vertex of $V(G_{k+i})$, then G is not connected. In the opposite case we build a graph $G_{|V|}$ with $V(G_{|V|}) = V$ and $E(G_{|V|}) \subseteq E$ by the above described procedure. Since $|V(G_{|V|})| = |E(G_{|V|})| = |V|$ we have a contradiction to the assumption $|E| = |V| - 1$.

(5) $(v) \Rightarrow (vi)$ The assumption that G is acyclic is common for both (v) and (vi). It remains to show that adding an edge $(u, v) \notin E$ to G for two distinct vertices $u, v \in V$ results in a cycle. This follows from Exercise 2.2.11 claiming that every graph H with $|E(H)| > |V(H)| - 1$ is cyclic.

(6) $(vi) \Rightarrow (i)$ According to Definition 2.2.12 the condition that G is acyclic is inherent in both statements (i) and (vi). From (vi) we know that any graph $G' = (V, E')$ with $E \subset E'$ is cyclic. We have to show that G is connected. Let u and v be arbitrary vertices of G. Let us assume that u and v are not adjacent, i.e., $\{u, v\} \notin E$. We add the edge $\{u, v\}$ to build $G' = (V, E \cup \{\{u, v\}\})$. Now G' has to contain a cycle involving $\{u, v\}$. But this is possible only if a path between u and v in G exists.

$\square$

Sometimes it is very useful to give some partial order for the vertices of a tree. The most common approach is to choose one vertex x (called the root later) and to assign a number to every vertex with respect to the distance from x. The formal definition follows.

Definition 2.2.14. *A **rooted tree** $T = (V, E)$ is a tree in which one vertex $v \in V$ is distinguished from the others. The distinguished vertex v is called the **root of T**. If the last edge on the path from the root v of T to a vertex x is $\{y, x\}$, then y **is the parent of** x, and x **is a child of** y. If two vertices* *have the same parent, they are* **siblings**. *A vertex with no child is a* **leaf**[1] *of T. A non-leaf vertex is an* **internal vertex of T**. *The* **depth of a vertex u** *of the rooted tree T,* **depth(u)**, *is the length of the path between the root v and u. For $i = 1, 2, \ldots$, the* **i-th level of T** *is the set* **$level_i(T) = \{x \in V \mid depth(x) = i\}$**. *The* **depth of T** *is the length of the longest path between the root and a leaf of T, i.e.,* **depth(T)** $= \max\{\mathrm{depth}(u) \mid u \in V\}$.

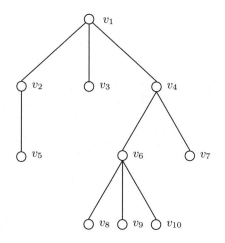

Fig. 2.7.

Figure 2.7 shows a rooted tree T with the root v_1. The vertex v_1 is the parent of the vertices v_2, v_3, v_4, i.e., the vertices v_2, v_3, v_4 are siblings. The vertex v_5 is the only child of v_2 (i.e., the vertex v_2 is the parent of v_5) and hence v_5 does not have any sibling. The vertices $v_3, v_5, v_7, v_8, v_9, v_{10}$ are leaves of T and all other vertices are internal vertices of T. For instance, depth(v_6) = 2 and depth(T) = 3.

Exercise 2.2.15. Draw all trees $T = (V, E)$ with $V = \{v_1, v_2, v_3, v_4\}$. Draw all rooted trees (V, E) with the root v_1.

Exercise 2.2.16. * Show that for every $n \geq 7$, there exists a tree of n non-labelled nodes such that picking each of the n nodes as a root results in a different rooted tree.

Definition 2.2.17. *Let $G = (V, E)$ be a graph. For every vertex $v \in V$, $\Gamma_G(v) = \{u \in V \mid \{u, v\} \in E\}$ is the* **neighborhood of v in G**. *The* **degree of a vertex $v \in V$**, **deg(v)**, *is the number of edges incident to v, i.e.,*

---

[1] If one considers a tree (i.e., an unrooted tree), then a leaf is any vertex with the degree 1.

$$\deg(v) = |\{\{v, u\} \in E \mid u \in V\}| = |\Gamma_G(v)|.$$

*The **degree of G** is*

$$\mathbf{deg}(G) = \max\{\deg(v) \mid v \in V\}.$$

*Any connected graph $G = (V, E)$ is called **regular** if, for all $v, u \in V$, $\deg(v) = \deg(u)$. G is **k-regular** for a $k \in \mathbb{N}$, if $\deg(v) = k$ for every $v \in V(G)$.*

The degree of the tree in Figure 2.7 is 4 because $\deg(v_6) = 4$, and all other vertices have a smaller degree. The complete graphs and (l, l)-complete bipartite graphs are regular graphs. The graph in Figure 2.2 is regular and the graphs in Figures 2.3, 2.4 and 2.5 are not regular.

Observation 2.2.18. *For every graph $G = (V, E)$,*

$$\Sigma_{v \in V} \deg(v) = 2 \cdot |E|.$$

Proof. Since every edge is incident to exactly two vertices, every edge is counted exactly twice in $\sum_{v \in V} \deg(v)$.

$\square$

Observation 2.2.19. *Every tree has at least two leaves.*

Proof. Let $T = (V, E)$ be a tree. First, we prove that T has at least a leaf. Let $u \in V$ be an arbitrary vertex of T. If u is a leaf, we are ready. If not, go to a vertex x that is adjacent to u. If x is not a leaf, then $\deg(x) \geq 2$ and we can continue to go to a neighbor y of x that is different from u. If we reach a vertex z different from all previous vertices and $\deg(z) \geq 2$, then we can use an edge different from the edge used to reach z in order to reach a new vertex w. Since T does not contain any simple cycle, w is different from all previous vertices visited. Since T is finite, the procedure must terminate and the only possibility to terminate is to reach a vertex of degree 1. Thus, T has at least one leaf.

Starting the same procedure from a leaf, the only possibility to terminate is to reach another leaf. Thus, T has at least two leaves.

$\square$

Observe that the paths in Figures 2.1 and 2.3 have exactly two leaves.

Exercise 2.2.20. Prove that every tree T different from a path has at least 3 leaves. What is the relation between $\deg(T)$ and the number of leaves of T?

Lemma 2.2.21. *For every graph G, the number of vertices of G with an odd degree is even.*

Proof. Let $G = (V, E)$, and let $X \subseteq V$ be the set of vertices of G with an odd degree. So, $X - V$ is the set of vertices of G with an even degree. Then, we have

$$\sum_{v \in V} \deg(v) = \sum_{x \in X} \deg(x) + \sum_{y \in V - X} \deg(y). \tag{2.1}$$

Since $2|E| = \sum_{v \in V} \deg(v)$, the left side of the equality (2.1) is an even number. The sum of even numbers is an even number, and hence $\sum_{y \in Y - X} \deg(y)$ is even. This implies that $\sum_{x \in X} \deg(x)$ must be even. Since $\deg(x)$ is odd for every $x \in X$, this is possible only if $|X|$ is even.

$\square$

Definition 2.2.22. *Two graphs* $G = (V, E)$ *and* $G' = (V', E')$ *are* **isomorphic** *if there exists a bijection* $f : V \to V'$ *such that*

$$\{u, v\} \in E \text{ if and only if } \{f(u), f(v)\} \in E'.$$

We say that a graph $G' = (V', E')$ *is a* **subgraph** *of* $G = (V, E)$ *if* $V' \subseteq V$ *and* $E' \subseteq E$. *Given a set* $V_0 \subseteq V$, *the* **subgraph of** G **induced by** V_0 *is the graph* $G_0 = (V_0, E_0)$, *where* $E_0 = \{\{u, v\} \in E \mid u, v \in V_0\}$.
Let $G = (V, E)$ *be a connected graph. A tree* $T = (V, E')$ *is a* **spanning tree of** G *if* $E' \subseteq E$.

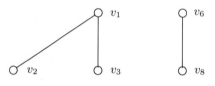

Fig. 2.8.

The three graphs in Figure 2.3 are isomorphic. These graphs are also subgraphs (even spanning trees) of the graph in Figure 2.2, but none of the graphs in Figure 2.3 is an induced subgraph of the graph in Figure 2.2. The graph in Figure 2.8 is the subgraph of the tree in Figure 2.7 that is induced by the set $\{v_1, v_2, v_3, v_6, v_8\}$.

Definition 2.2.23. *Let* $G = (V, E)$ *be a graph, and let* k *be a positive integer. Let there exist* $V_1, V_2, \ldots, V_k$ *such that*

(i) $V = \bigcup_{i=1}^{k} V_i$,
(ii) $V_i \cap V_j = \emptyset$ *for all* $i, j \in \{1, \ldots, k\}, i \neq j$,
(iii) for each $i \in \{1, \ldots, k\}$ *and all vertices* $u, w \in V_i$, *there is a simple path between* u *and* w *in* G, *and*
(iv) for any $x \in V_i$ *and any* $y \in V_j$, $i \neq j$ *implies that there is no path between* x *and* y *in* G.

Then, for each $i \in \{1, \ldots, k\}$, *the subgraph* G_i *induced by* V_i *is called a* **component** *of* G.

Observe that each component C of a graph G is a connected subgraph of G (see Definition 2.2.23(iii)) that is maximal with respect to connectivity (i.e., any subgraph of G that properly contains C is not connected according to Definition 2.2.23(iv)). The graph in Figure 2.4 has 4 components induced by the sets of vertices $V_1 = \{x_1, x_2, x_3, x_4, x_5\}$, $V_2 = \{x_6\}$, $V_3 = \{x_7, x_8\}$ and $V_4 = \{x_9, x_{10}, x_{11}, x_{12}\}$.

Lemma 2.2.24. *Each graph* $G = (V, E)$ *contains at least* $|V| - |E|$ *many components.*

Proof. We prove this assertion by induction according to the number $m = |E|$ of edges of G.

1. Let $m = 0$. Then G consists of $|V|$ isolated vertices which are the $|V|$ components of G.
2. Let the assertion of Lemma 2.2.24 be true for all graphs $G' = (V', E')$ with $|E'| \leq m$. We prove that it holds for graphs with $m + 1$ edges. Let $G = (V, E)$ with $|E| = m+1$ be an arbitrary graph. Consider $G_1 = (V, E_1)$ with $E_1 = E - \{\{u, v\}\}$, where $\{u, v\}$ is an edge from E. Following the induction hypothesis, G_1 has at least $|V| - m$ components. By adding the edge $\{u, v\}$ to G_1, the number of components either remains unchanged (if u and v are in the same component of G_1) or decreases at most by 1 (if u and v are from different components of G_1). Thus, G has at least $|V| - m - 1 = |V| - |E|$ components.

□

Definition 2.2.25. *A* **k-coloring** *of a graph* $G = (V, E)$ *is a function* $g : V \rightarrow \{0, 1, \ldots, k - 1\}$ *such that* $g(u) \neq g(v)$ *for every* $\{u, v\} \in E$. *A graph* G *is called* **k-colorable** *if a k-coloring of* G *exists.*

The graph in Figure 2.7 is 2-colorable. The color 0 can be assigned to v_1, v_5, v_6, v_7 and the color 1 can be assigned to the remaining vertices. The graph in Figure 2.4 is 3-colorable and is not 2-colorable. The complete graph K_n is n-colorable, but not $(n - 1)$-colorable.

Exercise 2.2.26. Prove that every tree is 2-colorable.

Lemma 2.2.27. *The following statements are equivalent.*

(i) G is bipartite.
(ii) G is 2-colorable.
(iii) G has no simple cycles of odd length.

Proof. We prove Lemma 2.2.27 by showing the three implications $(i) \Rightarrow (ii)$, $(ii) \Rightarrow (iii)$, and $(iii) \Rightarrow (i)$.

(1) **(i)** $\Rightarrow$ **(ii)** If $G = (V, E)$ is bipartite, then there exist $V_1, V_2 \subseteq V$, $V_1 \cup V_2 = V$, $V_1 \cap V_2 = \emptyset$, with $E \subseteq \{\{u, v\} \mid u \in V_1, v \in V_2\}$. Then the function $g : V \rightarrow \{0, 1\}$ defined by $g(v) = 0$ for every $v \in V_1$ and $g(u) = 1$ for every $u \in V_2$ is a 2-coloring of G.

(2) $(ii) \Rightarrow (iii)$ Let G be 2-colorable. This means that there exists a function $g : V \rightarrow \{0,1\}$ with $g(u) \neq g(v)$ for every $\{u,v\} \in E$. We will prove by contradiction that G does not contain any cycle of an odd length. Let k be an odd integer, $k \geq 3$, and let $x_1, x_2, \ldots, x_k, x_1$ be a cycle in G. Without loss of generality we assume $g(x_1) = 0$. Then $g(x_2) = 1$, $g(x_3) = 0, \ldots, g(x_{2i-1}) = 0$, $g(x_{2i}) = 1, \ldots, g(x_k) = 0$ because $\{\{x_1, x_2\}, \{x_2, x_3\}, \ldots, \{x_{k-1}, x_k\}\} \subseteq E$. Now, we have $g(x_1) = g(x_k) = 0$ and $\{x_1, x_k\} \in E$ which contradicts the fact that g is a 2-coloring of G.

(3) $(iii) \Rightarrow (i)$ Assume G has no cycle of odd length. We have to show that G is bipartite. Without loss of generality we assume that G is connected. Let $T = (V, E')$, $E' \subseteq E$ be a spanning tree of G. We choose an arbitrary vertex v of V and treat T as a rooted tree with root v. We set

$$V_1 = \{x \in V \mid \text{depth}(x) \text{ is even}\} \text{ and } V_2 = \{y \in V \mid \text{depth}(x) \text{ is odd}\}.$$

Obviously $V_1 \cap V_2 = \emptyset$.

We claim $E \subseteq \{\{u,v\} \mid u \in V_1, v \in V_2\}$. Clearly, this is true for every edge in E'. We observe that every edge $\{x,y\} \notin E'$ with both odd (even) depth(x) and depth(y) would be a part of a cycle of odd length in G^2. Thus, for every $\{x,y\} \in E - E'$, either depth(x) is even and depth(y) is odd (i.e., $x \in V_1$ and $y \in V_2$) or depth(x) is odd and depth(y) is even (i.e., $x \in V_2$ and $y \in V_1$).

□

Exercise 2.2.28. Prove that every graph G is $(\deg(G) + 1)$-colorable.

The notions defined above belong among the basic terms of graph theory. The notions defined in what follows are specific characteristics of graphs that are strongly related to communication facilities of interconnection networks.

Definition 2.2.29. *Let $G = (V, E)$ be a connected graph, and let $u, v \in V$. The **distance between u and v**, dist(u, v), is the length of the shortest path between u and v. The **diameter of G** is*

$$\textbf{diam}(G) = \max\{\text{dist}(u,v) \mid u, v \in V\}.$$

*The **radius**[3] **of a vertex v in G** is*

$$\textbf{rad}(v, G) = \max\{\text{dist}(v, x) \mid x \in V\}.$$

*The **radius of G** is*

$$\textbf{rad}(G) = \min\{\text{rad}(v, G) \mid v \in V\}.$$

---

[2] $\{x, y\}$ together with the path between x and y in T is a cycle of an odd length because the path between x and y in T is of an even length.

[3] Note that in graph theory one also uses the notion eccentricity of a vertex v instead of the radius of v.

Definition 2.2.30. *Let $G = (V, E)$ be a connected graph. A triple (E', V_1, V_2) is a* **cut of G** *if the following conditions hold:*

(i) $E' \subseteq E$,
(ii) $V = V_1 \cup V_2$, $V_1 \cap V_2 = \emptyset$, $V_1 \neq \emptyset$, $V_2 \neq \emptyset$, and
(iii) $E \cap \{\{u, v\} \mid u \in V_1, v \in V_2\} \subseteq E'$.

If $|E'| = 1$ for a cut (E', V_1, V_2), then the only element of E' is called a **bridge of G**. *A cut (E', V_1, V_2) is called* **almost balanced** *if $|V_i| \geq |V|/3$ for $i = 1, 2$. A cut (E', V_1, V_2) is called* **balanced** *if $-1 \leq |V_1| - |V_2| \leq 1$. A balanced cut is called an* **(edge) bisection of G**.
The **(edge) bisection width of a graph G** *is*

$$\mathbf{bisec}(G) = \min\{|H| \mid \text{ there are } U_1, U_2 \text{ such that } (H, U_1, U_2) \text{ is}$$
$$\text{a balanced cut of } G\}.$$

The diameter of a path of n vertices is $n - 1$ and the diameter of the cycle of n vertices is $\lfloor n/2 \rfloor$. The diameter of any complete graph is 1 and the diameter of any complete bipartite graph is 2. The radius of the path (cycle) of n vertices is exactly $\lfloor n/2 \rfloor$, and the radius of any complete bipartite graph is 2. Obviously, for any connected graph G,

$$\mathrm{rad}(G) \leq \mathrm{diam}(G) \leq 2 \cdot \mathrm{rad}(G).$$

Every edge of a tree T is a bridge of T. Cycles and complete graphs do not have any bridge.

The bisection width of any path is 1, and the bisection width of any cycle is 2. The bisection width of the complete graph K_{2n} [K_{2n+1}] is n^2 [$(n+1) \cdot n$].

Exercise 2.2.31. Let T be a tree of n vertices with $\deg(T) \leq 3$. Prove that $\mathrm{bisec}(T) \in O(\log_2 n)$.

Exercise 2.2.32. Estimate, for any even integer n, the bisection width of the complete bipartite graph $K_{n,n}$.

Definition 2.2.33. *Let $G = (V, E)$ be a graph. A* **matching H of G** *is a subset of E with the property: for every two edges $\{v_1, v_2\}$ and $\{u_1, u_2\} \in H$, $\{v_1, v_2\} \cap \{u_1, u_2\} = \emptyset$.*

A matching H is **maximal** *in G if $H \cup \{e\}$ is not a matching of G for any $e \in E - H$. A matching H is* **perfect** *if $|H| = |V|/2$ (i.e., all vertices of G are endpoints of edges in H).*

Note, that every perfect matching of G is a maximal matching of G, too, and that only graphs with an even number of vertices can have a perfect matching. The complete bipartite graph $K_{n,n}$ has a perfect matching for any positive integer n. The complete bipartite graph $K_{l,m}$ with $l \neq m$ does not have any perfect matching because every matching of $K_{l,m}$ has a cardinality of at most $\min\{l, m\}$.

We bring one's attention to the convention to use the notation (u, v) as an indication for an edge, rather than the set notation $\{u, v\}$. In this case (u, v) and (v, u) represent the same edge. In what follows we shall conveniently use both notations with a preference for (u, v).

Interconnection networks are always considered to be undirected, connected graphs, where processors of the network correspond to the vertices of the graph and physical links between processors correspond to the edges of the graph. For this reason we will always mean an undirected connected graph when referring to the notion graph, unless specified otherwise.

2.3 Combinatorics

The goal of this section is to present some basic notions and facts of combinatorics that are useful for counting objects with some given properties or when estimating the number of choice possibilities that one has in some situations.

Let $\log n$ denote $\log_2 n$ and $\ln n$ denote $\log_e n$ in what follows. Let, for any set A, $\mathrm{Pow}(A)$ denote the power set $\{B \mid B \subseteq A\}$ of A.

Definition 2.3.1. *A* **permutation of a finite set** S *is an ordered sequence of all the elements of* S. *The integer* $\boldsymbol{n!}$ *denotes the number of all permutations of a set of* n *elements.*

In what follows we frequently use the string notation $a_1 a_2 a_3 \ldots a_i$ for the sequence $a_1, a_2, a_3, \ldots, a_i$. For the set $S = \{a, b, c\}$, the strings abc, acb, bac, bca, cab, cba are the six permutations of S.

Lemma 2.3.2. *For every* $n \in \mathbb{N}$,

$$0! = 1$$
$$n! = n \cdot (n-1)! = n \cdot (n-1) \cdot (n-2) \ldots 2 \cdot 1.$$

Proof. If S is an empty set, the only sequence of elements of S is the empty sequence. If $|S| = n$, then there are n ways of choosing the first element of the sequence, and $(n-1)!$ ways of choosing an ordered sequence of the length $n-1$ from the remaining $n-1$ elements.

$\square$

Exercise 2.3.3. Let n be a positive integer. Estimate the number of perfect matchings in

(i) $K_{n,n}$, and
(ii) $K_{2,n}$.

Definition 2.3.4. *Let* S *be a finite, nonempty set, and let* k *be a positive integer. A* **k-permutation of** S *is an ordered sequence of* k *elements of* S, *with no element appearing more than once in the sequence.*

Lemma 2.3.5. *Let* $|S| = n$ *for a finite set* S, $n \geq 1$. *Let* k *be a positive integer,* $k \leq n$. *The number of* k-*permutations of* S *is*

$$n(n-1)(n-2) \cdot \ldots \cdot (n-k+1) = \frac{n!}{(n-k)!}.$$

Proof. There are n ways of choosing the first element, $n-1$ ways of choosing the second element, and so on until k elements are selected.

$\square$

Definition 2.3.6. *Let* k *and* n *be positive integers,* $k \leq n$. *A* k-**combination of a set** S *of* n *elements is a subset* $A \subseteq S$ *with* $|A| = k$. *The positive integer* $\binom{n}{k}$ *denotes the number of* k-*combinations of a set of* n *elements.*

Lemma 2.3.7. *For all* $k, n \in \mathbb{N}$, $n \geq k \geq 0$,

$$\binom{n}{k} = \frac{n!}{k!(n-k)!}.$$

Proof. For every k, $n \geq k \geq 0$, the integer $\frac{n!}{(n-k)!}$ is the number of k-permutations of S_n, where S_n is a set of n elements. For every k-combination A of S_n (a subset of k elements), there are exactly $k!$ k-permutations of S_n containing exactly the elements of A.

$\square$

Exercise 2.3.8. Prove that for all $k, n \in \mathbb{N}$, $n \geq k \geq 0$,

(i) $\binom{n}{k} = \binom{n}{n-k} = \frac{n}{k} \cdot \binom{n-1}{k-1} = \frac{n}{n-k}\binom{n-1}{k}$, and

(ii) $\binom{n}{k} = \binom{n-1}{k} + \binom{n-1}{k-1}$.

Observation 2.3.9. *For all* $k, n \in \mathbb{N}$, $n \geq k \geq 0$,

$$2^n = \sum_{k=0}^{n} \binom{n}{k}.$$

Proof. Let $S_n = \{x_1, x_2, \ldots, x_n\}$ be a set of n elements. Each subset $A \subseteq S_n$ can be identified with a vector $\alpha = (\alpha_1, \alpha_2, \ldots, \alpha_n) \in \{0,1\}^n$, where $\alpha_i = 1$ if and only if $x_i \in A$. On the other hand, each vector $\beta \in \{0,1\}^n$ unambiguously determines a subset of S_n. Thus, $|\{0,1\}^n|$ is equal to the number of the subsets of S_n. Obviously, $|\{0,1\}^n| = 2^n$. Now, we observe that $\sum_{k=0}^{n} \binom{n}{k}$ is the cardinality of the power set of S_n too, because $\binom{n}{k}$ is the number of subsets of S_n having exactly k elements for $k = 0, 1, \ldots, n$.

$\square$

Exercise 2.3.10. Prove that for all $k, n \in \mathbb{N}$, $n \geq k \geq 1$,

$$\binom{n}{k} \geq \left(\frac{n}{k}\right)^k.$$

Exercise 2.3.11. Prove that for every positive integer n,

$$\sum_{i=1}^{n} i = \binom{n+1}{2}.$$

As usual when dealing with functions we are interested in their asymptotic behavior. To do this, we need to define the standard O, Ω, and Θ notation.

Definition 2.3.12. *Let g be a function from $\mathbb{N}$ to $\mathbb{N}$.*

$$O(g) = \{f : \mathbb{N} \to \mathbb{N} \mid \exists c, n_0 \in \mathbb{N} \text{ such that}$$
$$\text{for all } n \in \mathbb{N}, n \geq n_0 : f(n) \leq cg(n)\}$$

is the set of all functions that asymptotically do not grow faster than the function g.

$$\Omega(g) = \{f : \mathbb{N} \to \mathbb{N} \mid \exists d, n_0 \in \mathbb{N} \text{ such that}$$
$$\text{for all } n \in \mathbb{N}, n \geq n_0 : f(n) \geq \frac{1}{d}g(n)\}$$

is the set of all functions that asymptotically do not grow slower than the function g.

$$\Theta(g) = \{f : \mathbb{N} \to \mathbb{N} \mid \exists c_1, c_2, n_0 \in \mathbb{N} \text{ such that for all}$$
$$n \in \mathbb{N}, n \geq n_0 : \frac{1}{c_1}g(n) \leq f(n) \leq c_2 \cdot g(n)\}$$

is the set of functions with the same asymptotic behavior as g.

If $f \in O(g)$ for some functions f and g we say that $\boldsymbol{g}$ is an asymptotic **upper bound** *of $\boldsymbol{f}$. If $f \in \Omega(g)$, we say that $\boldsymbol{g}$ is an asymptotic* **lower bound** *of $\boldsymbol{f}$. If $f \in \Theta(g)$, we say that $\boldsymbol{g}$ is an* **asymptotically tight bound** *for $\boldsymbol{f}$.*

In what follows, to indicate that a function f is an element of $O(g)$ $[\Omega(g), \Theta(g)]$ we also write

$$\boldsymbol{f(n) = O(g(n))} \; [\boldsymbol{f(n) = \Omega(g(n)), f(n) = \Theta(g(n))}].$$

Obviously, $f(n) = O(f(n))$ $[f(n) = \Omega(f(n)), f(n) = \Theta(f(n))]$ for every function $f : \mathbb{N} \to \mathbb{N}$.

Observation 2.3.13. *Let f and g be functions from $\mathbb{N}$ to $\mathbb{N}$. Then*

$$f(n) = O(g(n)) \text{ if and only if } g(n) = \Omega(f(n)).$$

Proof. If $f(n) = O(g(n))$, then according to the definition of $O(g)$ there exists $c, n_0 \in I\!N$ such that for all $n \geq n_0$, $f(n) \leq c \cdot g(n)$. This is equivalent to the claim that there exist $c, n_0 \in I\!N$ such that for all $n \geq n_0$, $g(n) \geq \frac{1}{c} f(n)$. But this is the definition of the fact $g(n) = \Omega(f(n))$.

$\square$

Observation 2.3.14. *For every function g from $I\!N$ to $I\!N$,*

$$\Theta(g) = O(g) \cap \Omega(g).$$

Proof. Similarly as in the proof of Observation 2.3.13, the equivalence of the claims $f(n) = \Theta(g(n))$ and $(f(n) = O(g(n)) \wedge f(n) = \Omega(g(n)))$ directly follows from Definition 2.3.12.

$\square$

To illustrate the above notations consider the function $f(n) = 2n^2 - 7n + 3$. Obviously $f(n) \leq 2n^2$ for all $n \geq 1$, i.e., $f(n) = O(n^2)$. On the other hand $f(n) \geq \frac{n^2}{2}$ for all $n \geq 5$, i.e., $f(n) = \Omega(n^2)$. Thus, we obtain that $f(n) = \Theta(n^2)$.

Exercise 2.3.15. Let $p(n) = a_1 n^k + a_2 n^{k-1} + \ldots + a_{k-1} n + a_k$ be a polynomial for some $k \in I\!N$, and real constants $a_1, a_2, \ldots, a_k$. Prove $p(n) = \Theta(n^k)$.

Exercise 2.3.16. Prove, for any real constants a, b, $b > 0$,

$$(n + a)^b = \Theta(n^b).$$

Exercise 2.3.17. Let a, b be positive real constants. For which a and b is $2^{a \cdot n + b}$ in $O(2^n)$?

As agreed upon above $f(n) = O(g(n))$ stands for $f \in O(g)$ (similarly for Ω and Θ). In what follows we allow that this asymptotic notation appears in a formula, like $n^3 + \Theta(\log_2 n)$. We interpret it as a representation for some anonymous function $f(n) \in \Theta(\log_2 n)$. For instance, we interpret the equality

$$f(n) = g(n) + \Theta(r(n))$$

to mean that a function $h(n) \in \Theta(r(n))$ exists such that

$$f(n) = g(n) + h(n).$$

If one writes $g(n) + \Theta(f(n)) = \Theta(r(n))$ for some functions g, f, and r, then it means that, for every $h \in \Theta(f)$, $g(n) + h(n) \in \Theta(r(n))$.

Finally we define the "little-oh" notation, enabling one to indicate that a function grows asymptotically faster than another one.

Definition 2.3.18. *Let g be a function from $I\!N$ to $I\!N$. We define*

$o(g) = \{f : I\!N \to I\!N \mid$ *for every positive constant* $c \in I\!N$,

there exists a positive integer $n_0 = n_0(c)$ *such that for all*

$$n \geq n_0 : \quad f(n) < \frac{1}{c}g(n)\}.$$

as the **set of functions growing asymptotically slower than** g. *We write* $f(n) = o(g(n))$ *to denote* $f \in o(g)$.

To prove $f(n) = o(g(n))$ for some functions $f, g : I\!N \to I\!N$ it is sufficient to prove $\lim_{n\to\infty} \frac{f(n)}{g(n)} = 0$. If $\lim_{n\to\infty} \frac{f(n)}{g(n)}$ exists and $\lim_{n\to\infty} \frac{f(n)}{g(n)} = 0$, it provides a convenient method to show that g grows asymptotically faster than f. We give two simple examples to illustrate this method. We can claim that $2^{cn} = o(2^n)$ for any real constant c, $0 < c < 1$, because

$$\lim_{n\to\infty} \frac{2^{c \cdot n}}{2^n} = \lim_{n\to\infty} \frac{1}{2^{(1-c) \cdot n}} = 0.$$

Similarly, we can claim $n^b = o(a^n)$ for any positive real constants a and b because

$$\lim_{n\to\infty} \frac{n^b}{a^n} = \lim_{n\to\infty} \frac{2^{\log_2(n^b)}}{2^{\log_2(a^n)}} = \lim_{n\to\infty} 2^{b \cdot \log_2 n - (\log_2 a) \cdot n} = 0.$$

Exercise 2.3.19. Prove or disprove:

(i)   For every positive constant k, $n^{(\log_2 n)^k} = o(2^n)$,
(ii) $n! = o(n^n)$,
(iii) $\left(\frac{n}{2}\right)^{\frac{n}{2}} = o(n!)$,
(iv) for every function $f : I\!N \to I\!N$, $f(n) + o(f(n)) = \Theta(f(n))$.

Exercise 2.3.20. Prove that for any three functions g, f, h from $I\!N$ to $I\!N$, $f(n) = o(g(n))$ and $g(n) = o(h(n))$ imply $f(n) = o(h(n))$.

Exercise 2.3.21. For which positive real constants c and d is 2^{cn} in $o(2^{d \cdot n})$?

Exercise 2.3.22. Prove that for every function f from $I\!N$ to $I\!N$,

$$o(f) \cap \Omega(f) = \emptyset.$$

Exercise 2.3.23. Prove that for every two real positive constants a and b,

$$(\log_2 n)^a = o(n^b).$$

One of the most useful formulas in the following combinatorial considerations is the **Stirling formula** that expresses $n!$ in terms of n^n.

Exercise 2.3.24. * **(Stirling's approximation).** Prove

$$n! = \sqrt{2\pi n} \left(\frac{n}{e}\right)^n \left(1 + \Theta\left(\frac{1}{n}\right)\right).$$

The Stirling formula can be used, for instance, to express $\binom{2n}{n}$ as an exponential function.

$$\binom{2n}{n} = \frac{(2n)!}{n!n!} = \frac{\sqrt{4\pi n} \cdot (\frac{2n}{e})^{2n} \cdot (1 + \Theta(\frac{1}{n}))}{(\sqrt{2\pi n} \cdot (\frac{n}{e})^n \cdot (1 + \Theta(\frac{1}{n})))^2}$$
$$= \frac{1}{\sqrt{\pi n}} \cdot 2^{2n}(1 + o(1)).$$

Observe that $\binom{2n}{n}$ is the largest number among $\binom{2n}{0}$, $\binom{2n}{1}, \ldots, \binom{2n}{n}$ and that the above estimation says that the number of subsets of n elements of a set of $2n$ elements is approximately

$$\frac{|\text{Pow}(\{1, 2, \ldots, 2n\})|}{\sqrt{\pi n}} = \Theta\left(\frac{2^{2n}}{\sqrt{n}}\right).$$

Exercise 2.3.25. Use Stirling's approximation to prove

(i) $n! = o(n^n)$,
(ii) $2^n = o(n!)$, and
(iii) $\log_2(n!) = \Theta(n \log n)$.

Another useful concept for analyzing the efficiency of communication algorithms is the concept of Fibonacci numbers.

Definition 2.3.26. *The* **Fibonacci numbers** F_i *for* $i = 0, 1, 2, \ldots$, *are defined by the following recurrence:*

$$F_0 = 0, \ F_1 = 1, \ and \ F_i = F_{i-1} + F_{i-2} \ for \ i \geq 2.$$

Exercise 2.3.27. Prove that $\log(F_n) = \Theta(n)$.

Definition 2.3.28. *The number* $\Phi = \frac{1+\sqrt{5}}{2} = 1.61803\ldots$ *is called the* **golden ratio**.

Exercise 2.3.29. Prove that $\Phi^2 = \Phi + 1$.

Exercise 2.3.30. * Prove that for every integer $i \geq 2$, $\Phi^{i-2} \leq F_i \leq \Phi^{i-1}$.

2.4 Communication Tasks and Algorithms

A lot of work has been done in recent years in the study of the properties of interconnection networks in order to find the best communication structures for parallel and distributed computing. An important feature characterizing the "quality" (suitability) of an interconnection network for parallel computing is the ability to effectively disseminate the information among its processors. Thus, the problem of dissemination of information has been investigated for most of the interconnection networks considered in parallel computing.

There are three main problems of information dissemination investigated in the current literature: *broadcasting*, *accumulation* and *gossiping*. For all three problems we may view any interconnection network as a connected undirected graph $G = (V, E)$, where the nodes in V correspond to the processors and the edges in E correspond to the communication links of the network. This abstraction is allowed because these three problems are purely communication problems, i.e., we do not need to deal with the computing actions of the processors.

Now, we will delve into the definitions of the broadcast problem, accumulation problem and gossip problem.

1. **Broadcast problem for a graph G and a node v of G**
   Let $G = (V, E)$ be a graph and let $v \in V$ be a node of G. Let v know a piece of information $I(v)$ which is unknown to all nodes in $V \setminus \{v\}$. The problem is to find a communication strategy (algorithm) such that all nodes in G learn this piece of information $I(v)$.

2. **Accumulation problem for a graph G and a node v of G**
   Let $G = (V, E)$ be a graph, and let $v \in V$ be a node of G. Let each node $u \in V$ know a piece of information $I(u)$, and let, for any $x, y \in V$, the pieces of information $I(x)$ and $I(y)$ be "disjoint" (independent[4]). The set

   $$I(G) = \{I(w) \mid w \in V\}$$

   is called the **cumulative message of G**. The problem is to find a communication strategy such that the node v learns the cumulative message of G.

3. **Gossip problem for a graph G**
   Let $G = (V, E)$ be a graph, and let $I(v)$ be a piece of information residing in v for all $v \in V$. The problem is to find a communication strategy such that each node from V learns the whole cumulative message.

As we have seen above, all these communication problems are natural for interconnection networks. The broadcast problem is to spread the knowledge of one processor to all other processors in the network, the accumulation problem is to accumulate the knowledge of all processors in one given processor, and the gossip problem is to accumulate the knowledge of all processors in each processor of the network. Obviously, the description above provides only an abstract characterization of broadcasting, accumulating, and gossiping. To make the characterization more precise, we have to explain the notion "communication strategy". The communication strategy means for us a communication algorithm (also called *communication scheme*) from an allowed set of communication algorithms. Each communication algorithm is a sequence of elementary communication steps called **communication rounds** (or simply

---

[4] This means that one cannot compute $I(u)$ from $I(v_1), I(v_2), \ldots, I(v_k)$ for any subset of nodes $\{v_1, \ldots, v_k\}$ that does not contain u. Thus, the only possibility to learn $I(u)$ is to get the message $I(u)$ from u.

rounds). To specify the set of allowed communication algorithms one defines
a so-called communication mode which precisely describes what may happen
in one communication round, i.e., the way in which the edges (communication
links) may or may not be used in one communication step (round). There are
several communication modes investigated in the literature. First, we present
the classical one-way and two-way modes. The investigation of these modes
was motivated by the first technologies used in telegraph networks and telephone networks.

- **One-way mode** (also called **telegraph communication mode**)
  In this mode each node may be active only via one of its adjacent edges
  either as sender or as receiver in a single round. This means that the information flow is one-way, i.e., one node sends a message to a given adjacent
  node. To specify the communication activity in a round, one has to determine which edges are active, and the direction of the communication flow
  has to be given for each active edge. Observe that the set of active edges
  in a round forms a matching of G.
  Formally, a **one-way communication algorithm for a graph $G=
  (V, E)$** is a sequence

  $$E_1, E_2, \ldots, E_m$$

  of sets $E_i \subseteq \overline{E}$, where $\overline{E} = \{(v \to u), (u \to v) \mid (u, v) \in E\}$, and if $(x_1 \to
  y_1), (x_2 \to y_2)$ are two distinct elements of E_i for some $i \in \{1, \ldots, m\}$,
  then $\{x_1, y_1\} \cap \{x_2, y_2\} = \emptyset$. For every $i \in \{1, \ldots, m\}$, E_i is called a **round
  (an elementary communication step) in the one-way mode**.

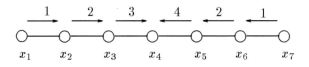

Fig. 2.9.

In Figure 2.9 an accumulation algorithm for the path of 7 nodes and
the node x_4 is depicted. In the first round the node x_1 sends its whole
knowledge to x_2, and x_7 sends its knowledge to x_6. In the second round
x_2 sends to x_3 and x_6 sends to x_5. In the third round x_3 sends to x_4,
and in the fourth round x_5 sends to x_4. Obviously, this communication
algorithm can be described as $\{(x_1 \to x_2), (x_7 \to x_6)\}, \{(x_2 \to x_3), (x_6 \to
x_5)\}, \{(x_3 \to x_4)\}, \{(x_5 \to x_4)\}$, and we see that the properties of the
one-way mode are satisfied.

We note that we shall use several distinct ways (e.g. Figure 2.9) to present
communication algorithms in this paper. But each of these ways will provide for each communication round the exact information which edges are
active (and in which direction they are active).

- **Two-way mode** (also called **telephone communication mode**)
  In this mode each node may be active only via one of its adjacent edges in a single round, and if it is active then it simultaneously sends a message and receives a message through the given, active edge (communication link). Phrased differently, if one edge is used for communication, the information flow is bidirectional.

  Thus, a **two-way communication algorithm for a graph $G= (V, E)$** is a sequence

  $$E_1, E_2, \ldots, E_r$$

  of some sets (matchings) $E_i \subseteq E$, where for each $i \in \{1, \ldots, r\}$, all $(x_1, y_1), (x_2, y_2) \in E_i$:

  $$\{x_1, y_1\} \neq \{x_2, y_2\} \text{ implies } \{x_1, y_1\} \cap \{x_2, y_2\} = \emptyset.$$

  For every $i \in \{1, \ldots, r\}$, E_i is called a **round in the two-way communication mode**.
  Figure 2.10 describes the following communication algorithm

  $$\{(x_1, x_3), (x_2, x_4)\}, \{(x_1, x_2), (x_3, x_4)\}$$

  for the cycle of four nodes.

Next, we formally specify what it means when we say that a communication algorithm solves a given communication task (broadcast, accumulation, gossip). The following definitions provide also the fundamental terms of a "formal language" needed for the analysis of the behavior of communication algorithms. Note that the following definition is independent of the communication mode considered.

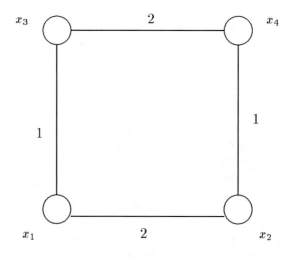

Fig. 2.10.

Definition 2.4.1. *Let $G = (V, E)$ be a graph, and let $A = A_1, A_2, \ldots, A_k$ be a communication algorithm for G. We assume that before the execution of A each node $v \in V$ knows a piece of information $I(v)$. For every $v \in V$ and every $i \in \{0, 1, \ldots, k\}$, we define the* **knowledge of v after the execution of i rounds of A** *as follows:*

$$I_0^A(v) = \{I(v)\},$$

and

$$\begin{aligned} I_i^A(v) = {} & I_{i-1}^A(v) \cup I_{i-1}^A(u) \text{ if the node } u \text{ sends its} \\ & \text{information to the node } v \text{ in the } i\text{-th round} \\ & (i.e., \text{ if } (u \to v) \in A_i \text{ in the one-way case and} \\ & (u, v) \in A_i \text{ in the two-way case), and} \end{aligned}$$

$$\begin{aligned} I_i^A(v) = {} & I_{i-1}^A(v) \text{ if } v \text{ does not receive any} \\ & \text{information in the } i\text{-th round.} \end{aligned}$$

For every $V' \subseteq V$ and every $i \in \{0, 1, \ldots, k\}$, we define

$$I_i^A(V') = \bigcup_{x \in V'} I_i^A(x)$$

as the **cumulative message of the set V' after i rounds.** *If $I_i^A(V') = I(G)$, then we say that V' is the* **cumulative set of nodes of G after i rounds of A.**

We define, for every node $v \in V$ and every $i \in \{1, \ldots, k\}$,

$$\mathbf{Broad}_0^A(v) = \{v\},$$

and

$$\begin{aligned} \mathbf{Broad}_i^A(v) = {} & \mathrm{Broad}_{i-1}^A(v) \cup \{u \in V \mid (x, u) \in A_i \ [(x \to u) \in A_i] \\ & \text{for some } x \in \mathrm{Broad}_{i-1}^A(v)\} \end{aligned}$$

as the **set of nodes knowing $I(v)$ after the execution of i rounds of A.**

Finally, we define, for every $i \in \{1, \ldots, k\}$,

$$\mathbf{Accu}_i^A = \{x \in V \mid I_i^A(x) = I(G)\}$$

as the **set of accumulation nodes of G after i rounds of A.**

The notation $I_i^A(v)$ is mainly used to analyze the performance of communication algorithms. Obviously, for every graph G, every communication algorithm $A = A_1, A_2, \ldots, A_k$ for G, and every $v \in V(G)$,

$$I_0^A(v) \subseteq I_1^A(v) \subseteq \ldots \subseteq I_{k-1}^A(v) \subseteq I_k^A(v).$$

A is a **broadcast algorithm** for G and v if and only if $I(v) \subseteq I_k^A(x)$ for every $x \in V(G)$.

The notation $I_i^A(V')$ for a $V' \subseteq V(G)$ is usually used to analyze accumulation and gossip algorithms. Clearly, A is an **accumulation algorithm for G and v** iff $I_k^A(\{v\}) = I(G)$. A is a **gossip algorithm for G** iff $I_k^A(\{x\}) = I(G)$ for all $x \in V(G)$.

For instance, for the communication algorithm

$$\{(x_1, x_3), (x_2, x_4)\}, \{(x_1, x_2), (x_3, x_4)\}$$

for the cycle in Figure 2.10 we have

$$I_0^A(x_i) = \{I(x_i)\} \text{ for } i \in 1, 2, 3, 4,$$

$$I_1^A(x_1) = \{I(x_1), I(x_3)\} = I_1^A(x_3),$$

$$I_1^A(x_2) = \{I(x_2), I(x_4)\} = I_1^A(x_4),$$

$$I_2^A(x_i) = I(G) = \{I(x_1), I(x_2), I(x_3), I(x_4)\} \text{ for all } i \in \{1, 2, 3, 4\}, \text{ and}$$

$$I_1^A(\{x_1, x_2\}) = I_1^A(\{x_3, x_4\}) = I(G).$$

Hence, the algorithm is a gossip algorithm (in the two-way mode) for the cycle of four nodes.

The notation $\mathrm{Broad}_i^A(v)$ is mainly used in the analysis of broadcast algorithms. This provides the following alternative definition of broadcast algorithms. A communication algorithm $A = A_1, A_2, \ldots, A_k$ for a graph G is a broadcast algorithm for G and a node $v \in V(G)$ iff $\mathrm{Broad}_k^A(v) = V(G)$.

Exercise 2.4.2. Let G be a graph, and let v be a node of $V(G)$. Prove that for every optimal broadcast algorithm $A = A_1, A_2, \ldots, A_k$ for G and v,

$$\mathrm{Broad}_0^A(v) \subset \mathrm{Broad}_1^A(v) \subset \cdots \subset \mathrm{Broad}_{k-1}^A(v) \subset \mathrm{Broad}_k^A(v).$$

The notation Accu_i^A is frequently used in the design and analysis of gossip algorithms. Obviously, a communication algorithm $A = A_1, A_2, \ldots, A_k$ for G is a gossip algorithm for G if and only if $\mathrm{Accu}_k^A = V(G)$.

Observation 2.4.3. *Let $A = A_1, A_2, \ldots, A_k$ be a communication algorithm for a graph G. Then, for every $i \in \{0, 1, \ldots, \min a(G) - 1\}$,*

$$\mathrm{Accu}_i^A = \emptyset.$$

The final topic of this section explains how the efficiency of communication algorithms is measured. We shall consider here one of the most popular possibilities – the number of communication rounds. This uniform measure does not deal with the length of the transmitted messages (with the amount of information exchanged). Thus, we assume that each node which is active as the sender in a given round sends its whole knowledge via the activated edge. The idea behind this is that one needs a lot of time to synchronize the network and to organize the information exchange in a given round, and the time needed for the direct information exchange via activated links is relatively small in comparison to the time for synchronization. This may be true in some cases, but there are also situations when one needs to measure the time required for any direct communication with respect to the length of the message. One may also be interested in measuring the physical work of the whole network during the communication algorithms. Such complexity measures will be discussed in later chapters. Next, we shall only deal with the number of rounds, which is one of the most commonly used complexity measures for communication algorithms.

Now, we give the definitions of the complexity measures investigated.

Definition 2.4.4. *Let $G = (V, E)$ be a graph, and let A be a one-way [two-way] communication algorithm. The (uniform)* **complexity of A, com(A),** *is the number of rounds in A.*
The **gossip complexity of G in one-way mode** *is*

$$\mathrm{r}(G) = \min\{\mathrm{com}(A) \mid A \text{ is a one-way communication}$$
$$\text{algorithm solving the gossip problem for } G\}.$$

The **gossip complexity of G in two-way mode** *is*

$$\mathrm{r_2}(G) = \min\{\mathrm{com}(B) \mid B \text{ is a two-way communication}$$
$$\text{algorithm solving the gossip problem for } G\}.$$

In what follows if A is a one-way [two-way] communication algorithm solving the gossip problem for a graph G we also say that A is a **one-way [two-way] gossip algorithm for G**.

Note that the fact that a graph G has the gossip complexity $\mathrm{r}(G)$ [$\mathrm{r_2}(G)$] means that there is a gossip algorithm for G with $\mathrm{r}(G)$ [$\mathrm{r_2}(G)$] rounds and there exists no gossip algorithm for G with fewer than $\mathrm{r}(G)$ [$\mathrm{r_2}(G)$] rounds. Thus, any one-way [two-way] gossip algorithm $A = A_1, A_2, \ldots, A_k$ for G is called **optimal** if $k = \mathrm{r}(G)$ [$k = \mathrm{r_2}(G)$].

Exercise 2.4.5. Give one-way and two-way communication algorithms solving the gossip problem for G, where

(i) G is a cycle of 8 nodes,
(ii) G is a path of 5 nodes.

Exercise 2.4.6. For every even $n \in \mathbb{N} - \{0\}$, estimate the two-way gossip complexity of G_n, where

(i) $G_n = (\{v_1, v_2, \ldots, v_n\}, \{(v_n, v_1), (v_1, v_2), \ldots, (v_{n-1}, v_n)\})$ is the cycle of n nodes, and
(ii) $G_n = (\{v_1, v_2, \ldots, v_n\}, \{(v_1, v_2), (v_2, v_3), \ldots, (v_{n-1}, v_n)\})$ is the path of n nodes.

Definition 2.4.7. *Let* $G = (V, E)$ *be a graph. For every node* $v \in V$, *we define the* **broadcast complexity for** G **and** v **in the one-way mode** *as*

$$\mathbf{b}(v, G) = \min\{\text{com}(A) \mid A \text{ is a one-way communication}$$
$$\text{algorithm solving the broadcast problem for } G \text{ and } v\},$$

and the **broadcast complexity for** G **and** V **in the two-way mode** *as*

$$\mathbf{b_2}(v, G) = \min\{\text{com}(A) \mid A \text{ is a two-way communication}$$
$$\text{algorithm solving the broadcast problem for } G \text{ and } v\}.$$

We define

$$\mathbf{b}(G) = \max\{\text{b}(v, G) \mid v \in V\} \text{ and } \mathbf{b_2}(G) = \max\{\text{b}_2(v, G) \mid v \in V\}$$

to be the **broadcast complexity of** G **in the one-way mode** *and* **in the two-way mode**, *respectively. We define*

$$\min \mathbf{b}(G) = \min\{\text{b}(v, G) \mid v \in V\}$$

as the **min-broadcast complexity of** G.

If A is a one-way [two-way] communication algorithm solving the broadcast problem for a graph G and a node $v \in V(G)$, then we also say that A is a **one-way [two-way] broadcast algorithm for** G **and** v. A one-way [two-way] broadcast algorithm $A = A_1, A_2, \ldots, A_k$ for G and v is called **optimal** if $k = \text{b}(v, G)$ [$k = \text{b}_2(v, G)$].

Definition 2.4.8. *Let* $G = (V, E)$ *be a graph. For every node* $v \in V$, *we define the* **complexity of the accumulation problem for** G **and** v **in the one-way mode** *as*

$$\mathbf{a}(v, G) = \min\{\text{com}(A) \mid A \text{ is a one-way communication algorithm}$$
$$\text{solving the accumulation problem for } G \text{ and } v\},$$

and the **complexity of the accumulation problem for** G **and** v **in the two-way mode** *as*

$$\mathbf{a_2}(v, G) = \min\{\text{com}(A) \mid A \text{ is a two-way communication algorithm}$$
$$\text{solving the accumulation problem for } G \text{ and } v\}.$$

We define

$$\mathbf{a}(G) = \max\{\text{a}(v, G) \mid v \in V\} \text{ and } \mathbf{a_2}(G) = \max\{\text{a}_2(v, G) \mid v \in V\}$$

to be the **accumulation complexity of** G **in the one-way mode** *and* **in the two-way mode**, *respectively. We define*

$$\min \mathbf{a}(G) = \min\{\text{a}(v, G) \mid v \in V\}$$

as the **min-accumulation complexity of** G.

If A is a one-way [two-way] communication algorithm solving the accumulation problem for a graph G and a node $v \in V(G)$, then we shortly say that A is a **one-way [two-way] accumulation algorithm for G and v**. A one-way [two-way] accumulation algorithm $A = A_1, A_2, \ldots, A_k$ for G and v is called **optimal** if $k = a(v, G)$ [$k = a_2(v, G)$].

Exercise 2.4.9. Let n be a positive, even integer. Estimate $a(G_n)$, $b(G_n)$, $\min a(G_n)$, and $\min b(G_n)$ for every G_n, where

(i) G_n is a path of n nodes,
(ii) G_n is a cycle of n nodes.

2.5 General Properties of the Complexity of Communication Algorithms

In this section we show that there is no difference between some of the complexity measures defined in the previous section, and so we show that it is sufficient to investigate only the broadcast problem in the one–way mode and the gossiping problem in both modes. First, we shall show that we do not need the complexity measures defined in Definition 2.4.8 because the accumulation problem is exactly as hard as the broadcast problem for the communication modes considered. Note that this may be wrong for other communication modes. We shall mention a large difference between broadcast complexity and accumulation complexity for some other modes in later sections.

Observation 2.5.1. *For every graph G and every node v of G,*

$$a_2(v, G) = b_2(v, G).$$

Proof. Let $A = A_1, A_2, \ldots, A_k$ be a broadcast algorithm for G and v in the two–way mode. We prove that $B = A_k, A_{k-1}, \ldots, A_1$ is an accumulation algorithm for G and v in the two–way mode. For this we show that, for every $i \in \{0, 1, \ldots, k\}$, $\text{Broad}_i^A(v)$ is the cumulative set of nodes of G after $k - i$ rounds of B. Obviously, this will complete our proof because it means for $i = k$ that $\text{Broad}_0^A(v) = \{v\}$ is the cumulative set of nodes of G after k rounds of B, i.e., that $I_k^B(\{v\}) = I(G)$.

We prove this fact by induction on the number j of rounds of B. For $j = 0$, it is obvious that $\text{Broad}_k^A(v) = V(G)$ is the cumulative set of nodes of G before the start of B. Let our induction hypothesis hold for all $j \in \{0, 1, \ldots, m\}$, $m < k$. We prove it for $m + 1$. So, we have to show that $\text{Broad}_{k-(m+1)}^A(v)$ is the cumulative set of nodes of G after $m + 1$ rounds of B. According to Definition 2.4.1, $\text{Broad}_{k-m}^A(v) = \text{Broad}_{k-(m+1)}^A \cup \{u \in V \mid (x, u) \in A_{k-m}$ for some $x \in \text{Broad}_{k-(m+1)}^A(v)\}$. Since $\text{Broad}_{k-m}^A(v)$ is the cumulative set of G after m rounds of B, and for every $u \in \text{Broad}_{k-m}^A(v) - \text{Broad}_{k-(m+1)}^A(v)$ there is an $x \in \text{Broad}_{k-(m+1)}^A(v)$ receiving the knowledge of u in the $(m + 1)$-st round of B (i.e., $(u, x) \in A_{k-m}$), we can conclude that $\text{Broad}^A_{k-(m+1)}(v)$ is the cumulative set of G after $m + 1$ rounds of B.

In the same way, a broadcast algorithm can be constructed from an accumulation algorithm. Obviously, the number of rounds does not change by this construction.

$\square$

Corollary 2.5.2. *For every graph G, $a_2(G) = b_2(G)$.*

Note that in what follows we usually omit the detailed proofs of obvious facts such as that $A_1, A_2, \ldots, A_k$ is a broadcast algorithm for a G and v iff $A_k, \ldots, A_2, A_1$ is an accumulation algorithm for G and v.

Observation 2.5.3. *For every graph G and every node v of G,*

$$a(v, G) = b(v, G).$$

Proof. Let $E_1, E_2, \ldots, E_s$ be a broadcast algorithm for G and v in the one–way mode. Let $R_i = \{(x \to y) \mid (y \to x) \in E_i\}$. Then $R_s, R_{s-1}, \ldots, R_2, R_1$ is an accumulation algorithm for G and v in the one–way mode.

A broadcast algorithm can be similarly constructed from an accumulation algorithm. Obviously, these constructions do not alter the number of rounds.

$\square$

Corollary 2.5.4. *For every graph G,*

$$a(G) = b(G) \ and \ \min a(G) = \min b(G).$$

We see that it is sufficient to deal only with the broadcast complexity for the considered communication modes because all results for broadcast complexity are also valid for accumulation complexity.

Now, we observe the intuitively obvious fact that the two-way mode cannot decrease the broadcast complexity in comparison with the one-way mode, because for broadcasting it is sufficient that the information flows in one direction from the source node to all other nodes.

Lemma 2.5.5. *For every graph $G = (V, E)$, and every node v of G,*

$$b(v, G) = b_2(v, G).$$

Proof. It is clear from the definition that $b(v, G) \geq b_2(v, G)$, because the one–way mode cannot be more powerful than the two–way mode.

To prove $b(v, G) \leq b_2(v, G)$, let $A = E_1, E_2, \ldots, E_s$ be a broadcast algorithm for G and v in the two–way mode. Note that $\text{Broad}^A_i(v)$, $i = 1, \ldots, s$, is the set of nodes receiving the piece of information $I(v)$ in the first i rounds (i.e., during the execution of the algorithm $E_1, E_2, \ldots, E_i$), and $\text{Broad}^A_0(v) = \{v\}$. Obviously, $\bigcup_{i=1}^{s} \text{Broad}^A_i(v) = V$. Let $V_i = \text{Broad}^A_i(v) \setminus \bigcup_{j=1}^{i-1} \text{Broad}^A_j(v)$ (see Figure 2.11). Thus, for $i = 0, 1, \ldots, s$, V_i is the set of nodes which receive $I(v)$

exactly in the i-th round and not before. Thus, $\bigcup_{i=0}^{s} V_i = V$ and $V_c \cap V_d = \emptyset$ for $c \neq d$, $c, d \in \{0, \ldots, s\}$.

Now, we remove the unnecessary edges (for example, (x, y) and (u, w) from E_4 in Figure 2.11) from the broadcast algorithm A in order to get the broadcast algorithm $A' = E_1', E_2', \ldots, E_s'$, where $E_i' = E_i \cap \left(\bigcup_{k=1}^{i-1} V_k \times V_i \right)$ for $i = 1, \ldots, s$. Obviously A' is a broadcast algorithm in the two–way mode with the property that each node from $V \setminus \{v\}$ receives $I(v)$ exactly once. The graphical representation of A' is the tree $T_{A'} = (V, \bigcup_{i=1}^{s} E_i')$ (see Figure 2.11).

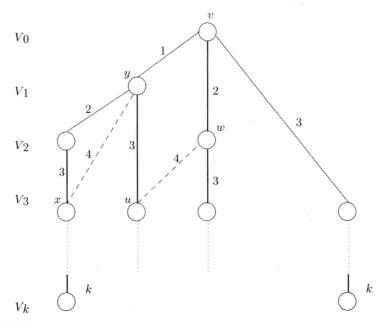

Fig. 2.11.

To get a broadcast algorithm in the one–way mode it suffices to direct the edges of A' in the direction from the root v to the leaves. Thus, $B = Z_1, Z_2, \ldots, Z_s$, where

$$Z_i = \{(x_1 \to x_2) \mid (x_1, x_2) \in E_i' \wedge x_1 \in \bigcup_{k=1}^{i-1} V_k \wedge x_2 \in V_i\}$$

for $i = 1, \ldots, s$. It is clear that B is a communication algorithm in the one–way mode. One can easily prove by induction that for $i = 1, \ldots, s$ all nodes in $\bigcup_{j=0}^{i} V_j = \text{Broad}_i^A(v) = \text{Broad}_i^B(v)$ know $I(v)$ after the i-th round of B (after $Z_1, Z_2, \ldots, Z_i$). Thus, B is a broadcast algorithm in the one–way mode with the same number of rounds as A.

$\square$

Corollary 2.5.6. *For every graph G, $b(G) = b_2(G)$.*

The proof of Lemma 2.5.5 shows that any broadcast algorithm A for G and v determines a spanning tree of G rooted at v. We call this tree together with the step labels on the edges a **broadcast tree of A**.

We see that it is sufficient to deal with the complexity measures r, r_2, b and $\min b$ because all others are identical with one of these four. In what follows we shall show that these four measures are really different. Hence we have to deal with all four measures.

Observation 2.5.7. *There exists a graph G such that*

$$r(G) = 2r_2(G).$$

Proof. Let us consider the cycle C_4 of 4 nodes as depicted in Figure 2.10. We see from Figure 2.10 that $r_2(C_4) = 2$. The following algorithms

$$\begin{aligned}
A_1 &= \{(x_1 \to x_3), (x_2 \to x_4)\}, \{(x_3 \to x_1), (x_4 \to x_2)\}, \\
&\quad \{(x_2 \to x_4), (x_1 \to x_2)\}, \{(x_4 \to x_2), (x_2 \to x_1)\}, \text{ and} \\
A_2 &= \{(x_1 \to x_3), (x_2 \to x_4)\}, \{(x_3 \to x_4)\}, \{(x_4 \to x_3)\}, \\
&\quad \{(x_3 \to x_1), (x_4 \to x_2)\}
\end{aligned}$$

are clearly gossip algorithms in the one–way mode. By checking all one–way communication algorithms with at most three rounds one can easily establish that $r(C_4) = 4$.

□

Thus $r(G)$ may be twice $r_2(G)$. As will be shown in the next lemma, this is the worst possible case.

Lemma 2.5.8. *For every graph G:*

$$\min b(G) \le b(G) \le r_2(G) \le r(G) \le 2r_2(G).$$

Proof. The inequalities $\min b(G) \le b(G) \le r_2(G) \le r(G)$ follow directly from the definitions. To show that $r(G) \le 2r_2(G)$ we consider $A = E_1, \ldots, E_r$ as a gossip algorithm for G in two–way mode. Then any $B = E_{11}, E_{12}, E_{21}, E_{22}, \ldots,$ E_{r1}, E_{r2}, where $E_{i1} \cup E_{i2} = \{(x \to y), (y \to x) \mid (x, y) \in E_i\}$ and E_{i1} and E_{i2} are defined such that B is a one–way communication algorithm, is a gossip algorithm for G in one–way mode.

□

Next we show that $\min b$ and b may be essentially different.

Example 2.5.9. Consider a path P_n of n nodes $x_1, x_2, \ldots, x_n$ (see Figure 2.12). We claim $\min b(P_n) = b(x_{\lceil \frac{n}{2} \rceil}, P_n) = \lceil \frac{n}{2} \rceil$ because

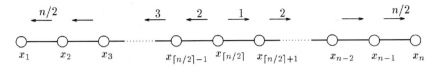

Fig. 2.12. Broadcasting in P_n for even n

$$\{(x_{\frac{n}{2}} \to x_{\frac{n}{2}+1})\}, \{(x_{\frac{n}{2}} \to x_{\frac{n}{2}-1}), (x_{\frac{n}{2}+1}, x_{\frac{n}{2}+2})\},$$
$$\ldots, \{(x_3 \to x_2), (x_{n-2} \to x_{n-1})\}, \{(x_2 \to x_1), (x_{n-1} \to x_n)\}$$

is a broadcast algorithm for $P_n, x_{\frac{n}{2}}$ and n even, and

$$\{(x_{\lceil\frac{n}{2}\rceil} \to x_{\lceil\frac{n}{2}\rceil+1})\}, \{(x_{\lceil\frac{n}{2}\rceil} \to x_{\lceil\frac{n}{2}\rceil-1}), (x_{\lceil\frac{n}{2}\rceil+1} \to x_{\lceil\frac{n}{2}\rceil+2})\},$$
$$\ldots, \{(x_{n-1} \to x_n), (x_3 \to x_2)\}, \{(x_2 \to x_1)\}$$

is a broadcast algorithm for P_n, $x_{\lceil\frac{n}{2}\rceil}$ and n odd. No algorithm with fewer rounds than these two algorithms exists because the distance between $x_{\lceil\frac{n}{2}\rceil}$ and x_1 is $\lceil\frac{n}{2}\rceil - 1$, the distance between $x_{\lceil\frac{n}{2}\rceil}$ and x_n is $\lfloor\frac{n}{2}\rfloor$, and $x_{\lceil\frac{n}{2}\rceil}$ can send $I(x_{\lceil\frac{n}{2}\rceil})$ in the first round only in one direction.

Clearly, $b(P_n) = b(x_1, P_n) = n - 1$ because the distance between x_1 and x_n is exactly $n - 1$.

We have seen that $b(G)$ may be almost twice $\min b(G)$. The next lemma shows that this is an upper bound.

Lemma 2.5.10. *For any graph G of at least two nodes:*

$$b(G) \le r(G) \le 2 \cdot \min b(G) \quad and \quad b(G) \le r_2(G) \le 2 \cdot \min b(G) - 1.$$

Proof. Let $G = (V, E)$ be a graph, and let $v \in V$ be a node with the property $b(v, G) = \min b(G)$. Let $A = E_1, E_2, \ldots, E_z$ for $z = \min b(G)$ be an optimal one–way broadcast algorithm for G and v. According to Observation 2.5.3 there exists a one–way accumulation algorithm $B = D_1, D_2, \ldots, D_z$ for G and v. Obviously, the concatenation of B and A,

$$B \circ A = D_1, D_2, \ldots, D_z, E_1, E_2, \ldots, E_z$$

is a one–way gossip algorithm for G. Hence

$$r(G) \le 2 \cdot \min b(G).$$

For the two-way mode, one round can be spared because $D_z = E_1$. Hence,

$$r_2(G) \le 2 \cdot \min b(G) - 1.$$

$\square$

To see that there exist graphs for which the equalities $r(G) = 2 \min b(G)$ and $r_2(G) = 2 \min b(G) - 1$ hold, it is sufficient to take the paths P_n for even n from Example 2.5.9. In this case, $\min b(G) = \frac{n}{2}$, $r(G) = n$, and $r_2(G) = n - 1$. These facts will be dealt with in detail in Chapter 4, where we will also discuss for which other graphs the equality $r(G) = 2 \min b(G)$ holds.

Observation 2.5.11. *For every even integer $n \geq 8$, there exists a graph G of n nodes such that $b(G) = r(G)$.*

Proof. A graph G for which $b(G) = r(G)$ is the graph D_n as displayed in Figure 2.13 for even $n \geq 8$.

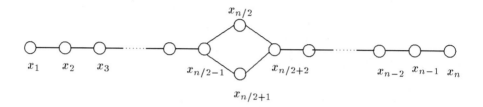

Fig. 2.13. The graph D_n

Clearly, $b(D_n) = n - 2$, because the distance between x_1 and x_n is $n - 2$ and broadcasting can be achieved in the same number of rounds. Also, $r(G) = r_2(G) = n - 2$, because $E_1, E_2, \ldots, E_{n-2}$ with

$$\begin{aligned}
E_1 &= \left\{ (x_1 \to x_2), (x_n \to x_{n-1}), (x_{\frac{n}{2}} \to x_{\frac{n}{2}-1}), (x_{\frac{n}{2}+1} \to x_{\frac{n}{2}+2}) \right\}, \\
E_2 &= \left\{ (x_2 \to x_3), (x_{n-1} \to x_{n-2}) \right\}, \ldots, \\
E_{\frac{n}{2}-2} &= \left\{ (x_{\frac{n}{2}-2} \to x_{\frac{n}{2}-1}), (x_{\frac{n}{2}+3} \to x_{\frac{n}{2}+2}) \right\}, \\
E_{\frac{n}{2}-1} &= \left\{ (x_{\frac{n}{2}-1} \to x_{\frac{n}{2}+1}), (x_{\frac{n}{2}+2} \to x_{\frac{n}{2}}) \right\}, \\
E_{\frac{n}{2}} &= \left\{ (x_{\frac{n}{2}} \to x_{\frac{n}{2}-1}), (x_{\frac{n}{2}+1} \to x_{\frac{n}{2}+2}) \right\}^5, \\
E_{\frac{n}{2}+1} &= \left\{ (x_{\frac{n}{2}-1} \to x_{\frac{n}{2}-2}), (x_{\frac{n}{2}+2} \to x_{\frac{n}{2}+3}) \right\}, \\
E_{\frac{n}{2}+2} &= \left\{ (x_{\frac{n}{2}-2} \to x_{\frac{n}{2}-3}), (x_{\frac{n}{2}+3} \to x_{\frac{n}{2}+4}), \right. \\
&\quad \left. (x_{\frac{n}{2}-1} \to x_{\frac{n}{2}+1}), (x_{\frac{n}{2}+2} \to x_{\frac{n}{2}}) \right\}, \ldots, \\
E_{n-1} &= \left\{ (x_3 \to x_2), (x_{n-2} \to x_{n-1}) \right\}, \\
E_{n-2} &= \left\{ (x_2 \to x_1), (x_{n-1} \to x_n) \right\}
\end{aligned}$$

is a gossip algorithm for D_n that takes $n - 2$ rounds.

$\square$

We conclude this section by showing how to obtain some straightforward lower bounds on the complexity of broadcasting and gossiping by simply investigating some basic properties of graphs.

Observation 2.5.12. *For every graph G:*

$$\mathrm{rad}(G) \leq \min \mathrm{b}(G).$$

---

[5] Note that after this $(\frac{n}{2})$-th round both vertices $x_{\frac{n}{2}-1}$ and $x_{\frac{n}{2}+2}$ know the cumulative message and they broadcast it to all other vertices in the remaining $\frac{n}{2} - 2$ rounds.

Proof. Obviously, for each $v, x \in V$, $b(v, G)$ must be at least $\text{dist}(v, x)$.

$\square$

The following observation follows also directly from the definitions.

Observation 2.5.13. *For every graph G,*

$$\text{rad}(G) \leq \text{diam}(G) \leq b(G).$$

Exercise 2.5.14. Prove, for any positive, even number n, that the path P_n of n nodes has the following properties:

(i) $\text{rad}(G) = \min b(G)$, and
(ii) $\text{diam}(G) = b(G)$.

Observation 2.5.15. *Let X stand for any of the complexity measures b, $\min b$, r, r_2. For any two graphs $G_1 = (V, E_1)$ and $G_2 = (V, E_2)$ such that $E_2 \subseteq E_1$,*

$$X(G_2) \geq X(G_1).$$

Exercise 2.5.16. Let G be a graph such that there are three pairwise distinct nodes u, x, y in $V(G)$ with the property $\text{dist}(u, x) = \text{dist}(u, y) = \text{diam}(G)$. Then

$$b(u, G) \geq \text{diam}(G) + 1.$$

We conclude this section by giving some general strategies for broadcasting in graphs having some special graph-theoretical properties.

Lemma 2.5.17. *For every graph G,*

(i) $\min b(G) \leq (\deg(G) - 1) \cdot \text{rad}(G) + 1$,
(ii) $b(G) \leq (\deg(G) - 1) \cdot \text{diam}(G) + 1$, *and*
(iii) $b(G) \leq \deg(G) \cdot \text{rad}(G)$.

Proof. Let $w \in V(G)$. First, we give the following general communication strategy for the broadcast problem for G and w.

1) w sends $I(w)$ to all its $d \leq \deg(G)$ neighbors in the first d rounds.
2) If a node $v \neq w$ learns the piece of information $I(w)$ in a round t, then v sends $I(w)$ to all its $k \leq \deg(G) - 1$ neighbors different from w in the next k rounds $t + 1, t + 2, \ldots, t + k$.

To remove the possible collisions we do the following agreements:

(i) If a node u knows $I(w)$ already, then no node sends $I(w)$ to u again.
(ii) If several nodes try to send $I(w)$ to the same node y in the same round, then only one of them is chosen to submit $I(w)$ to y and the remaining nodes send $I(w)$ to other neighbors (if any) in this round.

Clearly, this strategy results in a one–way communication algorithm A_w. We prove by induction the following assertion for $A = A_w$.

(1) After the execution of $r = (\deg(G) - 1) \cdot i + 1$ rounds of A,
$\{x \in V(G) \mid \text{dist}(w, x) \leq i\} \subseteq \text{Broad}_r^A(w)$ for every $i = 0, 1, \ldots, \text{com}(A)$.

1° If $i = 0$, then obviously $\text{Broad}_0^A(w) = \{w\} = \{x \in V(G) \mid \text{dist}(w, x) = 0\}$.
2° Let the induction hypothesis (1) hold for every $i \leq m$. We prove (1) for $m + 1$. By the induction hypothesis we have

$$\{x \in V(G) \mid \text{dist}(w, x) \leq m\} \subseteq \text{Broad}_{(\deg(G)-1) \cdot m + 1}^A(w).$$

Let $y \in V(G)$ and let $\text{dist}(w, y) = m + 1$. It is obvious that there is at least one node x with $\text{dist}(w, x) = m$ and $(x, y) \in E(G)$. So, every such x is in $\text{Broad}_{(\deg(G)-1) \cdot m + 1}^A(w)$. Since there are at most $\deg(G) - 1$ neighbors of A which do not know $I(w)$ after $(\deg(G) - 1) \cdot m + 1$ rounds of A, y definitely receives $I(w)$ in one of the next $\deg(G) - 1$ rounds of A. Hence, we have $y \in \text{Broad}_{(\deg(G)-1) \cdot (m+1) + 1}^A(w)$.

Now, we use (1) to prove (i), (ii), and (iii).

(i) Let $v \in V(G)$ be a node with the property $V(G) \subseteq \{x \in V(G) \mid \text{dist}(v, x) \leq \text{rad}(G)\}$. Thus, the above communication strategy A_v broadcasts in G from v in $(\deg(G) - 1) \cdot \text{rad}(G) + 1$ according to (1).
(ii) Let w be an arbitrary node of $V(G)$. Obviously, $V(G) \subseteq \{x \in V(G) \mid \text{dist}(w, x) \leq \text{diam}(G)\}$. The assertion (1) implies

$$V(G) \subseteq \text{Broad}_{(\deg(G)-1) \cdot \text{diam}(G) + 1}^{A_w}(w)$$

for every $w \in V(G)$. Thus $b(G) \leq (\deg(G) - 1) \cdot \text{diam}(G) + 1$.
(iii) To prove the inequality (iii) we need to modify the above communication strategy slightly. Let $v_0 \in V(G)$ be a node with $V(G) \subseteq \{x \in V(G) \mid \text{dist}(v_0, x) \leq \text{rad}(G)\}$. For every $w \in V(G)$ we define the following broadcast algorithm B_w for G and w:
1) Node w passes in at most $\text{rad}(G)$ rounds (exactly in $\text{dist}(w, v_0)$ rounds) $I(w)$ to v_0.
2) Node v_0 sends $I(w)$ to all its neighbors that are oblivious to $I(w)$ in at most $\deg(G) - 1$ next rounds.
3) The same strategy as the step 2 of the communication strategy A_{v_0} above.
Similarly, one can prove for B_w that, for every $i \in \{0, 1, \ldots, \text{rad}(G)\}$,

$$\{x \in V(G) \mid \text{dist}(w, x) \leq i\} \subseteq \text{Broad}_{\text{rad}(G) + (\deg(G)-1) \cdot i}^{B_w}(w).$$

For $i = \text{rad}(G)$ this yields

$$V(G) \subseteq \text{Broad}_{\text{rad}(G) \cdot \deg(G)}^{B_w}.$$

Thus, $b(G) \leq \text{rad}(G) \cdot \deg(G)$. $\square$

In the following exercise one has to search for graphs for which the communication strategies described in the proof of Lemma 2.5.17 provide optimal broadcast algorithms.

Exercise 2.5.18. Find some examples of graphs G which satisfy some of the following equalities:

(i') $\min b(G) = (\deg(G) - 1) \cdot \operatorname{rad}(G) + 1$,
(ii') $b(G) = (\deg(G) - 1) \cdot \operatorname{diam}(G) + 1$, or
(iii') $b(G) = \deg(G) \cdot \operatorname{rad}(G)$.

A direct consequence of Lemma 2.5.10 and Lemma 2.5.17 is the following assertion.

Corollary 2.5.19. *For every graph G,*

(i) $r(G) \leq 2(\deg(G) - 1) \cdot \operatorname{rad}(G) + 2$, and
(ii) $r_2(G) \leq 2(\deg(G) - 1) \cdot \operatorname{rad}(G) + 1$.

Exercise 2.5.20. Does a graph G exist for which (i) or (ii) of Corollary 2.5.19 is an equality?

2.6 Fundamental Interconnection Networks

In this section, we provide the definitions of the most studied networks, and we fix their notation for the rest of the book. Note that we fix not only the structures of these networks, but also the names (labels) of their nodes. For more information about these networks, we refer to [MS90].

The Path P_n. The *(simple) path of n nodes*, denoted by

$$P_n = (\{1, 2, \ldots, n\}, \{(i, i+1) \mid i = 1, \ldots, n-1\}),$$

is the graph whose nodes are all integers from 1 to n and whose edges connect each integer i with $i + 1$ for $1 \leq i \leq n - 1$.

P_n has n nodes, $n - 1$ edges, diameter $n - 1$, degree 2 and bisection width 1. An illustration of P_n is shown in Figure 2.14.

$$
\begin{array}{ccccccc}
1 & 2 & 3 & & & n-1 & n
\end{array}
$$

Fig. 2.14. The path P_n

The Cycle C_n. The *(simple) cycle of n nodes*, denoted by C_n with

$$V(C_n) = \{0, 1, \ldots, n-1\}$$

and

$$E(C_n) = \{(i, i+1) \mid i = 0, \ldots, n-2\} \cup \{(0, n-1)\},$$

is the graph whose nodes are all integers from 0 to $n-1$ and whose edges connect each integer i, $0 \le i \le n-1$, with $(i+1) \bmod n$.

C_n has n nodes, n edges, diameter $\lfloor \frac{n}{2} \rfloor$ and bisection width 2. It is also regular of degree 2. An illustration of C_4 is shown in Figure 2.15.

Fig. 2.15. The cycle C_4

Exercise 2.6.1. Estimate $b(C_n)$ for any positive integer n.

In what follows we will label the nodes not only by integers, but also through strings (words) over some alphabet. In typical cases we consider the alphabets

$$\Sigma_k = \{0, 1, \ldots, k-1\}$$

for every $k \in I\!N \setminus \{0\}$. A string over Σ_k is a finite sequence of symbols over Σ_k. For instance, 01100 is a string of symbols over Σ_2 and 3721177 is a string over Σ_{10}. For any string x, **the length of x**, denoted by $|x|$, is the number of symbols in x. Thus, $|001212| = 6$. The empty string, denoted by ε, has the length 0. For any alphabet Σ, Σ^* denotes the set of all finite strings over Σ. We define

$$\Sigma^m = \{x \in \Sigma^* \mid |x| = m\} \text{ and } \Sigma^{\le m} = \{x \in \Sigma^* \mid |x| \le m\}$$

for any alphabet Σ and $m \in I\!N$. For instance,

$$(\Sigma_2)^{\le 3} = \{\varepsilon, 0, 1, 00, 01, 10, 11, 000, 001, 010, 100, 011, 101, 110, 111\}.$$

Obviously, $|(\Sigma_k)^m| = k^m$.

Exercise 2.6.2. Prove that

$$|(\Sigma_k)^{\le m}| = \frac{k^{m+1} - 1}{k - 1}$$

for all $k, m \in I\!N \setminus \{0\}$.

The Complete k-ary Tree $T_k{}^m$. The *complete k-ary tree of depth m*, denoted by $T_k{}^m$, is the graph whose nodes are all k-ary strings of length at most m and whose edges connect each string α of length i, $0 \le i \le m-1$, with the strings αa, $a \in \{0, \ldots, k-1\}$, of length $i+1$. The node ε, where ε is the empty string, is the **root** of $T_k{}^m$ and a node α is at **level** i, $i \ge 0$, in $T_k{}^m$ if α is a string of length i. The nodes at level m are the *leaves* of the tree. For a node α at level i, $0 \le i \le m-1$, the nodes αa, $a \in \{0, \ldots, k-1\}$, are called the **children** of α. The node α is called the **parent** of αa. For any node α, the nodes αu, $u \in \{0, \ldots, k-1\}$, are called **descendants** of α, and α is called an **ancestor** of αu.

Formally,

$$V(T_k^m) = \bigcup_{i=0}^{m} \{0, 1, \ldots, k-1\}^i, \text{ and}$$

$$E(T_k^m) = \{(\alpha, \alpha i) \mid \alpha \in V(T_k^m), |\alpha| \le m-1, i \in \{0, 1, \ldots, k-1\}\}$$

$T_k{}^m$ has $(k^{m+1} - 1)/(k-1)$ nodes. Clearly, T_k^m has diameter $2m$, degree $k+1$ and bisection width $k/2$ for even k. The tree $T_2{}^3$ is depicted in Figure 2.16.

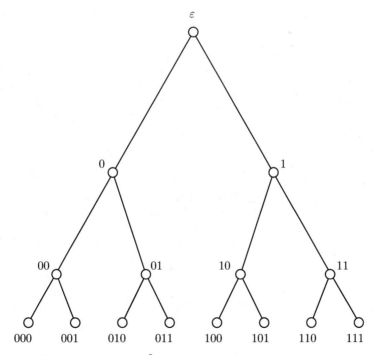

Fig. 2.16. The complete tree $T_2{}^3$

Exercise 2.6.3. Prove that the bisection width of T_k^m is in $\Theta(k \cdot m)$ for all odd k.

The Complete Graph K_n. The *complete graph of n nodes*, denoted by K_n, is the graph whose nodes are all integers from 1 to n and whose edges connect each integer i, $1 \leq i \leq n$, with each integer j, $1 \leq j \leq n$, $j \neq i$.
Formally,

$$V(K_n) = \{1, 2, \ldots, n\}, \text{ and}$$
$$E(K_n) = \{(i, j) \mid i, j \in \{1, \ldots, n\} \text{ and } i \neq j\}.$$

Obviously, K_n has n nodes, $\binom{n}{2}$ edges, diameter 1 and bisection width $\lfloor \frac{n}{2} \rfloor \cdot \lceil \frac{n}{2} \rceil$. It is also obvious that K_n is regular of degree $n - 1$. An illustration of K_4 is shown in Figure 2.17.

Fig. 2.17. The complete graph K_4

In all previous sections we have learnt that good communication structures must have a small diameter (radius) because the diameter provides a lower bound on the complexity of broadcasting, which is the simplest communication task. From this point of view, paths and cycles are not very suitable communication structures. The complete trees seem to be convenient for the fundamental communication tasks considered, because their diameters are logarithmic in the number of nodes. But there are also fundamental communication tasks (routing, for instance) that require an intensive exchange of messages between two disjoint parts of a network. This is the reason why one looks for communication structures with large bisection widths. The bisection width of a complete binary tree is 1 and so the edges incident to the root may become bottlenecks when many messages have to be exchanged between the left subtree and the right subtree. Obviously, the complete graph has a diameter equal to 1 and it has the maximal possible bisection width, i.e., the complete graph is the optimal communication structure. However, from the technological point of view, its creation is not realistic. First of all the cost of any network grows with the number of edges (wires) and the complete graph K_n has $\binom{n}{2} \in \Theta(n^2)$ edges. Secondly, the current technologies do not allow an arbitrary number of edges incident to a node. Usually, the number of wires leaving a node is bounded by a fixed constant independent of the number of nodes.

Summarizing the above considerations, one looks for communication structures with the following properties:

(i) the diameter (radius) is small (preferably logarithmic in the number of nodes[6]),
(ii) the bisection width is large, and
(iii) the degree is constant.

Moreover, so-called modularity would also be very much appreciated. Modularity means that small networks can be used as basic blocks for building large networks of the same kind.

Next, we introduce the hypercubes which belong amongst the most popular structures in parallel computing.

The Hypercube H_m. The *(binary) hypercube* of dimension m, denoted by H_m, is the graph whose nodes are all binary strings of length m and whose edges connect those binary strings which differ in exactly one position. For each i, $1 \leq i \leq m$, an edge

$$(a_1 a_2 \ldots a_{i-1} 0 a_{i+1} \ldots a_m, a_1 a_2 \ldots a_{i-1} 1 a_{i+1} \ldots a_m),$$

$a_1, a_2, \ldots, a_{i-1}, a_{i+1}, \ldots, a_m \in \{0,1\}$, is said to be in **dimension i**.
Formally,

$$V(H_m) = (\Sigma_2)^m, \text{ and}$$
$$E(H_m) = \{(a_1 \ldots a_m, b_1 \ldots b_m) \mid a_i, b_i \in \{0,1\} \text{ for } i = 1, \ldots, m, \text{ and}$$
$$\text{there exists exactly one } j \in \{1, \ldots, m\} \text{ such that } a_j \neq b_j\}.$$

H_m has 2^m nodes, $m \cdot 2^{m-1}$ edges, diameter m, and bisection width 2^{m-1}. It is regular of degree m. An illustration of H_3 is shown in Figure 2.18.

We see that hypercubes satisfy properties (i) and (ii). The following recursive definition of hypercubes shows that the property of modularity is also satisfied.

A 1-dimensional hypercube is an edge with one vertex labeled 0 and the other labeled 1. An $(m + 1)$-dimensional hypercube is constructed from two m-dimensional hypercubes, H_m^0 and H_m^1, by adding exactly one edge from each vertex in H_m^0 to the vertex in H_m^1 that has the same label and then by postfixing all of the labels in H_m^0 with a 0 and all of the labels in H_m^1 with a 1 (Figure 2.19).

Only property (iii) is not completely satisfied. The degree of the hypercube grows with its size, but only logarithmically in the number of nodes. Thus, at least hypercubes with a small dimension can be built. Next, we try to modify the structure of H_m in order to design networks that also satisfy property (iii).

---

[6] Note that a graph of a constant degree cannot have a diameter smaller than logarithmic in its size.

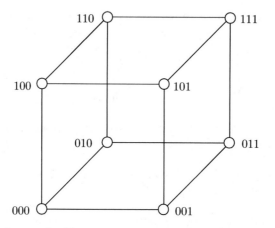

Fig. 2.18. The hypercube H_3

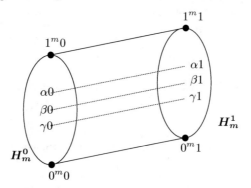

Fig. 2.19. Constructing H_{m+1} from two H_m's

The Cube-Connected-Cycles CCC_m. The *cube-connected-cycles network of dimension m*, denoted by CCC_m, has vertex-set

$$V(CCC_m) = \{0, 1, ..., m-1\} \times \{0, 1\}^m.$$

For each vertex $v = (i, \alpha) \in V(CCC_m)$, $i \in \{0, 1, ..., m-1\}, \alpha \in \{0, 1\}^m$, we call i the **level** and α the **position-within-level (PWL) string** of v. The edges of CCC_m are of two types. For each $i \in \{0, 1, ..., m-1\}$ and each $\alpha = a_0 a_1 ... a_{m-1} \in \{0, 1\}^m$, the vertex (i, α) on level i of CCC_m is connected

- by a **cycle-edge** with vertex $((i+1) \bmod m, \alpha)$ on level $(i+1) \bmod m$, and
- by a **cross-edge** with vertex $(i, \alpha(i))$ on level i.

Here, $\alpha(i) = a_0 ... a_{i-1} \bar{a}_i a_{i+1} ... a_{m-1}$, where $\bar{a}$ denotes the binary complement of a.

For each $\alpha \in \{0, 1\}^m$, the cycle

$$(0, \alpha), (1, \alpha), (2, \alpha), ..., (m-1, \alpha), (0, \alpha)$$

of length m will be denoted by $\boldsymbol{C_\alpha(m)}$ or $\boldsymbol{C_\alpha}$.

CCC_m has $m \cdot 2^m$ nodes, $3m \cdot 2^{m-1}$ edges, diameter $\lfloor 5m/2 \rfloor - 2$ for $m \geq 4$ and bisection width in $\Theta(2^m)$. It is also regular of degree 3. Graphical representations of CCC_3 are shown in Figures 2.20 and 2.21.

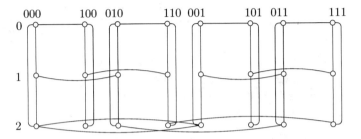

Fig. 2.20. The cube-connected-cycles CCC_3

The graphical representation of CCC_3 in Figure 2.20 shows in a transparent way that the design of cube-connected-cycles is based on the structure of the hypercube. The CCC_m is a modification of the hypercube H_m obtained by replacing each vertex of the hypercube with a cycle of m nodes. The i-th dimension edge incident to a node of the hypercube is then connected to the i-th node of the corresponding cycle of the CCC_m.

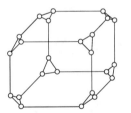

Fig. 2.21. The cube-connected-cycles CCC_3 as derived from H_3

Lemma 2.6.4. *The diameter of the cube-connected-cycles network of dimension m, CCC_m, is $2m - 1 + \max(1, \lfloor \frac{m-2}{2} \rfloor)$.*

Proof. Let $x = (i, x_1 x_2 \ldots x_m)$ and $y = (j, y_1 y_2 \ldots y_m)$ be two vertices of CCC_m. Let $i_1, i_2, \ldots, i_k$ be the coordinates in which x and y differ, in increasing order beginning with i (modulo m). Thus, i_1 is the first element of the sequence $i, i+1, i+2, \ldots, m-1, 0, 1, \ldots, i-1$ in which the coordinates differ, i_2 the second such element and so on.

To traverse a path from x to y means to go from x to the vertex $(i, x_1 x_2 \ldots x_{i_1} \ldots x_m)$, then change cycles via the vertex $(i, x_1 x_2 \ldots \bar{x}_{i_1} \ldots x_m)$ and so on. In this way, we arrive at $z = ((i-1) \bmod m, y_1 y_2 \ldots y_m)$ in at most $2m - 1$ steps. Another path, obtained by traversing the cycles in the opposite direction (i decreasing) takes us in $2m - 1$ steps to the vertex $z' = ((i + 1) \bmod m, y_1 y_2 \ldots y_m)$. From z or z', we can reach all the vertices $(j, y_1 y_2 \ldots y_m)$ with $j = i, i - 2, i + 2$ in one step, and the vertices of the cycle belonging to $y_1 y_2 \ldots y_m$ in at most $\lfloor \frac{m-2}{2} \rfloor$ steps. Thus, we have a path of length at most $2m - 1 + \max(1, \lfloor \frac{m-2}{2} \rfloor)$. This bound is achieved by considering the vertices $(0, 000)$ and $(0, 111)$ for $m = 3$ and the vertices $(0, 0^m)$ and $(\lfloor \frac{m}{2} \rfloor, 1^m)$ for $m \geq 4$.

$\square$

Exercise 2.6.5. Show that the distance in CCC_m is $2m - 1 + \max(1, \lfloor \frac{m-2}{2} \rfloor)$ between the vertices $(0, 000)$ and $(0, 111)$ for $m = 3$ and the vertices $(0, 0^m)$ and $(\lfloor \frac{m}{2} \rfloor, 1^m)$ for $m \geq 4$.

The Butterfly BF_m. The *butterfly network of dimension m*, denoted by BF_m, has vertex-set

$$V(BF_m) = \{0, 1, \ldots, m - 1\} \times \{0, 1\}^m.$$

For each vertex $v = (i, \alpha) \in V(BF_m)$, $i \in \{0, 1, \ldots, m - 1\}, \alpha \in \{0, 1\}^m$, we call i the **level** and α the **position-within-level (PWL) string** of v. The edges of BF_m are of two types. For each $i \in \{0, 1, \ldots, m - 1\}$ and each $\alpha = a_0 a_1 \ldots a_{m-1} \in \{0, 1\}^m$, the vertex (i, α) on level i of BF_m is connected

- by a **cycle-edge** with vertex $((i + 1) \bmod m, \alpha)$ and
- by a **cross-edge** with vertex $((i + 1) \bmod m, \alpha(i))$

on level $(i + 1) \bmod m$. Again, $\alpha(i) = a_0 \ldots a_{i-1} \bar{a}_i a_{i+1} \ldots a_{m-1}$, where $\bar{a}$ denotes the binary complement of a. For each $\alpha \in \{0, 1\}^m$, the cycle

$$(0, \alpha), (1, \alpha), \ldots, (m - 1, \alpha), (0, \alpha)$$

of length m will be denoted by $C_\alpha(m)$ or C_α.

BF_m has $m \cdot 2^m$ nodes, $m \cdot 2^{m+1}$ edges, diameter $\lfloor 3m/2 \rfloor$ and bisection width in $\Theta(2^m)$. It is also regular of degree 4. An illustration of BF_3 is shown in Figure 2.22. To obtain a clearer picture, level 0 has been replicated.

Similarly as for CCC_m, the design of BF_m is also based on the structure of H_m. The replacement of every hypercube node by a cycle of m nodes is the same for BF_m as for CCC_m. The only difference is in taking the cross-edges not between nodes at the same positions in the cycles as CCC_m, but by shifting the second end-point of the cross-edge to its cycle successor. In this way, every node of H_m gets two cross-edges.

Exercise 2.6.6. Show that the diameter of the butterfly network of dimension m, BF_m, is $\lfloor 3m/2 \rfloor$.

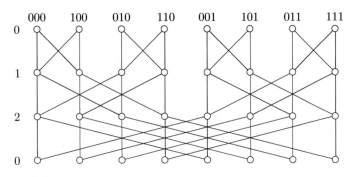

Fig. 2.22. The butterfly graph BF_3

The Shuffle-Exchange SE_m. The *shuffle-exchange network of dimension* m, denoted by SE_m, is the graph whose nodes are the binary strings of length m and whose edges connect each string αa, where α is a binary string of length $m-1$ and a is in $\{0,1\}$, with the string $\alpha\bar{a}$ and with the string $a\alpha$. An edge connecting αa with $\alpha\bar{a}$ is called an **exchange edge** and an edge connecting αa with $a\alpha$ is called a **shuffle edge**.

SE_m has 2^m nodes, diameter $2m-1$, degree 3 and bisection width $\Theta(\frac{2^m}{m})$. A graphical representation of SE_3 is shown in Figure 2.23.

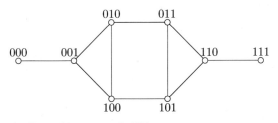

Fig. 2.23. The shuffle-exchange graph SE_3

Exercise 2.6.7. Show that the diameter of the shuffle-exchange network of dimension m, SE_m, is $2m-1$.

The (binary) deBruijn DB_m. The *deBruijn network of dimension* m, denoted by DB_m, is the graph whose nodes are the binary strings of length m and whose edges connect each string $a\alpha$, where $\alpha \in (\Sigma_2)^{m-1}$ and a is in $\{0,1\}$, with the strings αb, where b is a symbol in $\{0,1\}$. An edge connecting $a\alpha$ with αb, $a \neq b$, is called a **shuffle-exchange edge** and an edge connecting $a\alpha$ with αa is called a **shuffle edge**.

Formally,

$$V(DB_m) = \{0,1\}^m, \text{ and}$$
$$E(DB_m) = \{\{a\alpha, \alpha b\} \mid \alpha \in \{0,1\}^{m-1}, a, b \in \{0,1\}, a\alpha \neq \alpha b\}.$$

DB_m has 2^m nodes, diameter m, degree 4 and bisection width in $\Theta(\frac{2^m}{m})$. An illustration of DB_3 is shown in Figure 2.24.

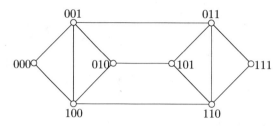

Fig. 2.24. The deBruijn graph DB_3

Exercise 2.6.8. Show that the diameter of the deBruijn network of dimension m, DB_m, is m.

The above introduced networks satisfy all properties (i), (ii) and (iii), but they are not modular. Only butterfly networks can be viewed as networks that partially fulfill the modularity property. The grid networks introduced in what follows are modular and easy to produce. But we pay for their simple structure by increasing the radius and decreasing the bisection width.

The Grid $Gr(a_1, a_2, \ldots, a_d)$. The *d-dimensional grid* of dimensions $a_1, a_2, \ldots, a_d$, denoted by $Gr(a_1, a_2, \ldots, a_d)$, is the graph whose nodes are *d*-tuples of positive integers $(z_1, z_2, \ldots, z_d)$, where $1 \leq z_i \leq a_i$, for all i, $1 \leq i \leq d$, and whose edges connect *d*-tuples which differ in exactly one dimension (coordinate) by one.

Formally,

$$V(Gr(a_1, \ldots, a_d)) = \{(z_1, \ldots, z_d\} \mid 1 \leq z_i \leq a_i \text{ for } i = 1, \ldots, d\}, \text{ and}$$
$$E(Gr(a_1, \ldots, a_d)) = \{\{(x_1, \ldots, x_d), (y_1, \ldots, y_d)\} \mid \exists j \in \{1, \ldots, d\} \text{ such that}$$
$$|x_j - y_j| = 1 \text{ and } x_i = y_i \text{ for all } i \in \{1, \ldots, d\} - \{j\}\}.$$

$Gr(a_1, a_2, \ldots, a_d)$ has $a_1 a_2 \ldots a_d$ nodes, diameter $a_1 + a_2 + \ldots + a_d - d$ and degree $2d$, if each a_i is at least three. An illustration of $Gr(3, 4)$ is shown in Figure 2.25.

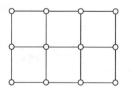

Fig. 2.25. The grid $Gr(3, 4)$

3

Broadcasting

*Not he who has reached an old age
has lived longer,
but he who has been feeling life more
inwardly.*

Jean-Jacques Rousseau

3.1 Introduction

Broadcasting from a node of a network seems to be the simplest communication task, and so one could assume that the design of optimal broadcast algorithms for fundamental networks is almost a routine matter. Unfortunately, the opposite is true for some fundamental networks for which one is unable to estimate the complexity of broadcasting. The hardness is not only in the communication algorithm design, but especially in proving lower bounds on the complexity of broadcasting, i.e., in proving the non-existence of broadcast algorithms with a bounded number of rounds. This is the typical problem with the combinatorial explosion. It causes that, already for small networks, the number of possible different communication strategies is so large that it is impossible to have a look at all of them in order to find the best one. Thus, one has to develop some more involved techniques than the total search in order to prove good lower bounds on the complexity of communication tasks. Because of this, in this chapter we focus on the communication algorithm as well as the combinatorial methods for proving lower bounds on the broadcast complexity.

This chapter is organized as follows. Section 3.2 presents some basic facts and elementary techniques for estimating the broadcast complexity. As an illustration of these techniques, we give an optimal broadcast algorithm for complete trees and some related networks. The analysis of the broadcast complexity for complete trees is not hard due to their regular structure and small number of edges (non-existence of cycles). The main aim of this presentation is to create a starting point for the development of more elaborated techniques used later. Section 3.3 is devoted to the design of broadcast algorithms for the hypercubes and hypercube-like networks such as cube-connected-cycles, shuffle-exchange, deBruijn and butterfly networks. The designed algorithms are optimal for hypercubes, cube-connected-cycles and shuffle-exchange networks because the complexity of the broadcast algorithms for them meets the trivial lower bound given by the diameter of these networks. For deBruijn networks and butterfly networks, we are far from that trivial lower bound and so we present in Section 3.4 a general technique for proving lower bounds on the broadcast complexity of graphs of bounded degree. Using this technique, one can improve the trivial lower bounds on the broadcast complexity of deBruijn networks and butterfly networks. The presented lower bound technique is robust, which means the positive fact that it can be successfully applied for many networks. But the negative side of this robustness is that it is often not fine enough to be able to prove an optimal lower bound for some specific communication structures. In Section 3.5, we give a short survey of the main known results and open problems related to broadcasting in common interconnection networks. Finally, in Section 3.6, we introduce the concept of approximation algorithms for the broadcast problem. Here, instead of designing individual broadcast schemes for specific networks, one is interested in constructing algorithms that, given a graph G and a node v in G, find/compute a good broadcast strategy.

3.2 Basic Facts and Techniques

As shown in Section 2.5, while investigating the complexity of broadcasting, we only have to consider the one-way communication mode.

Before going into more sophisticated results, let us start off with a simple general lower bound for the broadcast problem:

Observation 3.2.1. *Let G be a graph with n nodes, $n \in \mathbb{N}$. Then*

$$b(G) \geq \mathrm{minb}(G) \geq \lceil \log_2 n \rceil.$$

Proof. The fact $b(G) \geq \mathrm{minb}(G)$ is clear. To prove $\mathrm{minb}(G) \geq \lceil \log_2 n \rceil$, let $A(t)$ denote the maximum number of nodes which can know the message after t rounds. As the number of informed nodes can at most double during each time unit, we have the following recursive relation for $A(t)$:

$$A(0) = 1,$$
$$A(t+1) \leq 2 \cdot A(t) \quad \text{for all } t \geq 0.$$

It is easy to verify that the closed formula for $A(t)$ with $A(t+1) = 2 \cdot A(t)$ is

$$A(t) = 2^t.$$

Therefore, at most 2^t nodes are informed after t rounds. To inform all n nodes, the relation

$$2^t \geq n$$

must hold, hence $t \geq \lceil \log_2 n \rceil$. □

Now, the first question arising is whether there are graphs of n nodes satisfying the property $b(G) = \lceil \log_2 n \rceil$. Such graphs are called **minimal broadcast graphs**. We show that the complete graph and any graph having the hypercube as a spanning subgraph have this property. The crucial point is that in the complete graph and in the hypercube it is possible to double the number of informed nodes in each round.

Lemma 3.2.2. *For all positive integers n, m:*

    a) $\mathrm{minb}(K_n) = b(K_n) = \lceil \log_2 n \rceil$,

    b) $\mathrm{minb}(H_m) = b(H_m) = m$.

Proof. Observation 3.2.1 provides $\mathrm{minb}(K_n) \geq \lceil \log_2 n \rceil$ and $\mathrm{minb}(H_m) \geq m$. It remains to prove $b(K_n) \leq \lceil \log_2 n \rceil$ and $b(H_m) \leq m$. We do it separately.

a) Number the nodes of K_n from 0 to $n - 1$. Without loss of generality, let the originating node be 0. The following algorithm has the property that it doubles the number of informed nodes in each round.

    Algorithm BROADCAST-K_n

        <u>for</u> $t = 1$ <u>to</u> $\lceil \log_2 n \rceil$ <u>do</u>
            <u>for</u> all $i \in \{0, \ldots, 2^{t-1} - 1\}$ <u>do</u> in <u>parallel</u>
                <u>if</u> $i + 2^{t-1} \leq n$ <u>then</u>
                    i sends to $i + 2^{t-1}$;

It is easy to verify by induction on t that after t rounds of the algorithm, the nodes $0, 1, \ldots, \min\{2^t - 1, n - 1\}$ have been informed. Therefore, after $\lceil \log_2 n \rceil$ rounds, all nodes have received the information.

b) The algorithm for the hypercube H_m is exactly the same as for the complete graph K_n, where $n = 2^m$. Using the binary representation of the nodes, without loss of generality, the originating node is $00 \ldots 0$, and algorithm BROADCAST-K_n directly translates into

    Algorithm BROADCAST-H_m

        <u>for</u> $i = 1$ <u>to</u> m <u>do</u>
            <u>for</u> all $a_1, a_2, \ldots, a_{i-1} \in \{0, 1\}$ <u>do</u> in <u>parallel</u>
                $a_1 a_2 \ldots a_{i-1} 00 \ldots 0$ sends to $a_1 a_2 \ldots a_{i-1} 10 \ldots 0$;

In other words, in round i, each informed vertex sends the message in dimension i $(1 \leq i \leq m)$. From part a), we know that after m rounds all the nodes have received the information.

                                                           □

We note that there is no bounded-degree interconnection network G of n nodes having the property $b(G) = \lceil \log_2 n \rceil$, because a fast increase in the number of informed nodes is only possible if there are vertices in G of large degree. A detailed analysis of this situation is presented in Section 3.4.

Exercise 3.2.3. Let m be a positive integer. Design a graph G_m with $\mathrm{minb}(G_m) = m$, $|V(G_m)| = 2^m$, and $|E(G_m)|$ as small as possible.

Exercise 3.2.4. The same as in Exercise 3.2.3, but additionally $\mathrm{minb}(G_m) = \mathrm{diam}(G_m) = m$.

Another interesting problem is to find graphs having the property $\mathrm{b}(G) = \lceil \log_2 n \rceil$ and as few edges as possible. These graphs are called **minimum broadcast graphs**. This question has been investigated in several papers (for an overview see, for example, [HHL88, FL94]). Due to the technicality of its solution, we do not present it here.

From Chapter 2 (Observation 2.5.13) we know that

$$\mathrm{diam}(G) \leq \mathrm{b}(G) \quad \text{for all graphs } G,$$

i.e., the diameter of the graph is a trivial lower bound on the broadcast time. As $\mathrm{diam}(H_m) = m$, Lemma 3.2.2 shows that the hypercube H_m is another example of a graph G satisfying $\mathrm{diam}(G) = \mathrm{b}(G)$. In Section 3.3 we will find other graphs which are optimal (or near optimal) in this sense.

The diameter lower bound can be slightly improved in many cases as follows:

Observation 3.2.5. *Let G be a graph of diameter D. If there exist three different vertices u, v_1 and v_2 with both v_1 and v_2 at distance D from u, then*

$$\mathrm{b}(G) \geq D + 1.$$

Proof. Let S be a broadcasting scheme for G and v. By induction on i, we can see that in round i of the scheme, at most one vertex at distance i from the originator v can be informed. Therefore, to inform two nodes v_1 and v_2 at distance D, at least $D + 1$ rounds are needed.

$\square$

This observation will turn out to be quite useful for broadcasting in the cube-connected-cycles network in Section 3.3. A generalization of the idea contained in the proof of Observation 3.2.5 will lead to more powerful lower bounds on the broadcast time in Section 3.4.

Exercise 3.2.6. Construct a graph G_m with the following properties:

(i) $\mathrm{rad}(G_m) = m$,
(ii) there exists a vertex $v \in V(G_m)$ that has exactly 2^m vertices at distance m, and
(iii) $\mathrm{minb}(G_m) = 2m$.

Exercise 3.2.7. Construct a graph G_m with the properties (i), (ii) of Exercise 3.2.6, and

(iii') $\mathrm{minb}(G_m) = m - 1 + 2^m$.

We conclude this section by presenting two elementary but very instructive examples for broadcasting in certain types of networks. First, let us recall from Chapter 2 that any broadcast algorithm of a graph G and a node v determines a spanning tree of G rooted at v. This tree is called a **broadcast tree** of G and v. It turns out that this description of broadcast algorithms is quite useful for proving lower bounds on the broadcast time. This is demonstrated in the following instructive example determining the min-broadcast time of the k-ary tree:

Lemma 3.2.8. *For all integers $k \geq 2$, $m \geq 2$:*

$$\mathrm{minb}(T_k{}^m) = k \cdot m.$$

Proof. Let v_0 be the root of $T_k{}^m$. We show that

1. $\mathrm{b}(v_0, T_k{}^m) = k \cdot m$, and
2. $\mathrm{b}(v_0, T_k{}^m) \leq \mathrm{b}(v, T_k{}^m)$ for all $v \in V(T_k{}^m)$.

Statements 1. and 2. together imply the validity of the lemma. Now, we prove these statements separately.

1. The proof of $\mathrm{b}(v_0, T_k{}^m) = k \cdot m$.

First, we show that it is possible to broadcast from v_0 in $k \cdot m$ rounds. The algorithm works as follows:

Algorithm MINBROADCAST-$T_k{}^m$

    1. The root v_0 learns the message at time 0.
    2. Each non-leaf node receiving the message at time t sends it
       on to its k sons in the next k rounds.

We see that after $k \cdot i$ rounds ($1 \leq i \leq m$), each node at distance at most i from the root has received the information. Hence, after $k \cdot m$ rounds, each node in the tree has received the information.

Now, we show that any broadcast from v_0 takes at least $k \cdot m$ rounds. Let A be an arbitrary broadcast algorithm for $T_k{}^m$ and v_0. Without loss of generality, we may assume that every internal node broadcasts $I(v_0)$ to its k sons immediately in the next k rounds[1] after learning $I(v_0)$. Let T be a broadcast tree of $T_k{}^m$ determined by A. Label the edges of T as follows: Let v be any non-leaf vertex of T and let $v_1, \ldots, v_k$ be the sons of v. Suppose that the vertex v receives the message at time t and the vertex v_i receives the message at time $t + i$ from v ($i = 1, \ldots, k$). Then we label the edge connecting v with v_i by i (Figure 3.1) for $i = 1, \ldots, k$.

---

[1] If A does not have this property for some internal node v, then we can change A to an algorithm B that has this property. Since there is exactly one path in $T_k{}^m$ from v_0 to any other vertex, no conflict can be caused in this way. Obviously, the number of rounds cannot increase by this change.

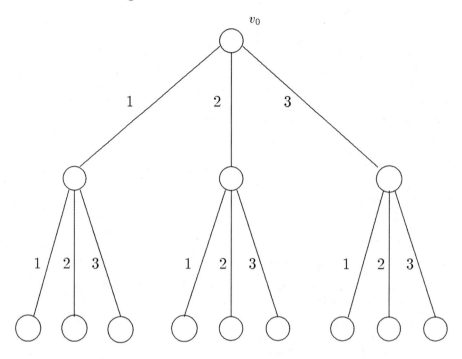

Fig. 3.1. Labeling of the edges of T

As $T_k{}^m$ is a complete tree, it is clear that there is a path of length m in T from the root v_0 to some leaf w which is only labeled with k's, and w is informed via this path. Hence, w receives the message at time $k \cdot m$.

2. The proof of $b(v_0, T_k{}^m) \leq b(v, T_k{}^m)$ for all $v \in V(T_k{}^m)$.

Denote the k subtrees rooted at the k sons of v_0 with $T_1, \ldots, T_k$ as in Figure 3.2.

Without loss of generality, one may assume that the originator v of the message is a node of the subtree T_1. To inform the subtrees $T_2, \ldots, T_k$, the message has to pass through the root v_0. Once v_0 has learnt the message, it has to broadcast it in the whole tree $T_k{}^m$ except for the subtree T_1. This subtree $\tilde{T}$ is depicted in Figure 3.3.

Using exactly the same arguments as for the lower bound in part 1., it can be seen that broadcasting in $\tilde{T}$ takes at least $k \cdot m - 1$ rounds. As at least one round is needed to inform v_0, we have

$$b(v, T_k{}^m) \geq k \cdot m = b(v_0, T_k{}^m).$$

$\square$

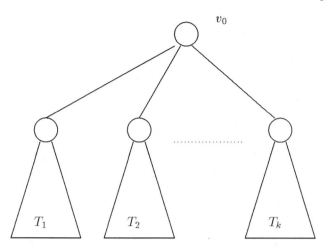

Fig. 3.2. Numbering of the subtrees of v_0

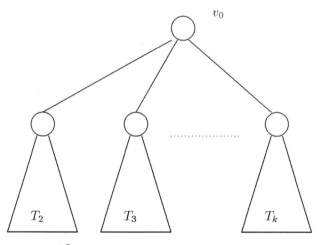

Fig. 3.3. The subtree $\tilde{T}$

Exercise 3.2.9. Consider a subclass of broadcast algorithms that must use only shortest paths from the source to all destinations. Find a graph U_m of m vertices such that

(i) $\mathrm{b}(U_m) \leq \lceil \log_2 m \rceil + 1$, and
(ii) every shortest-paths broadcast algorithm needs at least $m - 1$ rounds.

Exercise 3.2.10. Prove that, for all integers $k \geq 2$, $m \geq 2$,

$$\mathrm{b}(T_k{}^m) = (k + 1) \cdot m - 1 .$$

More sophisticated lower bound techniques will be presented in Section 3.4. There we look at graphs of bounded degree. Simple upper bounds on the broadcast time of these graphs were already established in Lemma 2.5.17:

Let G be a graph of degree d. Then

$$a) \quad \text{minb}(G) \le (d-1) \cdot \text{rad}(G) + 1,$$
$$b) \quad \text{b}(G) \le (d-1) \cdot \text{diam}(G) + 1,$$
$$c) \quad \text{b}(G) \le d \cdot \text{rad}(G).$$

We will see in Section 3.3 that the above-stated upper bounds for broadcasting in bounded-degree graphs are not very sharp in general. On the other hand, there are cases in which this simple algorithm already yields the best possible result.

Example 3.2.11. Let $\tilde{T}_k{}^m$ denote the k-ary tree consisting of a root v with k complete $(k-1)$-ary trees of depth $m-1$ as its sons (Figure 3.4).

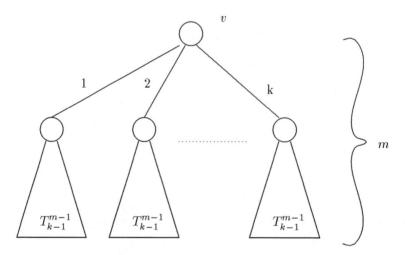

Fig. 3.4. The tree $\tilde{T}_k{}^m$

With the same techniques as used in the proof of Lemma 3.2.8 it can be shown that

$$\text{minb}(\tilde{T}_k{}^m) = (k-1) \cdot m + 1 = (\deg(\tilde{T}_k{}^m) - 1) \cdot \text{rad}(\tilde{T}_k{}^m) + 1 . \qquad (3.1)$$

$$\text{b}(\tilde{T}_k{}^m) = k \cdot m = \deg(\tilde{T}_k{}^m) \cdot \text{rad}(\tilde{T}_k{}^m) . \qquad (3.2)$$

Exercise 3.2.12. Prove equalities (3.1) and (3.2) for all positive integers $k, m \ge 2$.

3.3 Upper Bounds for Common Networks

In this section, we present upper bounds for broadcasting in common networks, namely cube-connected-cycles, butterfly, shuffle-exchange and deBruijn networks. A summary of results for these networks can be found in [FL94]. The simple upper bounds for bounded-degree graphs presented in Lemma 2.5.17 yield

1. $b(CCC_k) \leq 5k + O(1)$,
2. $b(BF_k) \leq 4.5k + O(1)$,
3. $b(SE_k) \leq 4k + O(1)$,
4. $b(DB_k) \leq 3k + O(1)$.

Exercise 3.3.1. Prove the above upper bounds by applying Lemma 2.5.17.

We will see that we can do a lot better than that for all these networks. We start by looking at the cube-connected-cycles network:

Theorem 3.3.2 ([LP88]). *For any integer $k \geq 4$:*

$$\left\lceil \frac{5k}{2} \right\rceil - 2 \leq \mathrm{minb}(CCC_k) = b(CCC_k) \leq \left\lceil \frac{5k}{2} \right\rceil - 1.$$

Proof.

(i) First, we prove that

$$\mathrm{minb}(CCC_k) \geq \left\lceil \frac{5k}{2} \right\rceil - 2.$$

To verify this, note that CCC_k has diameter $\lfloor 5k/2 \rfloor - 2$. If k is even, this implies that

$$\mathrm{minb}(CCC_k) \geq \left\lfloor \frac{5k}{2} \right\rfloor - 2 = \left\lceil \frac{5k}{2} \right\rceil - 2.$$

Let k be odd. Without loss of generality, the message originates at vertex $u = \langle 0, 00...0 \rangle$. There exist two nodes (namely $v_1 = \langle \lfloor k/2 \rfloor, 11...1 \rangle$ and $v_2 = \langle \lfloor k/2 \rfloor + 1, 11...1 \rangle$) at distance $\lfloor 5k/2 \rfloor - 2$ from u. From Observation 3.2.5, we obtain

$$b(CCC_k) \geq \mathrm{diam}(CCC_k) + 1 = \left(\left\lfloor \frac{5k}{2} \right\rfloor - 2 \right) + 1 = \left\lceil \frac{5k}{2} \right\rceil - 2.$$

As CCC_k is vertex-symmetric (i.e., for every node v in CCC_k, one can rename the nodes of CCC_k in such a way that v becomes $\langle 0, 00...0 \rangle$), it follows that $\mathrm{minb}(CCC_k) = b(CCC_k)$, hence

$$\mathrm{minb}(CCC_k) = b(CCC_k) \geq \left\lceil \frac{5k}{2} \right\rceil - 2.$$

(ii) Now, we present an algorithm which broadcasts in time $\lceil 5k/2 \rceil - 1$ from $v_0 = \langle 0, 00...0 \rangle$:

The algorithm consists of 2 phases. In the first phase, the message is sent "down" in the cycles C_α, alternating between cross-edges and straight-edges, thus distributing the information to all the cycles C_α. In the second phase, the message is distributed on the cycles C_α by using a broadcast algorithm for the cycle C_k. The first phase of the algorithm is illustrated for CCC_3 in Figure 3.5. (The label on an edge e denotes the round in which e is active.)

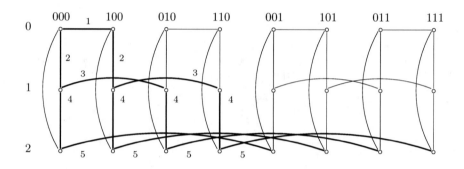

Fig. 3.5. Broadcasting in CCC_3

We now give a formal description of the algorithm.

Algorithm BROADCAST-CCC_k

1. $\langle 0, 00...0 \rangle$ sends to $\langle 0, 10...0 \rangle$;
   <u>for</u> $i = 1$ <u>to</u> $k - 1$ <u>do</u>
      <u>begin</u>
         <u>for</u> all $a_0, \ldots, a_{i-1} \in \{0,1\}$ <u>do in parallel</u>
            $\langle i - 1, a_0 \ldots a_{i-1} 00 \ldots 0 \rangle$ sends to $\langle i, a_0 \ldots a_{i-1} 00 \ldots 0 \rangle$;
         <u>for</u> all $a_0, \ldots, a_{i-1} \in \{0,1\}$ <u>do in parallel</u>
            $\langle i, a_0 \ldots a_{i-1} 00 \ldots 0 \rangle$ sends to $\langle i, a_0 \ldots a_{i-1} 10 \ldots 0 \rangle$;
      <u>end</u>;
2. <u>for</u> all $\alpha \in \{0,1\}^k$ <u>do in parallel</u>
      broadcast on the cycle $C_\alpha(k)$ from $\langle k - 1, \alpha \rangle$;

First, observe that Algorithm BROADCAST-CCC_k is a consistent algorithm in the one-way communication mode. Now, we prove by induction on i that after $2i - 1$ rounds of Phase 1 ($1 \leq i \leq k$), the nodes $\langle i - 1, a_0 \ldots a_{i-1} 00 \ldots 0 \rangle$, $a_0, \ldots, a_{i-1} \in \{0,1\}$ have received the information: In round 1, $\langle 0, 00...0 \rangle$ sends to $\langle 0, 10...0 \rangle$. Hence, after round 1, the nodes $\langle 0, 00 \ldots 0 \rangle$ and $\langle 0, 10 \ldots 0 \rangle$ have received the information. Assume that after $2i - 1$ rounds ($1 \leq i \leq k - 1$), the nodes

$\langle i-1, a_0 \ldots a_{i-1}00 \ldots 0 \rangle, a_0, \ldots, a_{i-1} \in \{0,1\}$ have received the information. In round $2i$, $\langle i-1, a_0 \ldots a_{i-1}00 \ldots 0 \rangle$ sends to $\langle i, a_0 \ldots a_{i-1}00 \ldots 0 \rangle$, and in round $2i+1$ $\langle i, a_0 \ldots a_{i-1}00 \ldots 0 \rangle$ sends to $\langle i, a_0 \ldots a_{i-1}10 \ldots 0 \rangle$. Hence, after $2i+1$ rounds, the nodes $\langle i, a_0 \ldots a_i 00 \ldots 0 \rangle, a_0, \ldots, a_i \in \{0,1\}$ have received the information.

Hence, we have proved that after $2k-1$ rounds, i.e., after Phase 1, all the nodes $\langle k-1, \alpha \rangle$, $\alpha \in \{0,1\}^k$ have received the message. In Phase 2, broadcasting on the cycles $C_\alpha(k)$, $\alpha \in \{0,1\}^k$ can be done in $\lceil k/2 \rceil$ rounds (see Example 2.5.9). Thus, overall the algorithm takes $\lceil 5k/2 \rceil - 1$ rounds. Finally, we observe that CCC_k is symmetric in the sense that for every node v of CCC_k, one can rename the nodes of CCC_k in such a way that v becomes $\langle 0, 00 \ldots 0 \rangle$. Thus, the complexity of broadcasting in CCC_k does not depend on the choice of the originating vertex.

$\square$

Note that the upper bound in Theorem 3.3.2 can be improved[2] to $\lceil \frac{5k}{2} \rceil - 2$. We omit the technicalities of the corresponding proof.

Theorem 3.3.3. *For any integer $k \geq 4$:*

$$\mathrm{minb}(CCC_k) = \mathrm{b}(CCC_k) = \left\lceil \frac{5k}{2} \right\rceil - 2.$$

Exercise 3.3.4.* Prove Theorem 3.3.3. [Hint: Design an algorithm which, like Algorithm BROADCAST-CCC_k, first distributes the message to all the cycles C_α (Phase 1), and then distributes the message on the cycles C_α (Phase 2). Unlike Algorithm BROADCAST-CCC_k, in Phase 1, the message now is distributed "downwards" as well as "upwards" around the cycles C_α. The difficulty is to overlay these two processes properly.]

Next, we investigate the shuffle-exchange network. In [HJM93] it is proved that $2k-1 \leq \mathrm{b}(SE_k) \leq 2k$. But changing the order of the shuffle and the exchange round, we obtain a broadcast algorithm of $2k-1$ rounds here. This improvement was pointed out in [DS93].

Theorem 3.3.5 ([HJM93]). *For any integer $k \geq 2$:*

$$\mathrm{minb}(SE_k) = \mathrm{b}(SE_k) = 2k - 1.$$

Proof. The lower bound comes from the fact that SE_k has diameter $2k-1$.

For the upper bound, the idea of the algorithm is to distribute the information by using alternatingly the exchange and the shuffle edges. The difficulty is to modify this general strategy such that it yields a valid communication algorithm in the one-way mode (i.e., each node is only active via one of its edges in one round), and to show that even with this modification (which

---

[2] Because no paper presents this improvement as its own achievement, this result is simply known as part of the folklore.

leaves some nodes not forwarding the information in some rounds) one can still inform all nodes in $2k - 1$ rounds. A further complication is that SE_k is not vertex-symmetric, hence we have to design a broadcast strategy working from an arbitrary source node in SE_k.

Now, we give a formal description of the algorithm. Let, for each word $w = a_1 a_2 \ldots a_k \in \{0,1\}^k$,

$$w_1 = a_1 \text{ and } w(t) = a_t a_{t+1} \ldots a_k$$

for $2 \leq t \leq k + 1$ ($w(k + 1) = \varepsilon$). Now, we shall describe the broadcast algorithm for an arbitrary source node $\alpha = a_1 a_2 \ldots a_k$ in SE_k.

Algorithm BROADCAST-SE_k

> $\alpha = a_1 a_2 \ldots a_{k-1} a_k$ sends to $a_1 a_2 \ldots a_{k-1} \bar{a}_k$ (exchange round);
> for $t = 1$ to $k - 1$ do
>   for all $\beta \in \{0,1\}^t$ do in parallel
>   begin if $\alpha(t) \notin \{\beta_1\}^+$
>   then $\alpha(t)\beta$ sends to $\alpha(t + 1)\beta a_t$ (shuffle round);
> $\alpha(t + 1)\beta a_t$ sends to $\alpha(t + 1)\beta \bar{a}_t$ (exchange round)
>   end;

For correctness, we need to prove the following two facts:

(1) There is no conflict in any of the $2k - 1$ rounds, i.e., the algorithm BROADCAST-SE_k works in the one-way mode (if a node is active in a round then it is active only via one edge in one direction).

(2) After $2r + 1$ rounds (r executions of the loop) all nodes $\alpha(r + 2)\beta, \beta \in \{0,1\}^{r+1}$, have learned the piece of information of α.

We prove these facts separately:

(1) There is no conflict in any exchange round because each sender has the last bit a_t and each receiver has the last bit $\bar{a}_t$. Let there be a conflict in a shuffle round, i.e., $\alpha(t)\beta = \alpha(t+1)\gamma a_t$ for some $\beta, \gamma \in \{0,1\}^t$. This implies $a_t \alpha(t + 1) = \alpha(t + 1)\gamma_1 \Rightarrow a_t = a_{t+1} = \cdots = a_k = \gamma_1 \Rightarrow \alpha(t) \in \{\gamma_1\}^+$. But this is a contradiction because we do not use the shuffle operation for $\alpha(t) \in \{\gamma_1\}^+$.

(2) We prove it by induction according to $r = t - 1$. One has to show that the nodes $\alpha(r + 2)\beta a_{r+1}$ with $\alpha(r + 1) \in \{\beta_1\}^+$ (which do not receive the information in the r-th execution of the loop) have got the information already in one of the previous rounds.

Clearly, our induction hypothesis [that all nodes $\alpha(r + 2)\beta$, for each $\beta \in \{0,1\}^{r+1}$ have learned the piece of information of α after r executions of the loop] is fulfilled before the first execution of the loop (i.e., when $r = 0$).

Now, let us consider the situation after r executions of the loop, $r \geq 1$. Clearly, if $\alpha(r+1) \notin \{\beta_1\}^+$, $\beta \in \{0,1\}^{r+1}$, then all $\alpha(r+2)\beta$ have learned the piece of information of α in the last execution of the loop according to the algorithm and the induction hypothesis.

If $\alpha(r+1) \in \{\beta_1\}^+$, $\beta \in \{0,1\}^{r+1}$, then $\alpha(r+2)\beta a_{r+1} = \alpha(r+1)\beta_1\beta a_{r+1}$ which already knows the piece of information of α according to the induction hypothesis.

□

The two previous results show that the upper bound algorithm for CCC_k and SE_k match the diameter lower bound. It turns out that this is not true for BF_k and DB_k.

Let us consider the butterfly network BF_k first. As the search for the best upper bound is still going on, we present an instructive yet very efficient algorithm by E. Stöhr [St91a] which needs $2k$ rounds for BF_k. This bound has been improved to $2k - 1$ by Klasing, Peine, Monien and Stöhr [KMPS92]. Refinements of these techniques [St92] show that an upper bound of $2k - \frac{1}{2}\log\log k + c$, for some constant c and all sufficiently large k, is also possible. But, for the sake of transparency, we have chosen to present the upper bound of $2k$. As for the lower bound, we first state the diameter lower bound. In Section 3.4, we will derive a non-trivial lower bound for broadcasting in BF_k.

Theorem 3.3.6 ([St91a]). *For any integer $m \geq 2$:*

$$\left\lfloor \frac{3m}{2} \right\rfloor \leq \mathrm{minb}(BF_m) = \mathrm{b}(BF_m) \leq 2m.$$

Proof. The lower bound comes from the fact that BF_m has diameter $\lfloor 3m/2 \rfloor$.

For the upper bound, first note that BF_m contains two isomorphic subgraphs F_0 and F_1. The subgraph F_0 has the vertex set

$$\{\langle l\,;\,\alpha0 \rangle \mid 0 \leq l \leq m-1,\ \alpha \in \{0,1\}^{m-1}\},$$

and the subgraph F_1 has the vertex set

$$\{\langle l\,;\,\alpha1 \rangle \mid 0 \leq l \leq m-1,\ \alpha \in \{0,1\}^{m-1}\}.$$

Obviously, the vertex sets of F_0 and F_1 do not intersect.

Then note that BF_m contains 2^m node-disjoint cycles $C_{\alpha i}$ of the length m, $\alpha \in \{0,1\}^{m-1}$, $i \in \{0,1\}$, of the form

$$(\langle 0\,;\,\alpha i \rangle, \langle 1\,;\,\alpha i \rangle, \ldots, \langle m-1\,;\,\alpha i \rangle, \langle 0\,;\,\alpha i \rangle).$$

Let $\alpha \in \{0,1\}^{m-1}$ be any string of length $m-1$. By $\sharp_1(\alpha)$ we denote the number of 1's in α and by $\sharp_0(\alpha)$ we denote the number of 0's in α. So from the definition we have $\sharp_1(\alpha) + \sharp_0(\alpha) = m-1$.

Consider the node $v_0 = \langle 0; 0 \ldots 0 \rangle$ of F_0. For every node $w_0 = \langle m-1; \alpha0 \rangle$ of F_0, $\alpha \in \{0,1\}^{m-1}$, there is a path in F_0 of length $m-1$ connecting v_0 and w_0. This path can easily be constructed as follows: the path traverses the straight edge between level i and level $i+1$ for every bit position i in which

α has a 0, and it traverses the cross-edge between level i and level $i+1$ for every bit position i in which α has a 1, $0 \le i \le m-2$.

Now consider $v_1 = \langle\, m-1\,;\, 0\ldots 01\,\rangle$, the level $m-1$ node of F_1. Similarly, there is a path in F_1 of length $m-1$ connecting v_1 with any level-0 node $w_1 = \langle\, 0\,;\, \alpha 1\,\rangle$.

Since the butterfly network is vertex symmetric (i.e., for every node v in BF_m, one can rename the nodes of BF_m in such a way that v becomes $\langle\, 0\,;\, 0\ldots 0\,\rangle$), we can assume that the message originates at vertex $v_0 = \langle\, 0\,;\, 0\ldots 0\,\rangle$, and the originator learns the message at time 0.

In the first step the node v_0 informs the neighbor $v_1 = \langle\, m-1\,;\, 0\ldots 01\,\rangle$. Now in F_0 as well as in F_1 one node is informed. Then broadcasting in F_0 and F_1 will be done in the following two phases:

Phase 1: In each cycle $C_{\alpha 0}$, $\alpha \in \{0,1\}^{m-1}$, we inform the node $w_0 = \langle\, m-1\,;\, \alpha 0\,\rangle$ and in each cycle $C_{\alpha 1}$, $\alpha \in \{0,1\}^{m-1}$, we inform the node $w_1 = \langle\, 0\,;\, \alpha 1\,\rangle$ in at most $\lfloor 3m/2 \rfloor$ rounds.

The broadcasting scheme we use is a little different in F_0 and F_1. In F_0 we prefer the straight edges. This means that any node $\langle\, l\,;\, \alpha 0\,\rangle$, $0 \le l \le m-2$, $\alpha \in \{0,1\}^{m-1}$, of F_0 that receives the message at time t informs its neighbor $\langle\, l+1\,;\, \alpha 0\,\rangle$ at time $t+1$ and its neighbor $\langle\, l+1\,;\, \alpha(l)0\,\rangle$ at time $t+2$.

In F_1 we prefer the cross-edges: from any node $\langle\, l\,;\, \alpha 1\,\rangle$, $1 \le l \le m-1$, $\alpha \in \{0,1\}^{m-1}$, of F_1 that receives the message at time t, the neighbor $\langle\, l-1\,;\, \alpha(l-1)1\,\rangle$ receives the message at time $t+1$ and the neighbor $\langle\, l-1\,;\, \alpha 1\,\rangle$ receives the message at time $t+2$.

Consider now any node $w_0 = \langle\, m-1\,;\, \alpha 0\,\rangle$, $\alpha \in \{0,1\}^{m-1}$, in F_0. The node gets the information from v_0 by broadcasting along the path in F_0 described above. This path traverses $\sharp_1(\alpha)$ cross-edges and $\sharp_0(\alpha)$ straight edges. Since in F_0 the straight edges are preferred, w_0 is informed at time $1 + 2\,\sharp_1(\alpha) + \sharp_0(\alpha) = m + \sharp_1(\alpha)$. (You have to add 1 since in the first round the node v_0 informs v_1.)

Similarly, by using the path in F_1 described above, for all $\alpha \in \{0,1\}^{m-1}$ the node $w_1 = \langle\, 0\,;\, \alpha 1\,\rangle$ is informed at time $1 + \sharp_1(\alpha) + 2\,\sharp_0(\alpha) = m + \sharp_0(\alpha)$.

Obviously for some $\alpha \in \{0,1\}^{m-1}$ the node w_0 or w_1 is informed in more than $\lfloor \frac{3m}{2} \rfloor$ rounds. For example, for $m = 3$ the node $\langle 0; 001 \rangle$ is informed in F_1 in round $1 + \sharp_1(00) + 2\,\sharp_0(00) = 5 > 4 = \lfloor \frac{3m}{2} \rfloor$.

In these cases we inform the nodes by using the cross-edges from level $m-1$ in F_0 to level 0 in F_1 or vice versa. In the example $m = 3$ we can inform the node $\langle 0; 001 \rangle$ by using the cross-edge from $\langle 3; 000 \rangle$ to $\langle 0; 001 \rangle$. Since $\langle 3; 000 \rangle$ is informed in round $1 + \sharp_0(00) + 2\,\sharp_1(00) = 3$, the node $\langle 0; 001 \rangle$ is informed in round $4 = \lfloor \frac{3m}{2} \rfloor$.

In general, we consider the following cases:

Case 1: m is odd.

    Case 1.1 : $\sharp_1(\alpha) < (m-1)/2$.

The node w_0 is informed from v_0 at time $m + \sharp_1(\alpha) < (3m - 1)/2 = \lfloor 3m/2 \rfloor$.

In the next round, w_0 sends the message to its neighbor w_1. So w_1 is informed at time at most $\lfloor 3m/2 \rfloor$.

Case 1.2: $\natural_0(\alpha) < (m - 1)/2$.

The node w_1 is informed from v_0 at time $m + \natural_0(\alpha) < (3m - 1)/2 = \lfloor 3m/2 \rfloor$.

So, the node w_0 which is adjacent to the node w_1 is informed at time at most $\lfloor 3m/2 \rfloor$.

Case 1.3: $\natural_0(\alpha) = \sharp_1(\alpha) = (m - 1)/2$.

w_0 is informed at time $m + \sharp_1(\alpha) = (3m - 1)/2 = \lfloor 3m/2 \rfloor$.

w_1 is informed at time $m + \natural_0(\alpha) = (3m - 1)/2 = \lfloor 3m/2 \rfloor$.

Case 2: m is even.

Case 2.1: $\sharp_1(\alpha) \leq (m - 2)/2$.

The node w_0 is informed from v_0 at time

$m + \sharp_1(\alpha) \leq 3m/2 - 1 < \lfloor 3m/2 \rfloor$.

So, the node w_1 is informed at time at most $\lfloor 3m/2 \rfloor$.

Case 2.2: $\natural_0(\alpha) \leq (m - 2)/2$.

The node w_1 is informed from v_0 at time $m + \natural_0(\alpha) \leq 3m/2 - 1 < \lfloor 3m/2 \rfloor$.

So, the node w_0 is informed at time at most $\lfloor 3m/2 \rfloor$.

Thus, after phase 1, for all $\alpha \in \{0, 1\}^{m-1}$ the nodes w_0 and w_1 received the message in at most $\lfloor 3m/2 \rfloor$ rounds.

Phase 2: Inform all nodes in the cycles $C_{\alpha i}$, $\alpha \in \{0, 1\}^{m-1}$, $i \in \{0, 1\}$.

From the informed node we can inform all other nodes of the cycle in $\lceil m/2 \rceil$ rounds (see Example 2.5.9). Hence, the broadcast time in the butterfly network is at most $\lfloor 3m/2 \rfloor + \lceil m/2 \rceil = 2m$.

$\square$

Now we investigate the deBruijn network DB_k. The best known upper bound so far was found by Bermond and Peyrat [BP88]. For the lower bound, we again state the diameter lower bound and refer to Section 3.4 for a nontrivial lower bound for broadcasting in DB_k.

Theorem 3.3.7 ([BP88]). *For any integer $d \geq 2$:*

$$d \leq \mathrm{minb}(DB_d) \leq \mathrm{b}(DB_d) \leq \frac{3}{2}(d + 1).$$

Proof. The lower bound comes from the fact that DB_d has diameter d.

For the upper bound, the idea of the broadcasting scheme is that any node broadcasts only to its right neighbors (i.e., $(y_1, y_2, \ldots, y_d)$ informs its neighbors $(y_2, \ldots, y_d, y_1)$ and $(y_2, \ldots, y_d, \overline{y_1})$). The order of broadcasting will be determined according to the 2-arity α of $(y_1, y_2, \ldots, y_d)$, that is $\alpha(y_1, \ldots, y_d) =$

$\left(\sum_{i=1}^{d} y_i\right) \bmod 2$. Note that $\alpha \in \{0, 1\}$. The node $(y_1, y_2, \ldots, y_d)$ will broadcast to its right neighbors in the order $(y_2, \ldots, y_d, \alpha), (y_2, \ldots, y_d, \overline{\alpha})$. (for example, in DB_6 for the node $(0, 0, 1, 1, 0, 0)$ the value of α is 0, and so the node informs at first $(0, 1, 1, 0, 0, 0)$ and then $(0, 1, 1, 0, 0, 1)$.) Now, consider the following two paths P_k, $k \in \{0, 1\}$, of length $d+1$ from $(y_1, \ldots, y_d)$ to any node $(z_1, \ldots, z_d)$:

$$P_k : \Big((y_1, \ldots, y_d), (y_2, \ldots, y_d, k), (y_3, \ldots, y_d, k, z_1), (y_4, \ldots, y_d, k, z_1, z_2),$$
$$\ldots, (y_d, k, z_1, \ldots, z_{d-2}), (k, z_1, \ldots, z_{d-1}), (z_1, \ldots, z_{d-1}, z_d) \Big).$$

(Note that the paths are not necessarily disjoint and not necessarily elementary.) Let $v_{0_i} = (y_i, \ldots, y_d, 0, z_1, \ldots, z_{i-2})$ $(v_{i_1} = (y_i, \ldots, y_d, 1, z_1, \ldots, z_{i-2}))$ be the i-th node of P_0 (P_1), $1 < i < d+2$. The nodes v_{0_i} and v_{1_i} differ just in one bit-position. Thus, we have

$$\alpha(y_i, \ldots, y_d, 0, z_1, \ldots, z_{i-2}) = \alpha(y_i, \ldots, \overline{y_d}, 1, z_1, \ldots, z_{i-2}) \in \{0, 1\}.$$

That means that the number of time units required to broadcast

from $(y_i, \ldots, y_d, 0, z_1, \ldots, z_{i-2})$ to $(y_{i+1}, \ldots, y_d, 0, z_1, \ldots, z_{i-2}, z_{i-1})$

is different from the number of time units required to broadcast

from $(y_i, \ldots, y_d, 1, z_1, \ldots, z_{i-2})$ to $(y_{i+1}, \ldots, y_d, 1, z_1, \ldots, z_{i-2}, z_{i-1})$.

Each time unit is either 1 or 2.

Let us have a look at the number of time units required to broadcast from $(y_1, \ldots, y_d)$ to $(z_1, \ldots, z_d)$ along the path P_k, $k \in \{0, 1\}$. The time t_k to broadcast the message via P_k is $t_k = t_{k_1} + t_{k_2} + \cdots + t_{k_{d+1}}$ with $t_{k_j} \in \{1, 2\}, 1 \leq j \leq d+1$. Thus, we have

$$\sum_{k=0}^{1} t_k = (d+1)(1+2) = 3(d+1).$$

Hence, the message will reach $(z_1, \ldots, z_d)$ on one of these paths at a time at most $3(d+1)/2$.

$\square$

3.4 Lower Bounds for Bounded-Degree Graphs

In this section, the overall goal is to improve the lower bounds for broadcasting in the butterfly and the deBruijn network. But in order to apply our proof techniques to other networks as well, we will concentrate on the general methods and arguments used and we will point out which properties of the graphs we are using.

The first property which helps us to improve the lower bound (at least for the deBruijn graph) is that the graph considered has degree d. This argument was developed by Liestman and Peters [LP88] for graphs of degree 3 and 4 and further refined and sharpened for general d in [BHLP92] and [CGV89]. As we are mainly interested in how the argument works, we only present the results for degree 3 and 4. The argument basically consists of finding an upper bound on the number of nodes which can be informed in t time steps.

Theorem 3.4.1 ([LP88]). *Let n be an integer, $n \geq 5$, and let G be a graph of n vertices.*

a) *If the degree of G is bounded by 3, then*
$$b(G) \geq minb(G) \geq 1.4404 \cdot \log_2 n - 3.$$

b) *If the degree of G is bounded by 4, then*
$$b(G) \geq minb(G) \geq 1.1374 \cdot \log_2 n - 2.$$

Proof. Let A be a broadcast algorithm for G and v_0 running in t rounds. For $i \in \{1, \ldots, t\}$, let $\mathrm{Rec}_i^A(v_0)$ denote the set of vertices (newly) receiving the information in round i, i.e.,

$$\mathrm{Rec}_i^A(v_0) = \mathrm{Broad}_i^A(v_0) \setminus \mathrm{Broad}_{i-1}^A(v_0).$$

Formally, we set $\mathrm{Rec}_0^A(v_0) = \{v_0\}$.

Our goal is to derive a suitable upper bound, $A(i)$, on $|\mathrm{Rec}_i^A(v_0)|$ for $i \in \{1, \ldots, t\}$ and thus on

$$|\mathrm{Broad}_i^A(v_0)| = \sum_{s=0}^{i} |\mathrm{Rec}_s^A(v_0)|.$$

As $|\mathrm{Broad}_t^A(v_0)| \geq n$ must hold for any broadcast algorithm A, this will lead to the desired lower bound on t.

Without loss of generality, we may assume that every internal node v in the broadcast tree representing A broadcasts $I(v_0)$ to its $\ell(v)$ sons immediately in the next $\ell(v)$ rounds. (Otherwise, we can change A to an algorithm B that has this property.)

Now, we prove the bounds a) and b) separately.

a) Since G has degree 3, $\ell(v_0) \leq 3$ and $\ell(v) \leq 2$ for all $v \neq v_0$. Hence, an upper bound, $A(i)$, on $|\mathrm{Rec}_i^A(v_0)|$ may be recursively defined as follows:

$$A(0) = 1, \ A(1) = 1, \ A(2) = 2, \ A(3) = 4,$$

$$A(i) = A(i-1) + A(i-2) \quad \text{for } i \geq 4.$$

We first prove by induction that $A(i) \leq 1.61804^i$ for any $i \geq 0$:

(1) We have
$$A(0) = 1 \leq 1 = 1.61804^0,$$
$$A(1) = 1 \leq 1.61804 = 1.61804^1,$$
$$A(2) = 2 \leq 2.61805 = 1.61804^2,$$
$$A(3) = 4 \leq 4.23612 = 1.61804^3.$$

(2) Let $i \geq 4$ and let $A(j) \leq 1.61804^j$ for any $j \leq i - 1$. Then
$$A(i) = A(i-1) + A(i-2)$$
$$\underset{induc.}{\leq} 1.61804^{i-1} + 1.61804^{i-2} \leq 1.61804^i,$$
as $1.61804 + 1 \leq 1.61804^2$.

Now, it follows that

$$n \leq |\text{Broad}_t^A(v_0)| = \sum_{i=0}^{t} |\text{Rec}_i^A(v_0)| \leq \sum_{i=0}^{t} A(i)$$

$$\leq \sum_{i=0}^{t} 1.61804^i = \frac{1.61804^{t+1} - 1}{1.61804 - 1} \leq 3 \cdot 1.61804^t ,$$

which yields $t \geq 1.4404 \cdot \log_2 n - 3$.

b) Since G has degree 4, $\ell(v_0) \leq 4$ and $\ell(v) \leq 3$ for all $v \neq v_0$. Hence, an upper bound, $A(i)$, on $|\text{Rec}_i^A(v_0)|$ may be recursively defined as follows:

$$A(0) = 1, \ A(1) = 1, \ A(2) = 2, \ A(3) = 4, \ A(4) = 8,$$

$$A(i) = A(i-1) + A(i-2) + A(i-3) \quad \text{for } i \geq 5.$$

A simple induction shows that

$$A(i) \leq 1.8393^i ,$$

hence

$$n \leq |\text{Broad}_t^A(v_0)| = \sum_{i=0}^{t} |\text{Rec}_i^A(v_0)| \leq \sum_{i=0}^{t} A(i) \leq 3 \cdot 1.8393^t ,$$

which yields $t \geq 1.1374 \cdot \log_2 n - 2$.

$\square$

Exercise 3.4.2. Fill in the technical details of the proof of Theorem 3.4.1 (b) (analogously to the proof of part (a)).

For the butterfly network BF_k, Theorem 3.4.1 yields a lower bound of

$$1.1374 \cdot k + 1.1374 \cdot \log_2 k - 2$$

which is worse than the diameter lower bound. But for the deBruijn network DB_k, we can improve the diameter lower bound by applying Theorem 3.4.1:

Corollary 3.4.3.

$$b(DB_k) \geq minb(DB_k) \geq 1.1374 \cdot k - 2 \quad \text{for all } k \in \mathbb{N}.$$

The technique of Liestman and Peters was extended by E. Stöhr [St91b]. She was the first to prove a non-trivial lower bound of $1.5621k$ on the broadcast time of the butterfly network BF_k. Her technique was again refined and extended in [KMPS92], where the lower bound was improved to $1.7417k$. In order to make things easier to understand, we prove a slightly weaker bound. The graph property which is needed for the improvement is the following:

There is a node from which many vertices have a large distance, where large distance means almost as large as the diameter.

The intuitive idea is that it is inherently difficult to send the message from the originating node to nodes very far away and to spread the information at the same time. This argument can be viewed as a generalization of Observation 3.2.5. It basically consists of finding an upper bound on the number of nodes which can be informed in t time steps at distance i. Taking also the distance into account is the difference to the technique of Liestman and Peters. As we will see, this makes the calculations much more difficult.

Let us start by stating the mentioned graph property more exactly for the butterfly network:

Lemma 3.4.4. *Let BF_m be the butterfly network of dimension $m \in \mathbb{N}$. Let $v_0 = \langle 0, 00...0 \rangle$. Let $\varepsilon > 0$ be any positive constant. Then there exist $2^m - o(2^m)$ nodes which are at a distance of at least $\lfloor 3m/2 - \varepsilon m \rfloor$ from v_0.*

Proof. Let

$$L = \{ \langle \lfloor m/2 \rfloor, \delta \rangle \mid \delta \neq \alpha 0^k \beta \text{ for some } k \geq \varepsilon m/2, \ \alpha 0^k \beta \in \{0,1\}^m \}$$

be the subset of the level-$\lfloor m/2 \rfloor$ vertices of BF_m. Then

$$|L| \geq 2^m - m 2^{m - \varepsilon m/2},$$

and the distance between any vertex $v \in L$ and v_0 is at least $\lfloor 3m/2 - \varepsilon m \rfloor$ (see Exercise 3.4.5).

$\square$

Exercise 3.4.5. Let BF_m be the butterfly network of dimension $m \in \mathbb{N}$. Let $v_0 = \langle 0, 00...0 \rangle$. Let $\varepsilon > 0$ be any positive constant. Let

$$L = \{ \langle \lfloor m/2 \rfloor, \delta \rangle \mid \delta \neq \alpha 0^k \beta \text{ for some } k \geq \varepsilon m/2, \ \alpha 0^k \beta \in \{0,1\}^m \} \ .$$

Prove that

(i) $|L| \geq 2^m - m 2^{m - \varepsilon m/2}$.
(ii) The distance between any vertex $v \in L$ and v_0 is at least $\lfloor 3m/2 - \varepsilon m \rfloor$.

Now, we are able to show the improved lower bound:

Theorem 3.4.6 ([KMPS92]).

$b(BF_m) = minb(BF_m) > 1.7396m$ *for all sufficiently large $m \in \mathbb{N}$.*

Proof. Suppose that broadcasting can be completed on BF_m in time $3m/2 + tm$, $0 \leq t < 1/2$. We prove a lower bound on t.

As in the proof of Theorem 3.3.6, we can assume that the message originates at vertex $v_0 = \langle 0, 00...0 \rangle$, and the originator learns the message at time 0.

Let T be a broadcast tree of v_0 and BF_m. Let $A_T(i,t)$ denote the number of nodes which are at distance i in T from the root v_0 and which receive the message in round t (along the broadcast tree T). Our first aim is to show an upper bound $A(i,t)$ on $A_T(i,t)$ which is independent of T.

Since BF_m has maximum degree 4, once a node has received the message it can only inform 3 additional neighbors in the next three rounds. Therefore, $A(i,t)$ can be recursively defined as follows:

$A(0,0) = 1,$

$A(1,1) = 1,$

$A(1,2) = 1, \; A(2,2) = 1,$

$A(1,3) = 1, \; A(2,3) = 2, \; A(3,3) = 1,$

$A(1,4) = 1, \; A(2,4) = 3, \; A(3,4) = 3, \; A(4,4) = 1,$

and

$A(i,t) = A(i-1,t-1) + A(i-1,t-2) + A(i-1,t-3)$ for $t \geq 5$.

It can easily be shown by induction (see [BP85]) that

$$A(n, n+l) \; \leq \; 2 \cdot \sum_{\substack{p+2q=l, \\ 0 \leq p,q \leq n}} \binom{n}{p+q} \cdot \binom{p+q}{q}$$

(see Exercise 3.4.7). Let $\varepsilon > 0$ be any positive constant. From Lemma 3.4.4, we know that for any broadcasting scheme

$$\sum_{n=3m/2-\varepsilon m}^{3m/2+tm} \; \sum_{l=0}^{3m/2+tm-n} \; A(n, n+l) \; \geq \; 2^m - o(2^m).$$

For ε tending towards 0, we have

$$2^m - o(2^m) \leq \sum_{n=3m/2}^{3m/2+tm} \sum_{l=0}^{3m/2+tm-n} A(n, n+l)$$

$$\leq \sum_{n=3m/2}^{3m/2+tm} \sum_{l=0}^{3m/2+tm-n} 2 \cdot \sum_{\substack{p+2q=l, \\ 0 \leq p,q \leq n}} \binom{n}{p+q} \cdot \binom{p+q}{q}$$

$$\leq 2 \cdot \sum_{n=3m/2}^{3m/2+tm} \sum_{0 \leq p+2q \leq 3m/2+tm-n} \binom{n}{p+q} \cdot \binom{p+q}{q}$$

$$\leq cm^3 \cdot \max_{\substack{3m/2 \leq n \leq 3m/2+tm, \\ 0 \leq p+2q \leq 3m/2+tm-n}} \binom{n}{p+q} \cdot \binom{p+q}{q}$$

for some constant c. It can easily be verified that the above maximum is obtained for $n = 3m/2$, $p + 2q = tm$ when $t < 1/2$ (see Lemma 3.4.8). Therefore,

$$\max_{\substack{3m/2 \leq n \leq 3m/2+tm, \\ 0 \leq p+2q \leq 3m/2+tm-n}} \binom{n}{p+q} \cdot \binom{p+q}{q} = \max_{0 \leq i \leq tm/2} \binom{3m/2}{tm-i} \cdot \binom{tm-i}{i}.$$

The latter term is maximized for $i = i_0 m$ where

$$i_0 = \frac{1}{4} + \frac{t}{2} - \sqrt{\left(\frac{1}{4} + \frac{t}{2}\right)^2 - \frac{t^2}{3}}$$

(see Lemma 3.4.8). For large m, an approximate expression for the factorial is given by Stirling's formula

$$m! \approx m^m e^{-m} \sqrt{2\pi m}.$$

Using Stirling's formula we obtain

$$\binom{3m/2}{tm - i_0 m} \cdot \binom{tm - i_0 m}{i_0 m}$$

$$\approx \frac{(3/2)^{3m/2}}{(3/2 - t + i_0)^{3m/2 - tm + i_0 m} (i_0)^{i_0 m} (t - 2i_0)^{tm - 2i_0 m}}.$$

Thus,

$$cm^3 \cdot \frac{(3/2)^{3m/2}}{(3/2 - t + i_0)^{3m/2 - tm + i_0 m} (i_0)^{i_0 m} (t - 2i_0)^{tm - 2i_0 m}} \geq 2^m - o(2^m).$$

Taking the m-th root on both sides, we have for large m

$$\frac{(3/2)^{3/2}}{(3/2 - t + i_0)^{3/2 - t + i_0} (i_0)^{i_0} (t - 2i_0)^{t - 2i_0}} \geq 2.$$

The latter inequality is not true for $t \leq 0.2396$ (see Lemma 3.4.8), and so we get the lower bound of the theorem. $\qquad\qquad\square$

Exercise 3.4.7. Let $A(i,t)$ as defined in the proof of Theorem 3.4.6. Prove that

$$A(n, n+l) \leq 2 \cdot \sum_{\substack{p+2q=l, \\ 0 \leq p,q \leq n}} \binom{n}{p+q} \cdot \binom{p+q}{q}.$$

Lemma 3.4.8.

(i) The maximum

$$\max_{\substack{3m/2 \leq n \leq 3m/2+tm, \\ 0 \leq p+2q \leq 3m/2+tm-n}} \binom{n}{p+q} \cdot \binom{p+q}{q}$$

is obtained for $n = 3m/2$, $p + 2q = tm$ when $t < 1/2$.

(ii) The term

$$\max_{0 \leq i \leq tm/2} \binom{3m/2}{tm-i} \cdot \binom{tm-i}{i}$$

is maximized for $i = i_0 m$ where

$$i_0 = \frac{1}{4} + \frac{t}{2} - \sqrt{\left(\frac{1}{4} + \frac{t}{2}\right)^2 - \frac{t^2}{3}}.$$

(iii) The inequality

$$\frac{(3/2)^{3/2}}{(3/2 - t + i_0)^{3/2 - t + i_0} (i_0)^{i_0} (t - 2i_0)^{t-2i_0}} \geq 2$$

where

$$i_0 = \frac{1}{4} + \frac{t}{2} - \sqrt{\left(\frac{1}{4} + \frac{t}{2}\right)^2 - \frac{t^2}{3}}$$

is not true for $t \leq 0.2396$.

Proof.

(i) The maximum

$$\max_{\substack{3m/2 \leq n \leq 3m/2+tm, \\ 0 \leq p+2q \leq 3m/2+tm-n}} \binom{n}{p+q} \cdot \binom{p+q}{q}$$

$$= \max_{\substack{3m/2 \leq n \leq 3m/2+tm, \\ 0 \leq p+2q \leq 3m/2+tm-n}} \frac{n!}{p! \cdot q! \cdot (n-p-q)!}$$

has to be calculated. It is clear that the above maximum is obtained for $n + p + 2q = 3m/2 + tm$.

[If $n + p + 2q < 3m/2 + tm$, set $n' := n + 1$, $p' := p$, $q' := q$. Then $n' + p' + 2q' \leq 3m/2 + tm$, and

$$\binom{n'}{p'+q'} \cdot \binom{p'+q'}{q'} = \binom{n+1}{p+q} \cdot \binom{p+q}{q}$$

$$\geq \binom{n}{p+q} \cdot \binom{p+q}{q} .]$$

Now, let $n + p + 2q = 3m/2 + tm$, $n \geq 3m/2$.

If $p > 0$, set $n' := n+1$, $p' := p-1$, $q' := q$. Then $n' + p' + 2q' = 3m/2 + tm$ and

$$\frac{n'!}{p'! \cdot q'! \cdot (n'-p'-q')!}$$

$$= \frac{(n+1) \cdot p}{(n-p-q+1) \cdot (n-p-q+2)} \cdot \frac{n!}{p! \cdot q! \cdot (n-p-q)!}$$

$$\leq \frac{n!}{p! \cdot q! \cdot (n-p-q)!}$$

as

$$\frac{(n+1) \cdot p}{(n-p-q+1) \cdot (n-p-q+2)} \leq \frac{(3/2+t)m \cdot tm}{((3/2-t)m)^2} \leq \frac{2 \cdot 1/2}{1 \cdot 1} \leq 1.$$

If $p = 0$ and $q > 0$, set $n' := n+2$, $p' := p$, $q' := q-1$. Then $n' + p' + 2q' = 3m/2 + tm$ and

$$\frac{n'!}{p'! \cdot q'! \cdot (n'-p'-q')!}$$

$$= \frac{(n+1) \cdot (n+2) \cdot q}{(n-p-q+1) \cdot (n-p-q+2) \cdot (n-p-q+3)}$$

$$\cdot \frac{n!}{p! \cdot q! \cdot (n-p-q)!}$$

$$\leq \frac{n!}{p! \cdot q! \cdot (n-p-q)!}$$

as

$$\frac{(n+1) \cdot (n+2) \cdot q}{(n-p-q+1) \cdot (n-p-q+2) \cdot (n-p-q+3)}$$

$$\leq \frac{((3/2+t)m)^2 \cdot tm/2}{((3/2-t)m)^3} \leq \frac{2 \cdot 2 \cdot 1/4}{1 \cdot 1 \cdot 1} \leq 1.$$

If $p = q = 0$, then

$$\binom{n}{p+q} \cdot \binom{p+q}{q} = \binom{3m/2+tm}{0} \cdot \binom{0}{0} = 1,$$

which cannot be the maximum.

The calculation above shows that

$$\max_{\substack{3m/2 \le n \le 3m/2+tm, \\ 0 \le p+2q \le 3m/2+tm-n}} \binom{n}{p+q} \cdot \binom{p+q}{q}$$

is obtained for $n = 3m/2$, $p + 2q = tm$. Hence, $tm - q = p + q$, $q \le tm/2$, and therefore

$$\max_{\substack{3m/2 \le n \le 3m/2+tm, \\ 0 \le p+2q \le 3m/2+tm-n}} \binom{n}{p+q} \cdot \binom{p+q}{q}$$

$$= \max_{0 \le i \le tm/2} \binom{3m/2}{tm-i} \cdot \binom{tm-i}{i}.$$

(ii) Let $n = 3m/2$, $k = tm$. The task now is to maximize

$$\binom{n}{k-i} \cdot \binom{k-i}{i} \qquad (*)$$

for $0 \le i \le k/2$. According to the properties of binomial coefficients, there is some $0 \le i_0 \le k/2$ such that
$(*)$ is increasing for increasing $i \le i_0$,
$(*)$ is decreasing for increasing $i > i_0$.
Hence, at i_0 we have

$$\binom{n}{k-i_0} \cdot \binom{k-i_0}{i_0} = \binom{n}{k-(i_0+1)} \cdot \binom{k-(i_0+1)}{i_0+1}$$

which is equivalent to the condition

$$\frac{(k-2i_0) \cdot (k-2i_0-1)}{(n-k+i_0+1) \cdot (i_0+1)} = 1 \,.$$

Solving this equality for i_0 (and neglecting low-order terms) yields

$$i_0 = \frac{n+3k}{6} \pm \sqrt{\left(\frac{n+3k}{6}\right)^2 - \frac{k^2}{3}} \,.$$

The value which is in the range $0 \le i_0 \le k/2$ is

$$i_0 = \frac{n+3k}{6} - \sqrt{\left(\frac{n+3k}{6}\right)^2 - \frac{k^2}{3}} \,.$$

Substituting $n = 3m/2$, $k = tm$, we obtain

$$i_0 = \left(\frac{1}{4} + \frac{t}{2} - \sqrt{\left(\frac{1}{4} + \frac{t}{2}\right)^2 - \frac{t^2}{3}}\right) \cdot m \,.$$

(iii) Let

$$f(t) = \frac{(3/2)^{3/2}}{(3/2 - t + i_0)^{3/2 - t + i_0} (i_0)^{i_0} (t - 2i_0)^{t - 2i_0}} .$$

Then f is strictly monotonically increasing for $0 \le t < 1/2$. Hence, for $t \le 0.2396$ we have

$$f(t) \le f(0.2396) = 1.9998486 < 2 .$$

□

For the deBruijn network, the proofs are completely analogous. First, we state the required graph property more exactly:

Lemma 3.4.9. *Let DB_m be the deBruijn network of dimension $m \in I\!N$. Let $v_0 = 00...0$. Let $\varepsilon > 0$ be any positive constant. Then there exist $2^m - o(2^m)$ nodes which are at a distance of at least $\lfloor m - \varepsilon m \rfloor$ from v_0.*

Proof. Let

$$L = \{ v \mid v \ne \alpha 0^k \beta \text{ for some } k \ge \varepsilon m, \ \alpha 0^k \beta \in \{0, 1\}^m \}$$

be the subset of vertices. Hence,

$$|L| \ge 2^m - m 2^{m - \varepsilon m}.$$

Let $v \in L$. As the longest sequence α of consecutive 0's in v has at most length $\lfloor \varepsilon m \rfloor$, the bit string $v_0 = 00...0$ has to be rotated at least $\lfloor m - \varepsilon m \rfloor$ times to change the 1's left and right of α. Therefore, the distance between any vertex v from L and v_0 is at least $\lfloor m - \varepsilon m \rfloor$.

□

Now, we are able to state the improved lower bound:

Theorem 3.4.10 ([KMPS92]).

$$b(DB_m) > 1.3042m \quad \text{for all sufficiently large } m \in I\!N.$$

Proof. The proof is similar to that of Theorem 3.4.6. We suppose that broadcasting can be completed on DB_m in time $m + tm$, $0 \le t < 1/3$. The node $v_0 = 00...0$ is taken as the originator of the message. The recursion formula for $A(i, t)$ is exactly the same as in the proof of Theorem 3.4.6. This time, the condition obtained from Lemma 3.4.9 is

$$\sum_{n=m}^{m+tm} \sum_{l=0}^{m+tm-n} A(n, n + l) \ge 2^m - o(2^m).$$

Similar estimations as before show that this cannot be true for $t \le 0.3042$ (see Exercise 3.4.11).

□

Exercise 3.4.11. Let $A(i,t)$ be defined as in the proof of Theorem 3.4.6. Show that the inequality

$$\sum_{n=m}^{m+tm} \sum_{l=0}^{m+tm-n} A(n, n+l) \geq 2^m - o(2^m)$$

cannot be true for $t \leq 0.3042$ by verifying the following:

(i) Prove that

$$\sum_{n=m}^{m+tm} \sum_{l=0}^{m+tm-n} A(n, n+l) \leq cm^3 \cdot \max_{\substack{m \leq n \leq m+tm, \\ 0 \leq p+2q \leq m+tm-n}} \binom{n}{p+q} \cdot \binom{p+q}{q}$$

for some constant c.

(ii) Show that the maximum

$$\max_{\substack{m \leq n \leq m+tm, \\ 0 \leq p+2q \leq m+tm-n}} \binom{n}{p+q} \cdot \binom{p+q}{q}$$

is obtained for $n = m$, $p + 2q = tm$ when $t < 1/3$.

(iii) Show that the term

$$\max_{0 \leq i \leq tm/2} \binom{m}{tm-i} \cdot \binom{tm-i}{i}$$

is maximized for $i = i_0 m$ where

$$i_0 = \frac{1}{6} + \frac{t}{2} - \sqrt{\left(\frac{1}{6} + \frac{t}{2}\right)^2 - \frac{t^2}{3}}.$$

(iv) Show that the inequality

$$\frac{1}{(1 - t + i_0)^{1-t+i_0} (i_0)^{i_0} (t - 2i_0)^{t-2i_0}} \geq 2$$

where

$$i_0 = \frac{1}{6} + \frac{t}{2} - \sqrt{\left(\frac{1}{6} + \frac{t}{2}\right)^2 - \frac{t^2}{3}}$$

is not true for $t \leq 0.3042$.

Exercise 3.4.12. Prove that

$$minb(DB_m) > 1.3042 \cdot m \quad \text{for all sufficiently large } m \in I\!N.$$

As we have already mentioned, the lower bounds in [KMPS92] are slightly better than the ones we proved, namely

$b(BF_k) \geq 1.7417k$ for all sufficiently large k,

$b(DB_k) \geq 1.3171k$ for all sufficiently large k.

To derive these bounds, a third property of the graph is used, namely:

Each edge is contained in a cycle of length at most 4.

This is true for the butterfly and the deBruijn networks. Using this additional property, the recursion for the number of informed nodes $A(i,t)$ is changed, and similar estimations as in the proofs above lead to the desired results. But the complete analysis is quite tedious and the improvement in the result is not very significant. Therefore, we omit this part here. We will rather give an overview on the broadcast time for small butterfly networks BF_k as displayed in Table 3.1.

Table 3.1. Broadcast time for small butterfly networks BF_k

k	lower bound	upper bound	no. processors
2	3	3	8
3	5	5	24
4	7	7	64
5	8	9	160
6	10	11	384
7	11	13	896
8	13	15	2048
9	15	17	4608
10	16	19	10240
11	18	21	22528
12	19	23	49152
13	21	25	106496
14	23	27	229376
15	24	29	491520
16	26	31	1048576
17	27	33	2228224
18	29	35	4718592

The upper bound is the $2k-1$ algorithm from [KMPS92], the lower bound comes from the exact evaluation of the $A(i,t)$'s in the proof of Theorem 3.4.6 and the exact computation of the distances in the butterfly network. The overall picture is that the upper and lower bound are very close together for small dimensions k. For $k \leq 4$, i.e., up to 64 nodes, they even coincide.

In the meantime, the lower bound method presented here has been further extended. In [Pe96], it is shown for the (binary) butterfly and deBruijn networks that, once a node has received the information, it basically does not help for the broadcast process to use the third outgoing edge. Using this additional argument leads to the improved lower bounds of

(i) $b(BF_m) > 1.7609m$ for all sufficiently large m,

(ii) $b(DB_m) > 1.4404m$ for all sufficiently large m.

The upper bound for broadcasting in BF_m has been improved to $2m - \frac{1}{2}\log\log m + c$, for some constant c and all sufficiently large m [St92]. The upper bound for broadcasting in DB_m is still $\frac{3}{2}(m+1)$ [BP88]. Empirically, the upper bound for broadcasting in BF_m has been improved by using genetic programming in [CG98]. The effort to close the gap between the upper and the lower bound for broadcasting in BF_m and DB_m is still going on.

With some more extensions, it is possible to apply the presented lower bound technique to networks of higher node degree, e.g., general butterfly networks as considered in [ABR90], or general deBruijn and Kautz networks as investigated in [BP88, HOS92].

3.5 Overview on Broadcasting in Concrete Networks

As a summary of this section, Table 3.2 contains an overview of the best currently known time bounds for broadcasting in common interconnection networks and the according references in this paper and in the literature.

Table 3.2. Broadcast times for common networks

graph	no. nodes	diameter	lower bound	upper bound
K_n	n	1	$\lceil\log_2 n\rceil$ Lemma 3.2.2	$\lceil\log_2 n\rceil$ Lemma 3.2.2
H_k	2^k	k	k Lemma 3.2.2	k Lemma 3.2.2
CCC_k	$k\cdot 2^k$	$\lfloor 5k/2\rfloor - 2$	$\lceil 5k/2\rceil - 2$ Theorem 3.3.2, [LP88]	$\lceil 5k/2\rceil - 2$ Theorem 3.3.2, [LP88]
SE_k	2^k	$2k - 1$	$2k - 1$ Theorem 3.3.5, [HJM93]	$2k - 1$ Theorem 3.3.5, [HJM93]
BF_k	$k\cdot 2^k$	$\lfloor 3k/2\rfloor$	$1.7609k$ [Pe96]	$2k - 1$ Theorem 3.3.6, [KMPS92]
DB_k	2^k	k	$1.4404k$ Theorem 3.4.10, [Pe96]	$\frac{3}{2}(k+1)$ Theorem 3.3.7, [BP88]

While improving the bounds for these networks is still of great interest, the search for new interconnection structures with better broadcasting capabilities has also been going on. For example in [BHLP92], new families of graphs are presented achieving broadcasting in time $1.8750\log_2 n$ and

$1.4167 \log_2 n$ for degree 3 and 4, respectively, further closing the gap towards the lower bounds of $1.4404 \log_2 n$ and $1.1374 \log_2 n$ from Theorem 3.4.1. For general fixed degree, other constructions of networks with efficient broadcasting schemes are given in [CGV89]. Broadcasting (and also gossiping) in the lately proposed star graph and pancake graph (as special cases of Cayley graphs, see [AK89, ABR90]) has been investigated in [MS92, BFP96, Go94], where efficient broadcast schemes are presented. The star graph and pancake graph have become very popular because they have a very regular interconnection pattern, and sublogarithmic degree and diameter.

In the meantime, in [FP01], families of bounded-degree graphs have been constructed in which the broadcast time matches the lower bound from Theorem 3.3.2 (for degree 3 and 4) and the generalized lower bound of [BHLP92] (for degree at least 5), thus showing that these bounds are optimal. (Moreover, it is shown in [FP01] that the lower bounds of [LP88, BHLP92] are even attained in the case of gossiping and systolic gossiping in the two-way mode.) Thus, the main open problem left is to find improved bounds for broadcasting in specific networks of bounded degree.

Open Problem 3.1. Improve the upper bound of $2m - \frac{1}{2} \log \log m + c$ or the lower bound of $1.7609m$ for broadcasting in BF_m.

Open Problem 3.2. Improve the upper bound of $\frac{3}{2}(m + 1)$ or the lower bound of $1.4404m$ for broadcasting in DB_m.

3.6 Approximation Algorithms

Instead of designing individual broadcast schemes for specific networks, in this section we are interested in constructing algorithms which, given a graph G and a node v in G, compute a "reasonable" broadcast strategy. More formally, we will investigate the following problem:

Minimum Broadcast Time (MBT)
INPUT: A graph $G = (V, E)$ and a node $v \in V$.
OUTPUT: An optimal broadcast strategy A from v, i.e., a broadcast tree for G rooted at v.
MEASURE: The number of rounds of A, i.e., $\text{com}(A)$.

The MBT problem in general graphs is NP-hard (see [GJ79]) and thus is unlikely to be solved exactly. In this section, we want to give an overview of several approaches coping with the MBT problem that have been considered in the literature. In particular, we will focus on approximation algorithms. An α-approximation algorithm is an algorithm that guarantees that the computed solution is at most a multiplicative factor of α away from the optimal solution. One has to add that the literature on approximation algorithms for the MBT problem is so vast that we can only present a small introduction here. In particular, we again focus on the presentation of some instructive underlying ideas and methods, before giving an overview of the bibliography of the research area and stating some open problems.

3.6.1 General Observations

We first start with some general observations.

Observation 3.6.1. *For every graph G and every node v of G,*

$$\mathrm{b}(v,G) \leq \mathrm{b}(G) \leq \mathrm{r}_2(G) \leq \mathrm{r}(G) \leq 2 \cdot \min \mathrm{b}(G) \leq 2 \cdot \mathrm{b}(v,G).$$

Exercise 3.6.2. Prove Observation 3.6.1 analogously to Lemma 2.5.8 and Lemma 2.5.10.

A direct consequence of Observation 3.6.1 is the following assertion.

Corollary 3.6.3. *Let A be an approximation algorithm for MBT with approximation ratio α. Then the following problems can be approximated within 2α:*

(a) The problem of computing $\min \mathrm{b}(G)$.
(b) The problem of computing $\mathrm{b}(G)$.
(c) The problem of computing $\mathrm{r}_2(G)$.
(d) The problem of computing $\mathrm{r}(G)$.

Exercise 3.6.4. Prove Corollary 3.6.3.

Corollary 3.6.3 shows that, in order to find good approximation algorithms for computing $\min \mathrm{b}(G)$, $\mathrm{b}(G)$, $\mathrm{r}_2(G)$, and $\mathrm{r}(G)$, it suffices to study the MBT problem.

A first observation is that the MBT problem can be solved in polynomial time on trees.

Theorem 3.6.5. *Let T be an arbitrary tree, let v be a node in T. Then the MBT problem can be solved in polynomial time for T and v.*

Proof. Consider the rooted tree T_v, obtained from T by specifying vertex v as the root of T_v. For any vertex u in T_v, let T_u denote the subtree of T_v rooted at u. Let $v_1, v_2, \ldots, v_k$ denote the children of u in T_u, and assume they are labeled so that $\mathrm{b}(v_1, T_{v_1}) \geq \mathrm{b}(v_2, T_{v_2}) \geq \ldots \geq \mathrm{b}(v_k, T_{v_k})$. Since $\mathrm{b}(v_i, T_{v_i})$ is the amount of time it will take to pass the information from v_i to the other vertices in T_{v_i}, an optimal calling sequence from u consists of first calling v_1, then v_2, then v_3, etc. A formal proof of this fact is left as an exercise to the reader.

$\square$

Exercise 3.6.6. Prove Theorem 3.6.5 formally.

A first simple approximation bound on the MBT problem in general graphs can be obtained as follows.

Lemma 3.6.7.

(a) MBT is approximable within $(n-1)/\lceil \log_2 n \rceil$.
(b) MBT is approximable within $(n-1)/\mathrm{rad}(G)$.

Proof. Every broadcast tree for G and v completes broadcasting in at most $n-1$ rounds. (In every round, at least one additional node learns the information.) As $\mathrm{b}(v, G) \geq \lceil \log_2 n \rceil$ (Observation 3.2.1) and $\mathrm{b}(v, G) \geq \mathrm{rad}(G)$ (Observation 2.5.12), the lemma follows.

$\square$

The following lemma can be proved analogously to Lemma 2.5.17.

Lemma 3.6.8. *For every graph G and every node v of G, there is an algorithm A that broadcasts from v in G in at most $\deg(G) \cdot \mathrm{rad}(v, G)$ rounds.*

Exercise 3.6.9. Prove Lemma 3.6.8.

As $\mathrm{rad}(v, G) \leq \mathrm{b}(v, G)$ (Observation 2.5.12), a direct consequence of Lemma 3.6.8 is the following assertion.

Corollary 3.6.10. *MBT is approximable within $\deg(G)$.*

Corollary 3.6.10 shows that MBT can be approximated in bounded-degree graphs within a constant factor. On the other hand, the case of bounded-diameter graphs is nearly as hard as the original problem, as testified by the following theorem. Moreover, the following theorem shows that, for studying the MBT problem in general graphs, it suffices to study the MBT problem in graphs of diameter 2.

Theorem 3.6.11. *Given a polynomial-time ρ-approximation algorithm for the MBT problem on graphs of diameter two, there is a polynomial-time $(4\rho + 2)$-approximation algorithm for the MBT problem on any graph.*

Proof. Suppose we are given a ρ-approximation algorithm for the minimum broadcast time problem for graphs of diameter two. Given an arbitrary input graph G with broadcast time $\mathrm{b}(G)$, we construct a graph H from G by adding a new node r with edges to all nodes in G. Clearly H has diameter two and the broadcast time of H is at most $\mathrm{b}(G) + 1$.

Claim (1). Any broadcast scheme for H starting from the node r and completing in C steps can be used to accomplish broadcast in G from any source node s in $2C + \mathrm{diam}(G)$ steps.

A ρ-approximation to the MBT problem in H from r outputs a scheme taking at most $\rho \cdot \mathrm{b}(H) \leq \rho \cdot (\mathrm{b}(G) + 1)$ steps. Using this scheme and Claim (1), we can obtain a scheme for broadcasting in G from any node s within $2\rho \cdot (\mathrm{b}(G)+1) + \mathrm{diam}(G)$ steps. Since $\mathrm{diam}(G) \leq \mathrm{b}(G) \leq 2 \cdot \mathrm{b}(s, G) - 1$, this is at most $(4\rho + 2) \cdot \mathrm{b}(s, G)$ steps, giving Theorem 3.6.11.

The Proof of Claim (1). We show that any broadcast scheme for H starting from the node r and completing in C steps can be used to accomplish broadcast from any source node s in G in $2C + \mathrm{diam}(G)$ steps. From the source s, we first send the message to all the nodes that the root r informs directly in the scheme for H. Call such nodes *primary* nodes. Since the whole broadcast in H takes at most C steps, there are at most C primary nodes. To send the message from s to the primary nodes, we use a breadth-first tree rooted at s pruned appropriately to contain all primary nodes and only the primary nodes as leaves. Thus this tree has at most C leaves. Furthermore, any path in this tree from s to a leaf has length at most $\mathrm{diam}(G)$ since we chose a breadth-first tree. It is easy to now verify that any greedy scheme that broadcasts using this tree takes time at most $C + \mathrm{diam}(G)$ to complete. The key observation is that the maximum time for the message to reach any of the leaves is at most the sum of the delays along the path from the source to this leaf. The delay due to the length of the path is at most $\mathrm{diam}(G)$, and the queuing delays along the path sum to at most the sum of the degree of all nodes. This latter sum is at most C, the total number of leaves in the tree. To summarize, all the primary nodes in H can be informed in G starting from any source s with $C + \mathrm{diam}(G)$ steps. Emulating the scheme for H in G for the next C steps ensures that all nodes in G are informed within the next $2C + \mathrm{diam}(G)$ steps in total. This completes the proof of Theorem 3.6.11.

□

3.6.2 Two Approximation Algorithms

The Heuristic Approach

Before presenting two approximation algorithms, we first demonstrate that some "natural" heuristics for approaching the MBT problem are inadequate in some simple cases. We demonstrate this for the so-called "matching-based" heuristics.

As defined, a broadcast algorithm for a graph G and a node v is a sequence of matchings

$$\{v\} = V_0, E_1, V_1, E_2, \ldots, E_k, V_k = V$$

such that, for any $i \in \{1, 2, \ldots, k\}$,

1. $V_i \subseteq V$ and $E_i \subseteq E$,
2. E_i is a (maximum) matching between V_{i-1} and $\Gamma(V_{i-1})$ where, for any set S of vertices, $\Gamma(S)$ denotes the set of all neighbors of the vertices in S in $V \setminus S$, and
3. $V_i = V_{i-1} \cup E_i(V_{i-1})$, where, for any set S of vertices and any matching M between S and $\Gamma(S)$, $M(S)$ denotes the set of vertices of $\Gamma(S)$ that are matched by M to vertices in S.

An optimal broadcast can always be transformed into one whose sequence of matchings are maximum matchings. This follows from Lemma 3.6.12 given below. (Here, $b(S, G)$ denotes the complexity of the broadcast problem for a graph G and a set $V \subseteq V(G)$, i.e., one assumes that the node set S already has the information and the problem is to find a communication strategy such that all nodes in G learn the piece of information.)

Lemma 3.6.12. *If $S_1 \subseteq S_2$, then $b(G, S_2) \leq b(G, S_1)$.*

Proof. The proof is left as an exercise to the reader.

$\square$

Exercise 3.6.13. Prove Lemma 3.6.12.

We therefore obtain for $b(G, S)$ the following recurrence relation

$$b(G, S) = 1 + \min_{M}\{b(G, S \cup M(S))\}$$

where the minimum is taken over all the matchings M between S and $\Gamma(S)$.

We now show that some "natural" ways of choosing the (maximum) matchings $E_1, E_2, \ldots, E_k$ are inappropriate for approximating the MBT problem. The example considered is a *wheel*, i.e., a cycle of $n - 1$ vertices numbered $1, 2, \ldots, n - 1$, arranged by increasing order of indices, with an extra vertex v_0 connected to all vertices in the cycle (see Figure 3.6).

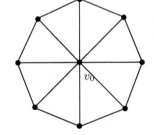

Fig. 3.6. The wheel of nine vertices

Let us first give tight bounds on the minimum broadcast time from v_0.

Lemma 3.6.14. *In the n-vertex wheel, $b(v_0) = \Theta(\sqrt{n})$.*

Proof. First let us show that $b(v_0) = \Omega(\sqrt{n})$. Assume that the broadcast takes k time units. The vertex v_0 can inform up to k different vertices. Therefore there is a vertex w whose distance on the ring from the vertices informed by v_0 is at least $\lfloor \lceil (n - 1)/k \rceil / 2 \rfloor$. Thus the time for informing w is at least $\lfloor \lceil (n - 1)/k \rceil / 2 \rfloor + 1$. The minimum broadcast time is achieved when $\lfloor \lceil (n - 1)/k \rceil / 2 \rfloor + 1 = k$, in which case $k \geq \sqrt{(n - 1)/2}$, yielding the desired lower bound on the broadcast time.

For proving that $b(v_0) = O(\sqrt{n})$, note that if v_0 informs all vertices whose index is congruent to 1 mod $\lceil \sqrt{n-1} \rceil$, and then these cycle vertices inform the other vertices, the total time for broadcasting is no more than $3 \cdot \lceil \sqrt{n-1} \rceil / 2 + 1$ time units.

$\square$

Now, consider the following three heuristics suggested in [SW84] for MBT. The first heuristic is based on defining for a set of vertices $V' \subseteq V$,

$$D_G(V') = \sum_{v \in V'} \deg(v).$$

At each round i, select from all possible maximum matchings between the set of informed vertices and the set of uninformed ones, a matching satisfying that the set $V'' \subseteq V \setminus V_i$ of "receiving" end vertices in the matching has maximal $D_G(V'')$. Let us describe a possible broadcast scenario. At the first round, v_0 delivers the message to the vertex 1. At round 2, v_0 calls $n-1$ and 1 calls 2. It is easy to see that starting from the third round one can choose a maximal matching that enlarges the set of informed vertices by 3 at each round. For example, at the third round v_0 may call $n-3$, $n-1$ calls $n-2$, and 2 calls 3, and so on. Note that the above scenario conforms with the given heuristic, since the degree of all the vertices in the wheel (except for v_0) is 3. Thus using this heuristic, it might take $\Omega(n)$ time units to complete the broadcasting.

A second heuristic approach suggested in [SW84] is the following. Define the *eccentricity* of a set $V' \subseteq V$ to be

$$E_G(V') = \sum_{v \in V'} \mathrm{rad}(v, G).$$

Choose, among all the possible maximal matchings, a matching for which the eccentricity of the set of newly informed vertices is maximal. In a similar way to the above it is easy to see that there are cases in which this approach leads to a broadcast time of $\Omega(n)$.

Exercise 3.6.15. Prove that the second heuristic suggested in [SW84] may lead to a broadcast time of $\Omega(n)$.

The last heuristic in [SW84] is a combination of the former two, and it, too, may lead to an $\Omega(n)$ broadcast scheme.

It follows that, for this example, all of these heuristics may yield a broadcast time that is $\Omega(\sqrt{n})$ times worse than optimal. We do not know whether this ratio is guaranteed by any of the three heuristics.

Decomposition into Clusters

In this subsection we consider an approximation scheme for broadcasting in general graphs. By Lemma 3.6.7, the scheme that chooses an arbitrary broadcast tree is at worst an $(n-1)/\lceil \log_2 n \rceil$ approximation scheme. We improve upon this approximation ratio, and present an algorithm which guarantees an $O(\sqrt{n})$ *additive* ratio.

A *cluster* in a graph G is a subset V' of the vertices such that the subgraph induced by V' is connected. Two clusters V', V'' are said to be *disjoint* if $V' \cap V'' = \emptyset$. Two disjoint clusters C_1 and C_2 are said to be *independent* if there is no edge connecting a vertex of C_1 to a vertex of C_2.

Definition 3.6.16. *Let $G = (V, E)$ be a graph and let $S \subseteq V$ be a subset of the vertices. A subtree $T = (V_1, E_1)$ of G rooted at a distinguished vertex v_0 is a* **shortest paths tree (SPT) leading from** v_0 **to** S *iff $S \subseteq V_1$, each path from v_0 to $v_i \in S$ in T is a shortest path in G, and every leaf of T belongs to S. Denote an SPT leading from a vertex v_0 to a set S by* **SPT(v_0, S)**.

The method used for the approximation is based on dividing the set of vertices into clusters of size $\lceil \sqrt{n} \rceil$, and broadcasting separately on those clusters. Let us start by describing the various tools used by the algorithm. At the heart of our scheme is the following decomposition lemma.

Lemma 3.6.17. *The set of vertices of any graph $G = (V, E)$ can be (poly-nomially) decomposed into sets of clusters $\mathcal{A}$ and $\mathcal{B}$, such that $|\mathcal{A}| \leq \sqrt{n}$, the clusters $\mathcal{A} \cup \mathcal{B}$ are pairwise disjoint, $(\bigcup \mathcal{A}) \cup (\bigcup \mathcal{B}) = V$, the size of each cluster $C' \in \mathcal{A}$ is $|C'| = \lceil \sqrt{n} \rceil$, the size of each cluster $C' \in \mathcal{B}$ is bounded by $|C'| \leq \sqrt{n}$, and the clusters in $\mathcal{B}$ are pairwise independent.*

Proof. The proof is left as an exercise to the reader.

$\square$

Exercise 3.6.18. Prove Lemma 3.6.17.

Our next tool involves broadcast on a tree. Recall that given a tree $T = (V_1, E_1)$ and a vertex $v \in V_1$ it is easy to compute the optimal scheme for broadcasting on T from v (Theorem 3.6.5). Let us call the optimal scheme for broadcasting in a tree the *OT scheme*.

In the next lemma we use a shortest paths tree $SPT(v, S)$ rooted at vertex v and leading to a set S of vertices (see Definition 3.6.16). Note that it is easy to construct such a tree in time polynomial in $|E|$ using a shortest path tree algorithm: simply construct a shortest path tree T spanning all the graph vertices, and iteratively exclude from it each leaf not belonging to S, until no such leaf exists.

Lemma 3.6.19. *Transmitting a message from a vertex v to a subset $V' \subseteq V$, $|V'| = \ell$ of the graph, can be performed in no more than $\ell - 1 + \text{rad}(v, G)$ time units.*

Proof. Construct an arbitrary tree $SPT(v, V')$ rooted at v and leading to the ℓ vertices of V'. Use the OT scheme to broadcast the message to all the members of the tree. Consider any leaf u. We would like to show that u gets the message within the specified time bounds. This is done by "charging" each time unit that elapses until u gets the message to a distinct vertex of the tree, and then bounding the number of charges.

Consider the situation immediately after an ancestor v' of u receives the message. The vertex v' is currently the lowest ancestor of u that knows the message. Thereafter v' starts delivering the message to its children. When v' delivers the message to a child whose subtree T' does not include u, choose arbitrarily a leaf in T' and charge this time unit to the leaf. When v' delivers the message to the ancestor of u, charge this time unit to v'.

Note that at every time unit we charge a single vertex, on account of u, and thus the total number of units charged is exactly the time before the message reaches u. Also note that no leaf is charged twice and that u is not charged. Finally note that every ancestor of u (except u itself) is charged once. Thus the time it takes the message to reach u is bounded by $\mathrm{rad}(v, G)$ plus the number of leaves in T besides u. Since each leaf is a member of V', the proof follows.

$\square$

We are now ready to combine the tools discussed in the above lemmas into an algorithm, named APPROX_MBT, for approximate broadcast on general graphs.

Algorithm APPROX_MBT

Input: A graph $G = (V, E)$ and a distinguished vertex $v_0 \in V$.
1. Decompose the vertices of V into two sets of clusters $\mathcal{A}$ and $\mathcal{B}$ using Procedure DECOMPOSE of Lemma 3.6.17.
2. Choose for each cluster C in $\mathcal{A}$ a single representative vertex v_C. Let R denote the set of representatives, $R = \{v_C | C \in \mathcal{A}\}$.
3. Transmit the message from v_0 to all the vertices of R by choosing an arbitrary tree $SPT(v_0, R)$ leading from v_0 to R, and applying the OT scheme to the tree.
4. Choose for each cluster $C \in \mathcal{A}$ an arbitrary spanning tree rooted at its representative v_C, and broadcast (in parallel) in the clusters of $\mathcal{A}$ according to the OT scheme.
5. Let $V_0 = (\bigcup \mathcal{A}) \cup \{v_0\}$. (Note that all the vertices in V_0 have received the message at this stage.) Transmit the message from V_0 to at least one vertex in each cluster in $\mathcal{B}$. (We omit the technical details.)
6. Choose for each cluster in $\mathcal{B}$ an arbitrary spanning tree rooted at an informed vertex and transmit the message in parallel to all the vertices in the clusters of $\mathcal{B}$ using the OT scheme in each cluster.

Theorem 3.6.20. *The broadcast time of Algorithm* APPROX_MBT *from a vertex* v_0 *in a graph* G *is bounded by* $3 \cdot \sqrt{n} + \mathrm{rad}(v, G) + \mathrm{b}(v_0)$ *time units.*

Proof. By Lemmas 3.6.19 and 3.6.17 the time it takes to complete stage 3 is bounded by $\sqrt{n} - 1 + \mathrm{rad}(v, G)$. The fact that each cluster in $\mathcal{A}$ has exactly

$\lceil \sqrt{n} \rceil$ vertices implies that stage 4 takes no more than $\sqrt{n}$ time units. In stage 5, a communication schedule taking no more than $b(v_0)$ time units can be calculated in polynomial time. (We omit the technical details, see, for example, [KP95].) Finally, the fact that the clusters in $\mathcal{B}$ are of size no larger than $\sqrt{n}$ implies that stage 6 takes no more than $\sqrt{n} - 1$ time units.

□

The ratio guaranteed by the theorem is $O(\sqrt{n})$-additive. Consequently, whenever the broadcast time of a network is $\Omega(\sqrt{n})$, e.g., in the wheel of n vertices, the scheme of Algorithm APPROX_MBT is a constant approximation scheme. For example, for each network whose diameter is at least $\sqrt{n}$, the above is a 5 approximation scheme. However, in the general case, the optimal broadcasting scheme may achieve time that is close to $\lceil \log n \rceil$. Thus in the general case our method is a $3 \cdot \sqrt{n}/\lceil \log n \rceil + 2$ approximation scheme and also an $O(\sqrt{n}/\mathrm{diam}(G))$ approximation scheme. In order to improve upon that, in the next subsection we will design a good approximation scheme for networks of "small" diameter.

Poise of a Graph

The next approximation algorithm we present is based on the idea of reducing the problem of finding a good broadcast scheme to that of finding a spanning tree of low degree and diameter. We first motivate the idea behind this approach. We can use the characterization of a broadcast scheme as a rooted spanning tree (also called *arborescence*) to derive a good lower bound on the broadcast time in the graph. The telephone model restriction dictates that the maximum out-degree of any node in the arborescence defined by a broadcast scheme is a lower bound on the broadcast time for this scheme. The maximum depth of any node from the root in this arborescence is also a lower bound on the broadcast time for this scheme since only neighbors can be informed in a time step. If A^* corresponds to the arborescence defined by the optimum broadcast scheme for a source node r, then the two observations above imply that

$$b(r, G) \geq \frac{1}{2} \cdot (\text{Max. out-degree}(A^*) + \text{Depth}(A^*))$$

$$\geq \frac{1}{2} \cdot \min_{\text{all arborescences } A} \{\text{Max. out-degree}(A) + \text{Depth}(A)\}.$$

Motivated by this we define the following problem: Given an undirected graph, find an undirected spanning tree in which the quantity (maximum degree of any node in the tree + diameter of the tree) is minimum over all the spanning trees. For any tree T, we call the quantity (maximum degree of any node in the tree + diameter of the tree) the *poise* of the tree[3] and denote this value by $P(T)$. The poise of a graph G, denoted by $P(G)$, is defined as the minimum

---

[3] This name is inspired by the fact that this quantity represents how well poised between width (the maximum degree) and height (the diameter) the tree is.

poise of any of its spanning trees. A tree on n nodes with the least poise is one with branching x at each internal node and with diameter roughly x. The value of x then obeys $x^x = n$ solving to $x = \Omega(\frac{\log n}{\log \log n})$. Thus the poise of any graph on n nodes is $x = \Omega(\frac{\log n}{\log \log n})$.

Exercise 3.6.21. Prove formally that for any graph G of n nodes, $P(G) = \Omega(\frac{\log n}{\log \log n})$.

We observed that the poise of a graph is a lower bound on the minimum broadcast time from any source in the graph. The next theorem shows that it is a good lower bound.

Theorem 3.6.22. *Let G be a graph of n nodes. Then*

$$\Omega(P(G)) \leq b(G) \leq O(P(G) \cdot \frac{\log n}{\log \log n}).$$

The proof of the second inequality in the above theorem is constructive: Given a tree with poise P, we demonstrate a broadcast scheme starting at any root in this tree which completes in $O(P \cdot \frac{\log n}{\log \log n})$ steps. This allows us to reduce the problem of finding a good broadcast scheme to that of finding a spanning tree of minimum poise, within a factor of $O(\frac{\log n}{\log \log n})$. As the poise of a graph can be approximated within $O(\log n)$ [Ra94], we obtain that the MBT problem is approximable within $O(\frac{\log^2 n}{\log \log n})$. Unfortunately, the proof that the poise of a graph can be approximated within $O(\log n)$ is well beyond the scope of this book. Here, we concentrate on proving Theorem 3.6.22 showing that the poise of a graph is a good approximation of the broadcast time from any source node in the graph.

Proof of Theorem 3.6.22. Let r be a node from which the minimum time to accomplish broadcasting is $b(G)$. Consider an optimum broadcast scheme finishing in time $b(G)$ starting at r. This defines an outward-directed arborescence rooted at r in a natural way as described in Chapter 2. Let T be the underlying undirected spanning tree defined by this arborescence rooted at r. Then $b(G)$ is at least as much as the maximum degree of any node in T minus one. Furthermore, $b(G)$ is also at least as much as the depth of the arborescence from which T is derived, and thus at least half the diameter of T. Since the poise of T is defined as the sum of the maximum degree and the diameter of T, we have that $b(G) = \Omega(P(G))$.

We prove the second inequality in Theorem 3.6.22 by showing that $b(G) = O(P(G) \cdot \frac{\log n}{\log P(G)})$, and applying the lower bound of $P(G) = \Omega(\frac{\log n}{\log \log n})$. To show the former claim, we exhibit a simple algorithm to complete broadcasting in this many time steps starting from any root r and using only edges in a spanning tree of G with poise $P(G)$. Let T be such a spanning tree. Let T_r denote the outward arborescence derived from T by rooting it at r and directing all the edges in T away from r. The maximum out-degree of any node in T is at most $P(G)$ and the longest directed path in T has length at most $P(G)$.

Note that given the tree T and the root r, the minimum time to accomplish broadcast from r using only edges of T can be determined in polynomial time by using dynamic programming and working bottom-up in the rooted tree T_r (Theorem 3.6.5). However, we wish to bound the broadcast time of the resulting scheme in terms of the poise of T, so we use a broadcast scheme that is simpler to analyze.

The scheme to broadcast the message from the root r is specified completely by specifying for each internal node v in T_r the order in which the children of v in T_r will be informed. We determine this order according to the number of nodes in the subtree of T_r rooted at this child. For any node v, let T_v denote the subtree of T_r rooted at node v and $|T_v|$ denote the number of nodes in T_v. We order the children of an internal node v of T_r as $v_1, v_2, \ldots, v_d$ where d is the out-degree of v and $|T_{v_i}| \geq |T_{v_{i+1}}|$ for $1 \leq i \leq d-1$. After the time at which v first receives the message from its parent in T_r, for the next d steps it sends the message to its children in the order $v_1, v_2, \ldots, v_d$. The broadcast is then accomplished recursively in each of the subtrees T_{v_i}.

Let $B(n, d, \delta)$ denote the time taken by this scheme to complete broadcast using a directed tree on n nodes with maximum out-degree d and depth δ. We have the recurrence

$$B(n, d, \delta) = \max_{1 \leq i \leq d'} \{B(n_i, d, \delta - 1) + i\}$$

where n_i is the number of nodes in the i-th subtree of the root, and d' is the out-degree of the root of this tree on n nodes. Since we ordered the subtrees in non-increasing order of size we have $n_i \leq \frac{n}{i}$.

Claim (2).

$$B(n, d, \delta) \leq d \cdot \frac{\log_2 n}{\log_2 d} + \delta.$$

Exercise 3.6.23. Prove by induction the validity of Claim (2).

Applying Claim (2) to the tree T_r and noting that the sum (maximum degree + depth) of this tree is $\Theta(P(G))$, we have that the broadcast completes in $O(P(G) \cdot \frac{\log n}{\log P(G)})$ steps. This completes the proof of Theorem 3.6.22.   □

We note that the bound of $O(P \frac{\log n}{\log P})$ on the broadcast time of a scheme derived from a tree of poise P is existentially tight in the following sense: For any value of P, there is a tree of poise P using which any broadcast scheme takes $\Omega(P \frac{\log n}{\log P})$ steps to complete. An example of such a tree is a complete P-ary tree of depth roughly $\log_P n = \frac{\log n}{\log P}$.

Exercise 3.6.24. Prove formally that for any value of P, there is a tree of poise P using which any broadcast scheme takes $\Omega(P \frac{\log n}{\log P})$ steps to complete.

The following exercise is a research problem.

Exercise 3.6.25. Derive improved bounds on the approximability and non-approximability of the MBT problem in general graphs as well as specific subclasses.

3.6.3 Bibliographical Remarks

General references on approximation algorithms can be found in the books [ACGK+99, Ho97, Hr03, Va01].

The NP-completeness of the MBT problem in general graphs and in certain restricted graph classes was shown in [FHMP79, GJ79, Mi93, JM95, JRS98]. Solvability of the MBT problem in polynomial time for trees (Theorem 3.6.5) was proved in [Pr81, SCH81]. Polynomial-time solvability in certain graph classes was shown in [JRS98, DLOY98].

There are numerous papers in the literature concerned with the approximation of the MBT problem. Approximation algorithms for the MBT problem in general graphs were first studied in [KP95], where the authors derived an $(O(\sqrt{n})$ *additive* ratio by using the concept of "Decomposition into Clusters". [Ra94] presented an algorithm with an approximation ratio of $O(\log^2 |V| / \log\log |V|)$ based on the concept of the "Poise of a Graph". This approximation ratio was later improved to $O(\log |V|)$ [BGNS00] and to $O(\log |V| / \log\log |V|)$ [EK03a].

Approximation algorithms for certain graph classes have been considered in the literature as well. [Ra94] shows that the MBT problem is approximable within $2B$ if the degree of G is bounded by a constant B (Corollary 3.6.10). Approximability within $O(\log |V|)$ is shown in [KP95], e.g., for all chordal or k-outerplanar graphs. MBT is also approximable within $O(\log |V| / \log\log |V|)$ if G has a bounded tree width [MRSR+95].

Hardness of approximation was considered in [BGNS00, Sc00a, Sc00b, EK02]. In [BGNS00] it is shown that broadcasting is NP-hard to approximate within a factor of $3 - \varepsilon$ in a slightly more restricted communication model. In [Sc00a, Sc00b] it is proved that broadcasting is NP-hard to approximate within a factor of $57/56 - \varepsilon$. Finally, [EK02] shows that broadcasting is NP-hard to approximate within a factor of $3 - \varepsilon$.

Heuristic approaches (that behave well in practice but have no approximation guarantee) like the "matching-based heuristics", the "coloring heuristics", genetic algorithms or simulated annealing are applied to broadcasting, gossiping and other structured communication problems, e.g., in [SW84, FV96, FV97a, FV97b, FV99, BS00, CG98]. (General references on heuristics can be found, for example, in [Hr03].) Randomized algorithms for the MBT problem are developed in [FPRU90], whereas broadcasting in random graphs is considered in [KP95].

Approximation algorithms for various structured communication problems are studied in [BF97, BF00]. Approximation algorithms in the EDP/VDP modes are investigated in [CF00, CFM99, CFM02, Fr00, Fr01a, Fr01b]. Approximation algorithms for the directed telephone multicast problem are presented in [EK02, EK03b]. Inapproximability results for the radio broadcast problem are derived in [EK03c].

4

Gossiping

*Life is sweet and full of joy for him
who is capable of loving someone and
who has a clear conscience.
Grow spiritually and help others to grow.
That is the sense of life.*

Lev Nikolayevich Tolstoy

4.1 Introduction

This chapter is devoted to the gossip problem. Unlike the previous section we have to consider both one-way and two-way modes here. We shall mainly deal with the one-way mode because the two-way mode is so powerful in some cases that one can get optimal gossip algorithms in the two-way mode in a straightforward way by reducing the gossip problem to the broadcast problem (for instance, gossiping examples with graphs G for which $r(G) = 2 \cdot \min b(G)$ or $r(G) = r_2(G)$ or $r_2(G) = b(G)$, etc.). On the other hand, finding optimal one-way gossip algorithms is rather technical for most of the fundamental interconnection networks. We will see that even for very simple graph structures (such as a cycle) a lot of effort is necessary to find an optimal one-way broadcast algorithm and especially to prove its optimality.

It is impossible to provide the proofs of all important and interesting results for gossiping here. Hence, we have chosen only some which represent nice, principle proof ideas with possibly broader applications than only deriving the results presented.

Following the concept to start with simple techniques and to finish with more advanced methods, this chapter is structured as follows. In Section 4.2 we present some general lower and upper bounds holding for any graph and show a new way of describing gossip algorithms. This way is especially useful for proving lower bounds on gossiping in specific networks. In Section 4.3 we give some optimal algorithms for graphs with a small bisection width (paths, trees, rings, etc.). Here we also find several graphs G with the property $r(G) = 2 \cdot \min b(G)$ or $r(G) = r_2(G)$. Section 4.4 is concerned with gossiping in the hypercube-like interconnection networks of constant degree. Some efficient algorithms are presented which nicely use the optimal algorithms for gossiping in the ring as a subroutine. Section 4.5 is devoted to optimal gossiping in complete graphs. The design essentially uses the concept of the

Fibonacci number. The proof of its optimality is based on a very elegant algebraic technique whose potential is not restricted to the results presented here. As usual, the last section is devoted to bibliographical remarks and discusses open problems.

4.2 General Observations

First, we present a trivial lower bound on the gossip complexity of any graph.

Lemma 4.2.1. *Let $G = (V, E)$ be a graph with n nodes, $n \in I\!N$. Then*

$$\mathrm{r}(G) \geq \mathrm{r}_2(G) \geq \begin{cases} \lceil \log_2 n \rceil & \text{for even } n, \\ \lceil \log_2 n \rceil + 1 & \text{for odd } n. \end{cases}$$

Proof. From the definitions of communication problems, we have $\mathrm{r}(G) \geq \mathrm{r}_2(G) \geq \mathrm{b}(G)$. According to Observation 3.2.1, $\mathrm{b}(G) \geq \lceil \log_2 n \rceil$ holds for any n.

Let n be odd. To see that $\mathrm{r}_2(G) \geq \lceil \log_2 n \rceil + 1$, we first prove the following fact:

> Let $E_1, E_2, \ldots, E_i$ be an arbitrary two–way communication algorithm of i rounds for a graph G. Then after the execution of this algorithm none of the nodes of G knows more than 2^i pieces of information originally distributed in G.

We prove this fact by induction on i.

1. For $i = 0$ (no executed round) each node knows exactly $1 = 2^i$ pieces of information originally residing in it.
2. Let the fact (hypothesis) hold for $k \leq m$. Let us show it also holds for $m + 1$. Let E_{m+1} be the $(m + 1)$-st round of a communication algorithm $E_1, E_2, \ldots, E_m, E_{m+1}, \ldots, E_s$. Following the induction hypothesis each node knows at most 2^m pieces of information after the m-th round. Thus, no node can get more than 2^m new pieces of information from another one in the $(m + 1)$-st round. Since $2^m + 2^m = 2^{m+1}$, each node knows at most 2^{m+1} pieces of information after the $(m + 1)$-st round.

Using the hypothesis above, we see that after $i = \lceil \log_2 n \rceil - 1$ rounds, each node of G knows at most $2^i = 2^{\lceil \log_2 n \rceil - 1}$ pieces of information. As the number n of nodes is odd, there is a node v_0 which cannot take part in any communication in round $\lceil \log_2 n \rceil$. Hence, v_0 has at most $2^{\lceil \log_2 n \rceil - 1} < n$ pieces of information after $\lceil \log_2 n \rceil$ rounds.

$\square$

Exercise 4.2.2. * Prove, for any graph G of n nodes,

$$\mathrm{r}(G) \geq \log_2 n + \Omega(\log_2 n).$$

Now, the question appears whether there exist graphs of n nodes with the property $r_2(G) = \lceil \log_2 n \rceil$. These graphs are called **minimal gossip graphs**. The answer is "yes". To show this we have to find a graph and a communication strategy which enables us to double the knowledge of each node in each round. We show that any graph of 2^m nodes having the hypercube H_m as a spanning subgraph has this property. Later we will give more examples of graphs having this property.

Lemma 4.2.3. *For any positive integer m:*

$$r_2(H_m) = m.$$

Proof. Clearly, $E(H_m)$ has exactly $m \cdot 2^{m-1}$ edges which can be distributed into m equal-sized disjoint sets $E_1, \ldots, E_m$, where

$$E_i = \{((\alpha_1\alpha_2 \ldots \alpha_{i-1}0\alpha_{i+1} \ldots \alpha_m), (\alpha_1\alpha_2 \ldots \alpha_{i-1}1\alpha_{i+1} \ldots \alpha_m))$$
$$\mid \alpha_j \in \{0,1\} \text{ for } j = 1, 2, \ldots, i-1, i+1, \ldots, m\}$$

for $i = 1, \ldots, m$. The sequence $E_1, \ldots, E_m$ (or any other permutation of this sequence) is a two-way gossip algorithm for H_m.

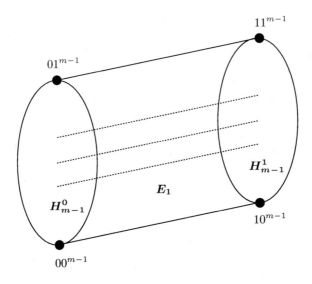

Fig. 4.1.

To see this it is sufficient to realize that by removing the edges in E_1 from H_m one gets two isolated $(m-1)$-dimensional hypercubes H_{m-1}^0 and H_{m-1}^1, and that $(V(H_m), E_1)$ is a bipartite graph of degree 1. Hence, after the first round E_1, the nodes in H_{m-1}^k (for $k = 0, 1$) together know the whole cumulative message of H_m, and to complete the task it suffices to gossip in the $(m-1)$-dimensional hypercubes H_{m-1}^0 and H_{m-1}^1 separately. $\square$

We note that there is no constant-degree interconnection network G of n nodes having the property $r_2(G) = \lceil \log_2 n \rceil$ because the doubling of the information is only possible if each node is active in each round via another edge. Thus, the degree $\lceil \log_2 n \rceil$ is necessary for each G with the property $r_2(G) = \lceil \log_2 n \rceil$.

Now, we give a general upper bound on the gossip complexity.

Lemma 4.2.4. *For every graph* $G = (V, E)$ *with* $|V| = n$,

(i) $r(G) \leq 2n - 2$, *and*
(ii) $r_2(G) \leq 2n - 3$.

Proof. As we have observed in Section 3.3, $\min b(G) \leq n-1$ for every graph G of n nodes. Following Lemma 2.5.10 claiming $r(G) \leq 2 \min b(G)$ and $r_2(G) \leq 2 \min b(G) - 1$, we obtain (i) and (ii).

□

To see that the above upper bounds cannot be improved, we consider the tree $T_k^1 = (\{v_0, v_1, \ldots, v_k\}, \{(v_0, v_i) \mid i = 1, \ldots, k\})$.

Observation 4.2.5. *For every positive integer* k,

(i) $r(T_k^1) = 2k$, *and*
(ii) $r_2(T_k^1) = 2k - 1$.

Proof. The upper bounds $r(T_k^1) \leq 2k$ and $r_2(T_k^1) \leq 2k-1$ follow from Lemma 4.2.4. To obtain the lower bounds we observe that every communication round in T_k^1 can contain at most one edge. In every one-way gossip algorithm for T_k^1 every node from the nodes $v_1, v_2, \ldots, v_k$ has to be the receiver and the sender[1] at least once. Since T_k^1 does not contain any edge (v_i, v_j) for $i, j \in \{1, \ldots, k\}$, $2k$ rounds are necessary to arrange it[2]. So, $r(T_k^1) \geq 2k$.

The proof of $r_2(T_k^1) \geq 2k - 1$ is left as an exercise to the reader.

□

Exercise 4.2.6. Prove, for every $k \in \mathbb{N} \setminus \{0\}$,

$$r_2(T_k^1) \geq 2k - 1.$$

We conclude this section with some definitions which enable us to see gossip algorithms from a point of view other than as a sequence of communication rounds. This point of view is especially useful if one wants to prove lower bounds on the complexity of the gossip problem in some graphs.

---

[1] This is obviously true for every gossip algorithm A for any graph G because each node has to give its information to other nodes and each node has to obtain at least one message during the execution of A.

[2] Every edge (v_0, v_i) has to be used twice. Once as $(v_i \rightarrow v_0)$ and once as $(v_0 \rightarrow v_i)$.

Definition 4.2.7. *Let* $G = (V, E)$ *be a graph, and let* $X = x_1, \ldots, x_m$ *be a simple path (i.e.,* $x_i \neq x_j$ *for* $i \neq j$*) in* G. *Let* $A = E_1, \ldots, E_s \; [R_1, \ldots, R_s]$ *be a communication algorithm in two-way [one-way] mode. Let* $T = t_1, \ldots, t_{m-1}$ *be an increasing sequence of positive integers such that* $(x_i, x_{i+1}) \in E_{t_i}$ *[*$(x_i \to x_{i+1}) \in R_{t_i}$*] for* $i = 1, \ldots, m-1$. *We say that* $X[t_1, \ldots, t_{m-1}]$ *is a* **time-path of** A *(Figure 4.2)) because it provides the information flow from* x_1 *to* x_m *in* A. *If* $t_{i+1} - t_i - 1 = k_i \geq 0$ *for some* $i \in \{1, \ldots, m-2\}$ *then we say that* $X[T] = X[t_1, \ldots, t_{m-1}]$ *has a* k_i**-delay** *at the node* x_{i+1}.
The **global delay of** $X[t_1, \ldots, t_{m-1}]$ *is*

$$\mathbf{delay}(X[t_1, \ldots, t_{m-1}]) = t_1 - 1 + \sum_{i=1}^{m-2} k_i.$$

The **global time of** $X[t_1, \ldots, t_{m-1}]$ *is*

$$\mathbf{time}(X[T]) = m - 1 + \mathrm{delay}(X[t_1, \ldots, t_{m-1}]).$$

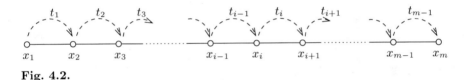

Fig. 4.2.

Observe that a delay at a node x_i is the waiting (passive) time[3] of the node x_i from the point of view of the time-path $X[T]$. The global delay, delay$(X[T])$, for $X = x_1, \ldots, x_m$ is the sum of the delays at the nodes x_i for $i = 2, \ldots, m-1$ plus $t_1 - 1$ (the number of rounds before the communication via the time-path $X[T]$ starts).

The global time, time$(X[T])$, is none other than t_{m-1}, i.e., the round in which the communication along $X[T]$ has finished.

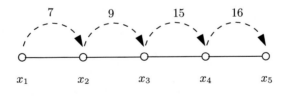

Fig. 4.3.

---

[3] The node x_i may be active in the rounds $t_{i-1} + 1, t_{i-1} + 2, \ldots, t_i - 1$ but it is passive from the point of view of $X[T]$ in these rounds because the message received by x_i in round t_{i-1} will be submitted to x_{i+1} in round t_i and not before with respect to T.

The time-path $x_1, x_2, x_3, x_4, x_5[7, 9, 15, 16]$ in Figure 4.3 has a 1-delay at the node x_2, a 5-delay at the node x_3 and a 0-delay at the node x_4. Since the communication starts in the round 7 on this time-path, the global delay is $6 + 1 + 5 + 0 = 12$. The global time of this time-path is $4 + 12 = 16 = t_4$.

The existence of a time-path (of a communication algorithm A) starting in a node v and finishing in a node u assures us that u will learn the message $I(v)$ in A.

Thus, the necessary and sufficient condition for a communication algorithm A to be a gossip algorithm in a graph G is the existence of time-paths of A among all pairs of nodes in G, i.e., one can view the gossip algorithm for a graph G as a set of time-paths between any ordered pair of nodes.

From this point of view, the complexity (the number of rounds) of a communication algorithm can be measured as

$$\max\{\text{length of } X[T] + \text{global delay of } X[T] \mid X[T] \text{ is a time-path of } A\}$$
$$= \max\{\text{time}(X[T]) \mid X[T] \text{ is a time-path of } A\}.$$

Viewing a gossip algorithm as a set of time-paths is mainly helpful for proving lower bounds. A collision of two time-paths (the meeting of two time-paths at the same node at the same time) causes some delays on these time-paths (because of the restriction given by the communication modes). Too many unavoidable collisions mean too many delays, and hence one can derive much better lower bounds for gossiping in some graphs G than the trivial lower bounds $\text{diam}(G)$ and $\text{rad}(G)$ provide. A combinatorial analysis providing lower bounds by analyzing the numbers of collisions and delays requires a precise definition and use of the terms "collision" and "delay". We introduce some definitions of these terms here. These specifications are suitable for our purposes but the reader is reminded that one is free to modify these definitions to suit the problem under analysis.

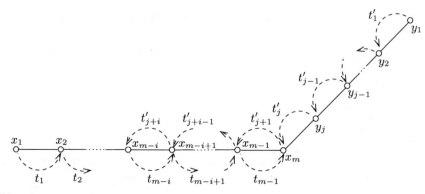

Fig. 4.4.

Definition 4.2.8. *Let* $G = (V, E)$ *be a graph, and let* $A = E_1, \ldots, E_s$ *be a communication algorithm in the two-way mode. Let*

$$X[T] = x_1, \ldots, x_m[t_1, \ldots, t_{m-1}]$$

and

$$Y[T'] = y_1, y_2, \ldots, y_j, x_m, x_{m-1}, \ldots, x_{m-i+1}, x_{m-i}[t'_1, \ldots, t'_{j+i}]$$

be two time-paths of A *for some* $i \in \{0, \ldots, m-1\}$, *and* $y_l \neq x_r$ *for all* $l \in \{1, \ldots, j\}, r \in \{1, \ldots, m\}$ *(Figure 4.4).*

We say that $X[T]$ and $Y[T']$ have a collision in a node x_k *for some* $k = m - d$ *(see Figure 4.5) if*

(i) $t_{k-2} = t'_{j+d-1}$ *or* $\max\{t_{k-2}, t'_{j+d-1}\} < \min\{t_{k-1}, t'_{j+d}\}$

  {*The condition (i) predetermines the conflict in the node* x_k, *because (i) ensures that messages flowing via the time-paths* $X[T]$ *and* $Y[T']$ *have reached the nodes* x_{k-1} *and* x_{k+1} *resp. before any of them has reached the node* x_k.}

(ii) $t_{k-1} < t'_{j+d}$ *or* $t'_{j+d} < t_{k-1}$.

  {*These two conditions describe all possible solutions of the conflict predetermined by (i).*}

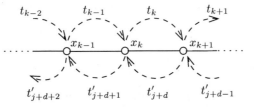

Fig. 4.5.

If $t_{k-1} < t'_{j+d}$ *we say that* **the collision causes at least a 1-delay on the time-path** $Y[T']$. *If* $t'_{j+d} < t_{k-1}$ *we say that* **the collision causes at least a 1-delay on the time-path** $X[T]$.

Exercise 4.2.9. For which n's one can be sure that every two-way optimal gossip algorithm for P_n contains at least one delay on some of its paths?

Next we define the collisions for one-way communication algorithms. It can easily be observed that a collision in this communication mode causes at least one 2-delay on one of the two time-paths in the collision or a 1-delay on each of these two paths.

Definition 4.2.10. *Let* $G = (V, E)$ *be a graph, and let* $A = E_1, \ldots, E_s$ *be a communication algorithm in the one-way mode. Let* $X[T] = x_1, \ldots, x_m$ $[t_1, \ldots, t_{m-1}]$, *and* $Y[T'] = y_1, y_2, \ldots, y_j, x_m, x_{m-1}, \ldots, x_{m-i}$ $[t'_1, \ldots, t'_{j+i}]$ *be two time-paths of* A *for some* $i \in \{1, \ldots, m-1\}$ *and* $y_l \neq x_r$ *for all* $l \in \{1, \ldots, j\}, r \in \{1, \ldots, m\}$. *We say that* $X[T]$ *and* $Y[T']$ **have a collision in a node** x_k *for some* $k = m - d$ *(see Figure 4.6) if*

(i) $t_k > t'_{j+d+1} > t_{k-1} > t'_{j+d}$, *or*
(ii) $t'_{j+d+1} > t_k > t'_{j+d} > t_{k-1}$, *or*
(iii) $t_k > t'_{j+d+1} > t'_{j+d} > t_{k-1}$, *or*
(iv) $t'_{j+d+1} > t_k > t_{k-1} > t'_{j+d}$.

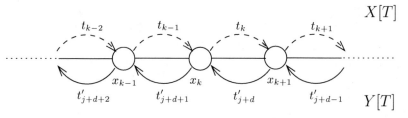

Fig. 4.6.

If (i) or (ii) occurs, then we say that **the collision causes a 1-delay on** $X[T]$ *and* **at least a 1-delay on** $Y[T']$. *If (iii) occurs, then we say that* **the collision causes at least a 2-delay on** $X[T]$. *If (iv) occurs, then we say that* **the collision causes at least a 2-delay on** $Y[T']$.

We note that our Definitions 4.2.8 and 4.2.10 are not the only possibilities for formalizing the notions of collisions and delays caused by distinct information flows. If some combinatorial analysis requires a definition of collisions which covers more conflicts appearing in communication algorithms, then one can redefine this in several distinct ways. One such broader definition will be later considered in the proof of the precise lower bound for gossiping in rings. Since we prefer to give some basic proof ideas here rather than to present too detailed, technical proofs, Definitions 4.2.8 and 4.2.10 will be sufficient for most purposes here.

4.3 Gossiping in Graphs with Small Bisection Width

We shall investigate the gossip problem for graphs with small bisection width in this section. The reason for doing this first is that these graphs are relatively simple (there are not many disjoint paths between the pairs of nodes in such graphs) and hence the gossip problem in such communication structures is easier to analyze. We start with paths which are the simplest communication structure.

Theorem 4.3.1.

(1) $r_2(P_n) = n - 1$ *for any even integer* $n \geq 2$,
(2) $r_2(P_n) = n$ *for any odd integer* $n \geq 3$,
(3) $r(P_n) = n$ *for any even integer* $n \geq 2$, *and*
(4) $r(P_n) = n + 1$ *for any odd integer* $n \geq 3$.

Proof. We prove claims *(1),(2),(3)* and *(4)* separately.

(1) The obvious relation $r_2(P_n) \geq b(P_n) \geq \text{diam}(P_n) = n - 1$ provides the lower bound.

To prove the upper bound $r_2(P_n) \leq n - 1$ we state two gossip algorithms. Let $V(P_n) = \{x_1, \ldots, x_n\}$. The following gossip algorithm

$$\{(x_1, x_2), (x_{n-1}, x_n)\}, \{(x_2, x_3), (x_{n-2}, x_{n-1})\}, \ldots,$$
$$\{(x_{\frac{n}{2}-1}, x_{\frac{n}{2}}), (x_{\frac{n}{2}+1}, x_{\frac{n}{2}+2})\}, \{(x_{\frac{n}{2}}, x_{\frac{n}{2}+1})\},$$
$$\{(x_{\frac{n}{2}-1}, x_{\frac{n}{2}}), (x_{\frac{n}{2}+1}, x_{\frac{n}{2}+2})\}, \ldots, \{(x_1, x_2), (x_{n-1}, x_n)\}$$

works in $n - 1$ rounds. Another gossip algorithm working in $n - 1$ rounds is $A = E_1, E_2, \ldots, E_{n-1}$, where

$$E_i = \{(x_1, x_2), (x_3, x_4), \ldots, (x_{n-1}, x_n)\}$$

for odd i, and

$$E_j = E(P_n) - E_1$$

for all even j.

(2) The upper bound $r_2(P_n) \leq n$ follows from the following gossip algorithm $A' = E'_1, E'_2, \ldots, E'_{n-1}, E'_n$, where

$$E'_i = \{(x_1, x_2), (x_3, x_4), \ldots, (x_{n-2}, x_{n-1})\}$$

for i odd, and

$$E'_j = E(P_n) - E'_1$$

for all even j.

To prove the lower bound $r_2(P_n) \geq n$, we consider the paths

$$X = x_1, x_2, \ldots, x_n \text{ and } X^R = x_n, x_{n-1}, \ldots, x_1.$$

Obviously each gossip algorithm for P_n must contain two time-paths $X[T]$ and $X^R[T']$ for some $T = t_1, \ldots, t_{n-1}$ and $T' = t'_1, \ldots, t'_{n-1}$. Thus it is sufficient to prove that at least one of the two time-paths $X[T]$ and $X^R[T']$ has a global delay of at least 1 (note that the length of X is $n - 1$). Let us assume that there is no positive delay on the time-paths $x_1, \ldots, x_{\lceil \frac{n}{2} \rceil - 1}[t_1, \ldots, t_{\lceil \frac{n}{2} \rceil - 2}]$ and $x_n, x_{n-1}, \ldots, x_{\lceil \frac{n}{2} \rceil + 1}[t'_1, \ldots, t'_{\lceil \frac{n}{2} \rceil - 2}]$ (otherwise, we are done), i.e., that $t_i = t'_i = i$ for $i = 1, \ldots, \lceil \frac{n}{2} \rceil - 2$.

Following Definition 4.2.8 we see that there must be a collision between $X[T]$ and $X^R[T']$ in the node $x_{\lceil \frac{n}{2} \rceil}$ (Figure 4.5). Thus at least one of $X[T]$ and $X^R[T']$ has a positive global delay, i.e., the maximum of the global time of $X[T]$ and the global time of $X^R[T']$ is at least n.

(3)   The gossip algorithms showing $r(P_n) \leq 2 \cdot \lceil \frac{n}{2} \rceil$ for any $n \geq 2$ can be derived by combining Example 2.5.9 and Lemma 2.5.10.

Any one-way gossip algorithm for P_n must contain the time-paths

$$X[T] = x_1, \ldots, x_n[t_1, \ldots, t_{n-1}] \text{ and } X^R[T'] = x_n, \ldots, x_1[t'_1, \ldots, t'_{n-1}].$$

Because these time-paths are going in the opposite direction on the same path $x_1, \ldots, x_n$ there must be a collision between $X[T]$ and $X^R[T']$ at some node x_i. Obviously this implies $r(P_n) \geq n$ for any $n \geq 2$, which completes the proof of claim 3.

(4)   Now, we prove $r(P_n) \geq n + 1$ for any odd n. Again, a collision occurs at some node x_i. If $x_i \neq x_{\lceil \frac{n}{2} \rceil}$, one of the time-paths already has a delay before the collision, and we are done because any collision in the one-way mode requires either a 2-delay for some time-path or a 1-delay for each of the two paths in collision. Assume $x_i = x_{\lceil \frac{n}{2} \rceil}$ and $t_j = t'_j = j$ for $j = 1, 2, \ldots, \lceil \frac{n}{2} \rceil - 2$ (see Definition 4.6). Then, the collision in $x_{\lceil \frac{n}{2} \rceil}$ is one of the 4 types (i), (ii), (iii), (iv) of Definition 4.2.10. Obviously (iii) and (iv) cause a 2-delay on one of the time-paths which completes the proof. Now, we consider case (i) (case (ii) is analogous), where $\lceil \frac{n}{2} \rceil - 1 = t_{\lceil \frac{n}{2} \rceil - 1} < t'_{\lceil \frac{n}{2} \rceil - 1} < t_{\lceil \frac{n}{2} \rceil} < t'_{\lceil \frac{n}{2} \rceil}$. Obviously, $t'_{\lceil \frac{n}{2} \rceil - 1} - t'_{\lceil \frac{n}{2} \rceil - 2} > 1$ and $t'_{\lceil \frac{n}{2} \rceil} - t'_{\lceil \frac{n}{2} \rceil - 1} > 1$. Thus, this collision causes at least two positive delays on the path $X^R[T']$.

$\square$

Now, we estimate the gossip complexity for complete k-ary trees. To do so we start with a technical lemma that provides an important property of all trees.

Lemma 4.3.2. *For any tree T,*

$$r(T) = 2 \min b(T).$$

Proof. From Lemma 2.5.10, we have $r(T) \leq 2 \cdot \min b(T)$ for any tree T. To show $r(T) \geq 2 \min b(T)$, let $A = E_1, \ldots, E_s$ be an arbitrary one-way gossip algorithm for T, and let t_A be the first round after which at least one node of T knows the cumulative message (i.e., $\text{Accu}_{t_A}^A \neq \emptyset$ and $\text{Accu}_{t_A - 1}^A = \emptyset$). Obviously $t_A \geq \min b(T)$. Note that $\text{Accu}_{t_A}^A$ contains all nodes having the cumulative message after t_A rounds. We show via contradiction that

$$|\text{Accu}_{t_A}^A| = 1.$$

Let $u, v \in \text{Accu}_{t_A}^A$, $u \neq v$. Because T is a tree there exists exactly one path $u, y_1, \ldots, y_k, v$ (k may be 0) between u and v (Figure 4.7).

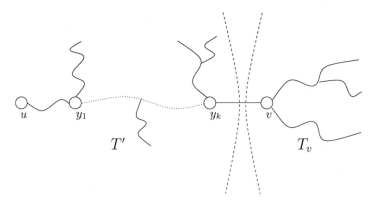

Fig. 4.7.

Let T_v be the subtree rooted at v excluding the edge (v, y_k) and the subtree rooted at y_k. Let T' be the subtree rooted at y_k excluding the edge (y_k, v) and the tree T_v (Figure 4.7). Let $t \le t_A$ be the last round in which (y_k, v) was used for communication in the first t_A rounds of A. We distinguish between two possibilities depending on the direction of the communication.

1. Let $(y_k \to v) \in E_t$. Then y_k must already know the cumulative message of T' after the $(t-1)$-th round (if not then v cannot know the cumulative message of T after t_A rounds because all pieces of information originally residing in T' can flow to v only via y_k). Since $(y_k \to v)$ is the last use of the edge (y_k, v) in $E_1, E_2, \ldots, E_{t_A}$ (i.e., no information from T_v will flow to u later in the rounds $E_t, \ldots, E_{t_A}$), any piece of information originally distributed in T_v can flow to u only via y_k, and u must know the cumulative message of T_v after t_A rounds of A, and y_k must already know the cumulative message of T_v before the t-th round (after the last round containing $(v \to y_k)$). Thus, y_k already knows the whole cumulative message of T (union of the cumulative messages of T' and T_v) before the t-th round. But this is a contradiction to the assumption that no node knows the cumulative message after $t_A - 1$ rounds.
2. Let $(y_k \leftarrow v) \in E_t$. In the same way as with the first case, it can be shown that v must already learn the cumulative message of T before the t_A-th round, which is again in contradiction with the fact $\text{Accu}_{t_A - 1}^A = \emptyset$.

Now, we may assume $\text{Accu}_{t_A - 1}^A = \{w\}$ for a node w in T. Let us view w as the root of T with k sons $w_1, w_2, \ldots, w_k$ (Figure 4.8) for some positive integer k.

Let T_i denote the subtree rooted at w_i for $i = 1, \ldots, k$. Since w knows the cumulative message after t_A rounds, each w_i knows the cumulative message of T_i (each information flowing from a node in T_i to w must flow via w_i), and no node in T_i knows a piece of information which is unknown to w_i (each piece of information flowing to a node in T_i from a node outside of T_i must

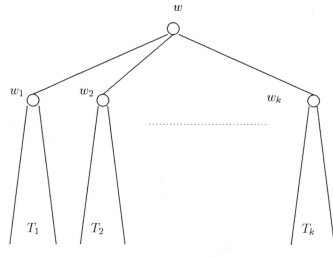

Fig. 4.8.

flow via w_i) for any $i \in \{1, \ldots, k\}$. On the other hand, none of the nodes $w_1, w_2, \ldots, w_k$ knows the cumulative message of T after t_A rounds. Thus, for each $i \in \{1, \ldots, k\}$, there is a piece of information $p(i)$ which is unknown to every node in T_i. We see that to complete A after the t_A-th round, the remaining rounds $E_{t_A+1}, \ldots, E_s$ must contain the broadcast of $p(1)$ from w in T_1, the broadcast of $p(2)$ from w in T_2, etc. Obviously, this cannot be easier than broadcasting from w in T. Hence,

$$s = t_A + (s - t_A) \geq \min b(T) + \min b(T) \geq 2 \min b(T).$$

$\square$

Following Lemma 4.3.2 and Lemma 3.2.8, we get the following result.

Theorem 4.3.3.

$$r\left(T_k^m\right) = 2 \min b\left(T_k^m\right) = 2 \cdot k \cdot m \quad \text{for all integers } m \geq 1, k \geq 2. \qquad \square$$

How Lemma 4.3.2 can be extended to the two-way mode is formulated in the following exercise.

Exercise 4.3.4. Prove, for any tree T, $r_2(T) = 2 \min b(T) - 1$.

Now, the question arises whether there exist non-tree graphs G with $r(G) = 2 \min b(G)$. The answer is "yes" and the next technical lemma enables us to find such graphs. The proof for this lemma is a generalization of the idea used in the proof for Lemma 4.3.2.

Lemma 4.3.5. *Let G be a graph with a bridge (v, u) whose removal (from G) divides G into two components G_1 and G_2. Then*

$$r(G) \geq \min b(G) + 1 + \min\{\min b(G_1), \min b(G_2)\}.$$

Proof. Let A be a one-way gossip algorithm for G. Let t be the time unit in which at least one node of G has learned the whole cumulative message (i.e., all pieces of information distributed in G), and no node of G knows the cumulative message at time $t-1$ (i.e., after $t-1$ rounds). Let $G = (V, E)$, $G_1 = (V_1, E_1)$, $G_2 = (V_2, E_2)$, $u \in V_1$, $v \in V_2$. Note that Accu_t^A is the set of all nodes that know the whole cumulative message after t rounds. We shall prove by contradiction that

either $\text{Accu}_t^A \subseteq V_1$ or $\text{Accu}_t^A \subseteq V_2$.

Let there exist two nodes

$v_1 \in V_1 \cap \text{Accu}_t^A$ and $v_2 \in V_2 \cap \text{Accu}_t^A$.

Since $v_1[v_2]$ knows all pieces of information distributed in G_2 $[G_1]$ after t rounds, and the whole information exchange between G_1 and G_2 flows through the edge (v, u), the whole cumulative message has flown through the edge (v, u) in the first t rounds. Thus, the nodes v and u belong to Accu_t^A. But this is impossible because if the last information exchange between u and v was from u (v) to v (u) in a round $t' \leq t$ then u (v) has already learned the cumulative message before this information exchange (i.e., before the round t) [Figure 4.9].

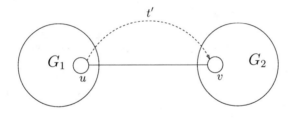

Fig. 4.9.

We have proved that either $\text{Accu}_t^A \subseteq V_1$ or $\text{Accu}_t^A \subseteq V_2$. Without loss of generality we can assume that $\text{Accu}_t^A \subseteq V_1$. Since the nodes in $\text{Accu}_t^A \subseteq V_1$ know all pieces of information distributed in G_2, node $v \in V_2$ must also know all pieces of information distributed in G_2. Since $v \notin \text{Accu}_t^A$, v does not learn at least one of the pieces of information distributed in G_1 in the first t rounds. Thus, we need at least $1 + b(v, G_2)$ rounds to distribute this piece of information in G_2.

Clearly, in the case $\text{Accu}_t^A \subseteq V_2$ we need at least $1 + b(u, G_1)$ rounds to finish the gossiping after t rounds. Since $t \geq \min b(G)$ we obtain the claimed inequality.

□

Now, we show that Lemma 4.3.5 provides optimal lower bounds for gossiping on some infinite class of graphs. Consider two cycles R_1 and R_2, each

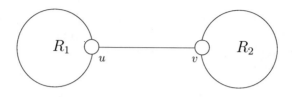

Fig. 4.10.

with n nodes, n even, connected by one edge (u, v) (Figure 4.10) as a graph G_n.

Using Example 2.5.9, $\min b(R_1) = \min b(R_2) = \frac{n}{2}$, and $\min b(G_n) = \frac{n}{2} + 1$. Applying Lemma 4.3.5 we obtain

$$r(G_n) \geq n + 2.$$

An optimal gossip algorithm for G_n first concentrates the cumulative information in u in $\min b(G_n) = \frac{n}{2} + 1$ rounds and then it disseminates the cumulative message from u to all nodes in G in $\min b(G_n)$ rounds.

Thus, for two connected cycles of the same size we have proved $r(G) = 2 \min b(G)$, i.e., we have found graphs different from trees with the property that gossiping is exactly two times harder than broadcasting. We note that Lemma 4.3.5 provides optimal lower bounds for $r(G)$ of several further graphs (see, for example, some trees, cycles connected by one simple path, etc.). Many of them also have the property $r(G) = 2 \min b(G)$. (To be more precise, all graphs with the structure of Figure 4.10 and for which $\min b(G) = b(u, G_1) + 1 = b(v, G_2) + 1$.)

Exercise 4.3.6. Find, for any integer $k \geq 3$, an infinite class of graphs $\{G_n\}_{n=1}^{\infty}$ such that $r(G_n) = 2 \min b(G_n)$ and G_n contains at least k distinct cycles.

Now, we present a version of Lemma 4.3.5 providing lower bounds for the two-way communication mode.

Lemma 4.3.7.   *Let G be a graph with a bridge (v, u) whose removal divides G into two components G_1 and G_2. Then*

$$r_2(G) \geq \min b(G) + \min\{\min b(G_1), \min b(G_2)\}.$$

Sketch of the Proof: Similarly as in the proof of Lemma 4.3.5 it can be proved that either $\mathrm{Accu}_t^A \subseteq V_i$ for some $i \in \{1, 2\}$ or that $\mathrm{Accu}_t^A = \{v, u\}$. $\mathrm{Accu}_t^A = \{v, u\}$ holds exactly in the case when v and u make an information exchange in the t-th round.

Clearly, if $\mathrm{Accu}_t^A \subseteq V_i$ for some $i \in \{1, 2\}$ then

$$r_2(G) \geq t + 1 + \min b(G_j) \geq \min b(G) + \min b(G_j) + 1$$

for $j \in \{1,2\} - \{i\}$. If $\mathrm{Accu}_t^A = \{v, u\}$ then

$$r_2(G) \geq t + \max\{\min b(G_1), \min b(G_2)\}.$$

Thus,

$$r_2(G) \geq \min b(G)$$
$$+ \min\{1 + \min b(G_1), 1 + \min b(G_2), \max\{\min b(G_1), \min b(G_2)\}\}.$$

$\square$

Exercise 4.3.8. Give a detailed proof of Lemma 4.3.7.

Exercise 4.3.9. Let G be a graph with a node v whose removal divides G into two components G_1 and G_2. Prove an assertion analogous to the assertion of Lemmas 4.3.5 and 4.3.7 for such graphs.

Considering the graph G_n consisting of two connected cycles (Figure 4.10), Lemma 4.3.7 implies an optimal lower bound $r_2(G_n) \geq n + 1$. Thus G_n is an interesting example because the two-way mode decreases the complexity of gossiping only by 1 (note that both upper bounds of Lemma 2.5.10, $r(G_n) = 2 \cdot \min b(G_n)$ and $r_2(G_n) = 2 \cdot \min b(G_n) - 1$ are satisfied).

Next, we shall establish the exact values for $r(C_n)$ and $r_2(C_n)$. This is of importance because in Section 4.4 we shall design algorithms for gossiping in some prominent interconnection networks whose efficiency strongly depends on a subroutine that performs the gossiping in cycles. While finding an optimal two-way gossip algorithm for the cycle C_n is a simple task, the one-way version of this task is already hard. Since the lower bound proof for $r(C_n)$ is very technical we do not start this detailed combinatorial analysis here. First we illustrate the proof idea based on the analysis of collisions by proving a weaker lower bound in a transparent way. Then we show how to improve the considerations for the weaker lower bound to obtain the exact estimation on $r(C_n)$. First of all we present the estimation of $r_2(C_n)$ that can be easily proved.

Theorem 4.3.10.

$$r_2(C_k) = \frac{k}{2} \text{ for even } k \geq 4, \text{ and } r_2(C_k) = \lceil \frac{k}{2} \rceil + 1 \text{ for odd } k \geq 3.$$

Proof. We provide the proof only for even k. The case for k odd is left as an exercise for the reader.

Obviously, $\mathrm{rad}(C_k) = \frac{k}{2}$ and so $r_2(C_k) \geq \frac{k}{2}$. Let $V(C_k) = \{x_1, \ldots, x_k\}$. The gossip algorithm for C_k is $A = E_1, E_2, \ldots, E_{\frac{k}{2}}$, where

$$E_i = \{(x_1, x_2), (x_3, x_4), \ldots, (x_{k-1}, x_k)\}$$

for all odd i, and

$$E_j = \{(x_2, x_3), (x_4, x_5), \ldots, (x_{k-2}, x_{k-1}), (x_k, x_1)\}$$

for all even j. To see that A is a two-way gossip algorithm it is sufficient to realize that after i rounds each node knows exactly $2i$ pieces of information.

□

Exercise 4.3.11. Prove by induction on the number of rounds that the algorithm A given in the proof of Theorem 4.3.10 is a two-way gossip algorithm.

Exercise 4.3.12. Give a detailed proof of the fact $r_2(C_k) = \lceil \frac{k}{2} \rceil + 1$ for every odd $k \geq 3$.

Next, we present the optimal one-way gossip algorithm in cycles of even length.

Theorem 4.3.13.

$$r(C_n) \leq \frac{n}{2} + \lceil \sqrt{2n} \rceil - 1 \text{ for each even } n > 3, \text{ and}$$

$$r(C_n) \leq \lceil \frac{n}{2} \rceil + \lceil 2\sqrt{\lceil \frac{n}{2} \rceil} \rceil - 1 \text{ for each odd } n \geq 3.$$

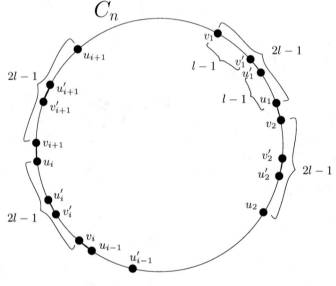

Fig. 4.11.

Proof. To explain the idea we first give the algorithm for $n = 2l^2$, where l is even. Then we extend the algorithm for any positive integer n. We divide the cycle C_n into l disjoint paths of $2l$ nodes (i.e., of length $2l - 1$), the i-th path starting with v_i and ending with u_i, as depicted in Figure 4.11. Let v'_i [u'_i] be a node between v_i and u_i with the distance $l - 1$ from v_i [u_i]. The algorithm works in the following two phases.

<u>The first phase</u> For each $i \in \{1, \ldots, l\}$:

there is a time-path of length $\frac{n}{2}$ from v_i to $v_{(i+\frac{l}{2}-1) \bmod l+1}$ going through u_i, and

there is a time-path of length $\frac{n}{2} - 1$ from u_{i-1} to $v_{(i+\frac{l}{2}-1) \bmod l+1}$ going through v_{i-1}.

Observe that the time-paths starting in v_i's go in an opposite direction to the time-paths starting in u_i's and that both time-paths starting in u_{i-1} and v_i finish in the same node (Figure 4.12).

Thus, after the first phase all nodes v_i already know the cumulative message because for each v_i there are two time-paths that finish in v_i: one from $v_{(i+\frac{l}{2}-1) \bmod l+1}$ to v_i and the second one from $u_{(i+\frac{l}{2}-1-1) \bmod l+1}$ to v_i.

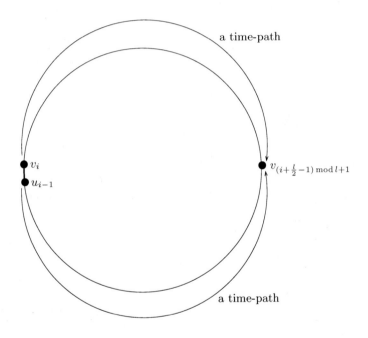

a time-path

v_i

u_{i-1}

$v_{(i+\frac{l}{2}-1) \bmod l+1}$

a time-path

Fig. 4.12.

<u>The second phase</u> For each $i \in \{1, \ldots l\}$:

v_i sends the cumulative message to u_{i-1},

v_i sends the cumulative message to v_i',

u_i sends the cumulative message to u_{i-1}'.

In this second phase the node v_i broadcasts the cumulative message in the cycle part from u_{i-1}' to v_i' by the strategy depicted in Figure 4.13. Clearly, each node knows the cumulative message after the second phase.

Now, let us count the number of rounds. Each time-path starting in a v_i in the first phase has length $\frac{n}{2}$ and exactly $l - 1$ collisions with $l - 1$ time-paths going in the opposite direction. When the collision of two time-paths is solved in such a way that the collision causes a 1-delay for each time-path, then the first phase uses $\frac{n}{2} + l - 1$ rounds. Since the distance between v_i and v_i' is $l - 1$ and the distance between u_{i-1} and u_{i-1}' is $l - 1$, the second phase uses l rounds (Figure 4.13). One can simply see that $\frac{n}{2} + 2l - 1 = \frac{n}{2} + \lceil \sqrt{2n} \rceil - 1$.

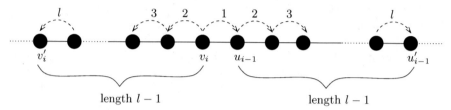

Fig. 4.13.

Now, we give an algorithm for even n. For each even, positive integer $n > 3$ there is a positive integer l such that

$$2l^2 \le n < 2(l + 1)^2 = 2l^2 + 4l + 2.$$

Thus $n = 2l^2 + 2i$ for some $i \in \{0, 1, \ldots, 2l\}$. If $1 \le i \le l$ then we divide the cycle into l parts $P_1, \ldots, P_l$ such that i parts have $2(l + 1)$ nodes and $l - i$ parts have $2l$ nodes. For each i, part P_i starts with the node v_i and ends with the node u_i (similarly as in Figure 4.11, only the distances between v_i and u_i may be different for distinct i's). The first phase of the generalized algorithm executes the time-paths of length $\frac{n}{2}$ from all nodes v_i, and the time-paths of length $\frac{n}{2} - 1$ from all nodes u_i. Again, the first phase can be performed in $\frac{n}{2} + l - 1$ rounds. After the first phase we have exactly l cumulative points (nodes that know the cumulative message after the first phase). Note that these points may be different from the v_i's and u_i's in this case, and the distance between two neighboring cumulative points is at most $2(l + 1)$. Thus, the second phase of the distribution of the cumulative message can be executed in $l + 1$ rounds.

Since

$$2 \cdot \sqrt{\frac{n}{2}} = 2\sqrt{l^2 + i} < 2(l + \frac{1}{2}) \text{ for } 1 \le i \le l,$$

the total number of rounds is

$$\frac{n}{2} + 2l = \frac{n}{2} + (2l + 1) - 1 = \frac{n}{2} + \lceil 2 \cdot \sqrt{\lceil \frac{n}{2} \rceil} \rceil - 1.$$

If $l < i \le 2l$ then the cycle is divided into l parts, where $i - l$ parts have length $2(l+2)$ and $l - (i - l) = 2l - i$ parts have length $2(l+1)$. The algorithm for gossiping works exactly in the way described above, the only difference is that the second phase uses $l+2$ rounds instead of $l+1$ rounds. Thus the total number of rounds of the algorithm is $\frac{n}{2} + 2l + 1$. Since

$$2 \cdot \sqrt{\frac{n}{2}} = 2\sqrt{l^2 + i} > 2(l + \frac{1}{2}) \text{ for } l < i \leq 2l$$

we have

$$\frac{n}{2} + 2l + 1 = \frac{n}{2} + (2l + 2) - 1 = \frac{n}{2} + \lceil 2\sqrt{\frac{n}{2}} \rceil - 1.$$

In the case that $n > 1$ is an odd positive integer one can use the algorithm for C_{n+1} to design the algorithm for C_n. The number of rounds of this algorithm is at most

$$r(C_{n+1}) = \frac{n+1}{2} + \lceil 2\sqrt{\frac{n+1}{2}} \rceil - 1 = \lceil \frac{n}{2} \rceil + \lceil 2\sqrt{\lceil \frac{n}{2} \rceil} \rceil - 1.$$

$\square$

In what follows we deal with the lower bound. Because the complete proof of the optimal lower bound requires too many specific considerations, we first present the following weaker lower bound:

$$r(C_n) \geq \frac{n}{2} + \frac{\sqrt{2n}}{4} - O(1).$$

The proof of this lower bound will be sufficient for learning more about the combinatorial lower bound proof technique based on the investigation of collisions. For the reader interested in learning the full applicability of this technique for rings, the full technical proof of the optimal lower bound is given later.

Lemma 4.3.14. *For every positive integer n*

$$r(C_n) \geq \frac{n}{2} + \frac{\sqrt{2n}}{4} - O(1).$$

Proof. Since the optimal algorithm for gossiping in C_n in the two-way mode uses at least $\lceil \frac{n}{2} \rceil$ rounds (Theorem 4.3.10) we may assume that $r(C_n) = \lceil \frac{n}{2} \rceil + f(n)$ for some function f from positive to nonnegative integers. Next we shall show that $f(n) \geq \frac{\sqrt{2n}}{4} - O(1)$.

Let A be an arbitrary optimal algorithm for gossiping in C_n in one-way mode. Let A work in $t(A) = \lceil \frac{n}{2} \rceil + f(n)$ rounds. From the upper bound on $r(C_n)$ we know $f(n) \leq \lceil \sqrt{2n} \rceil$. Thus, each time-path of A has a global time of at most $\lceil \frac{n}{2} \rceil + f(n)$, i.e., there is no time-path in A for paths longer than $\lceil \frac{n}{2} \rceil + f(n)$. It implies that any two nodes x and y lying at distance $\lfloor \frac{n}{2} \rfloor - f(n) - 1$ must have two time-paths in A, one leading from x to y and

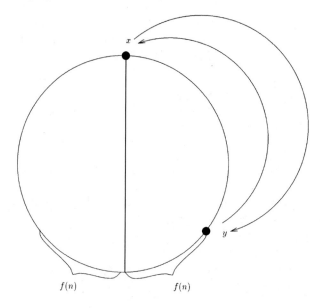

Fig. 4.14.

another one going from y to x, both performed on the shortest path between x and y (Figure 4.14).

We now consider the sets of time-paths

$$S_1 = \left\{ x_1, x_2, \ldots, x_{\lfloor \frac{n}{2} \rfloor - f(n)}[T_1]; x_2, x_3, \ldots, x_{\lfloor \frac{n}{2} \rfloor - f(n) + 1}[T_2]; \ldots; \right.$$
$$\left. x_{\lfloor \frac{n}{2} \rfloor - 3f(n) - 2}, x_{\lfloor \frac{n}{2} \rfloor - 3f(n) - 1}, \ldots, x_{2\lfloor \frac{n}{2} \rfloor - 4f(n) - 3} \left[T_{\lfloor \frac{n}{2} \rfloor - 3f(n) - 2} \right] \right\}$$

and

$$S_2 = \left\{ x_{\lfloor \frac{n}{2} \rfloor - f(n)}, x_{\lfloor \frac{n}{2} \rfloor - f(n) - 1}, \ldots, x_1[T_1']; x_{\lfloor \frac{n}{2} \rfloor - f(n) + 1}, \ldots, x_2[T_2']; \ldots; \right.$$
$$\left. x_{2\lfloor \frac{n}{2} \rfloor - 4f(n) - 3}, x_{2\lfloor \frac{n}{2} \rfloor - 4f(n) - 2}, \ldots, x_{\lfloor \frac{n}{2} \rfloor - 3f(n) - 2} \left[T'_{\lfloor \frac{n}{2} \rfloor - 3f(n) - 2} \right] \right\}.$$

Obviously, $|S_1| = |S_2| = \lfloor \frac{n}{2} \rfloor - 3f(n) - 2$.

First, we observe that each time-path from $S_1 \cup S_2$ can contain at most $2(f(n) + 1)$ delays because A finishes in $\lceil \frac{n}{2} \rceil + f(n)$ rounds and the distances between the endnodes in these time-paths are $\lfloor \frac{n}{2} \rfloor - f(n) - 1$. Thus, the sum of all delays on the time-paths in $S_1 \cup S_2$ is at most

$$2\left(f(n) + 1\right) \cdot (|S_1| + |S_2|) \leq 2 \cdot (f(n) + 1) \cdot 2 \left(\lfloor \frac{n}{2} \rfloor - 3f(n) - 2 \right). \quad (4.1)$$

On the other hand, each time-path from S_1 must have a collision with each time-path from S_2 because of the fact

$$j - i = 2f(n) - 2$$

(Figure 4.15). Furthermore, at most $2f(n) + 3$ distinct time-paths from S_1 [S_2] going in the same direction can use the same edge at the same time. If at least $2f(n) + 4$ time-paths use the same edge at the same time, then at least one of these time-paths must already have the delay $2f(n) + 3$ which is impossible for time-paths in $S_1 \cup S_2$. Obviously, if k time-paths from S_1 collide with m time-paths from S_2 in the same node in the same round, then the number of delays caused in this collision is at least $\min\{k + m, 2k, 2m\}$.

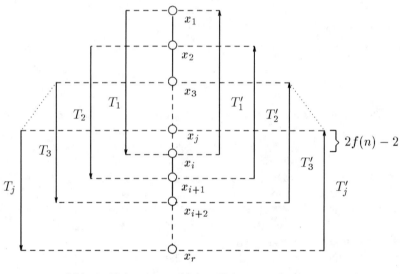

Fig. 4.15. $j = \lfloor \frac{n}{2} \rfloor - 3f(n) - 2$, $i = \lfloor \frac{n}{2} \rfloor - f(n)$, $r = j + i - 1$

Thus, each

$$2f(n) + 2$$

collisions between time-paths from S_1 and S_2 cause at least 2 delays (the worst cases: $k = 1, m = 2f(n) + 2$, and one time-path is waiting while the m time-paths continue without any delay, or $k = 2f(n) + 2, m = 1$). Based on the facts stated above we obtain that the sum of all delays on the time-paths from $S_1 \cup S_2$ is at least

$$2 \cdot \frac{(|S_1| \cdot |S_2|)}{2f(n) + 2} = \frac{\left(\lfloor \frac{n}{2} \rfloor - 3f(n) - 2\right)^2}{f(n) + 1}. \tag{4.2}$$

Comparing (4.1) and (4.2) we get

$$\frac{\left(\lfloor\frac{n}{2}\rfloor - 3f(n) - 2\right)^2}{f(n) + 1} \leq 2\left(f(n) + 1\right) \cdot 2\left(\lfloor\frac{n}{2}\rfloor - 3f(n) - 2\right)$$

$$2\left(f(n) + 1\right)^2 \geq \frac{\left(\lfloor\frac{n}{2}\rfloor - 3f(n) - 2\right)^2}{2\left(\lfloor\frac{n}{2}\rfloor - 3f(n) - 2\right)}$$

$$= \frac{1}{2} \cdot \left(\lfloor\frac{n}{2}\rfloor - 3f(n) - 2\right),$$

which provides the lower bound $f(n) \geq \frac{\sqrt{2}}{4} \cdot \sqrt{n} - O(1)$.

$\square$

In what follows we give the proof of the (almost) precise lower bounds on the one-way gossip complexity in cycles.

Theorem 4.3.15. *

$$r(C_n) \geq \frac{n}{2} + \lceil\sqrt{2n}\rceil - 1 \text{ for } n \text{ even, and}$$

$$r(C_n) \geq \lceil\frac{n}{2}\rceil + \lceil\sqrt{2n} - \frac{1}{2}\rceil - 1 \text{ for } n \text{ odd.}$$

Proof. Since the optimal algorithm for gossiping in an n-node cycle in two-way mode uses at least $\lceil\frac{n}{2}\rceil$ rounds we may assume that $r(C_n) = \lceil\frac{n}{2}\rceil + f(n)$ for some function f from positive integers to nonnegative integers. We shall show by detailed analysis that $f(n) \geq \sqrt{2n} - \frac{3}{2}$.

Let A be an arbitrary optimal algorithm for gossiping in C_n in one-way mode. This algorithm A can be considered as a sequence of rounds $A_1, A_2, \ldots, A_{t(A)}$, where $t(A) = \lceil\frac{n}{2}\rceil + f(n)$ is the number of rounds in A.

To prove our lower bound we use the fact that there are exactly two simple paths between any two nodes v and u in C_n.

We shall denote by $\boldsymbol{L(v, u)}$ $[\boldsymbol{R(v, u)}]$ the path from v $[u]$ to u $[v]$ in the clockwise [anticlockwise] direction of the cycle. The clockwise direction in the cycle is the direction leading from the right-most node to the left-most node through the lower half of the cycle. We define the **left distance** $\boldsymbol{d_l(v, u)}$ from v to u as the length of the path $L(v, u)$ and the **right distance** $\boldsymbol{d_r(v, u)}$ from v to u as the length of the path $R(v, u)$. Obviously, $d_l(v, u) + d_r(v, u) = n$. The rough idea of the proof is based on the following observation (Figure 4.16).

Observation 4.3.16. *For each node v of C_n there exists an edge (l_v, r_v) such that:*

(a) The distances $d_l(l_v, v), d_r(r_v, v) \in [\lceil\frac{n}{2}\rceil - f(n) - 1, \lceil\frac{n}{2}\rceil + f(n)]$

(b) For each node u on the path $L_v = L(l_v, v)$ which does not go through the edge (l_v, r_v), the piece of information $I(u)$ (originally distributed in u) is submitted to v through the path which does not involve the edge (l_v, r_v). (Note that the submission of $I(u)$ also through (l_v, r_v) is not forbidden.)

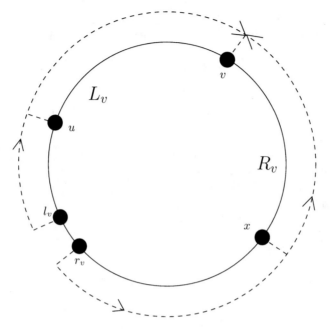

Fig. 4.16. Accumulation of the distributed information pieces to node v

(c) *For each node x on the path $R_v = R(r_v, v)$ which does not go through the edge (l_v, r_v), the piece of information $I(x)$ is submitted to v through the path which does not involve the edge (l_v, r_v).*

Let us give an informal idea of the proof. We shall divide the nodes in C_n into two disjoint subsets B_A (that contains all nodes which have learned the cumulative message in one round from a neighboring node which already knew the whole cumulative message) and its complement C_A according to the algorithm A. For each node v of C_n, let t_v^A be the minimal number of rounds after v has learned the cumulative message in the algorithm A. We define

$$\mathbf{C_A} := \{v| \text{ no message submitted to } v \text{ in any round } t \leq t_v^A$$
$$\text{was the cumulative message}\},$$

and

$$\mathbf{B_A} := \{v| \text{ the message submitted to } v \text{ in the round } t_v^A$$
$$\text{is the cumulative message}\}.$$

One can view the algorithm A as a process in which the nodes in C_A first concentrate the whole cumulative message of C_n and then the nodes in C_A distribute the cumulative message to the nodes in B_A. Note that these two parts of the process may be non-disjoint in time, i.e., in the same round one node in B_A can receive the cumulative message from a neighbor, and one node in C_A can receive partial information in order to complete its knowledge about the cumulative message. Let l_A denote the cardinality of C_A, and let $C_A = \{u_1, u_2, \ldots, u_{l_A}\}$.

For each $i \in \{1, \ldots, l_A\}$ we define

$$\mathbf{B(u_i)} := \{u_i\} \cup \{v \in B_A| \text{ there is a time-path } u_i, \ldots, v[t_1, \ldots, t_v^A]$$
$$\text{from } u_i \text{ to } v \text{ for some } t_1 > t_{u_i}^A \}.$$

Clearly $\bigcup_{i=1}^{l_A} B(u_i)$ includes all nodes of the cycle C_n iff A is a gossip algorithm for C_n. Obviously $|B(u_i)| \le 2(t(A) - t_{u_i}^A) = 2(\lceil \frac{n}{2} \rceil + f(n) - t_{u_i}^A)$.

Thus, if A is a gossip algorithm the following inequality must be true

$$n \le \sum_{i=1}^{l_A} |B(u_i)|$$

$$\le 2 \cdot \sum_{i=1}^{l_A} (\lceil \frac{n}{2} \rceil + f(n) - t_{u_i}^A)$$

$$= 2 \cdot l_A(\lceil \frac{n}{2} \rceil + f(n)) - 2 \sum_{i=1}^{l_A} t_{u_i}^A. \tag{4.3}$$

To use (4.3) for obtaining a lower bound for $f(n)$ we need to obtain a lower bound on $\sum_{i=1}^{l_A} t_{u_i}^A$. To do so we need the idea of collisions and delays. We shall investigate the collisions among the paths bringing the pieces of information to the nodes in C_A in what follows.

We consider only the paths leading to the nodes in C_A. In this way, we are able to secure that no two distinct collisions take place simultaneously at the same place and in the same round. Thus, all collisions among the paths leading to nodes in C_A will be "disjoint" in this sense, and we will have no problem counting them.

First we give the formal definition of collisions between two time-paths going in opposite directions. For the purpose of this proof we extend the definition of collisions as follows.

Let $X[T] = x_1, \ldots, x_m[t_1, \ldots, t_{m-1}]$ and $Y[T'] = y_1, \ldots, y_k[t_1', \ldots, t_{k-1}']$ be two time-paths of A going in opposite directions, and let $x_m \ne y_k$. We say that $X[T]$ **and** $Y[T']$ **have a collision in a node** v iff $(x_1 = y_1 = v)$ or

(1°) $\exists i \in \{1, \ldots, m-1\}$ and $\exists j \in \{1, \ldots, k-1\}$ $(i,j) \ne (1,1)$, such that $x_i = y_j = v$, and

(2°) $(t_{i-1} < t_{j-1}' < t_i)$ or $(t_{j-1}' < t_{i-1} < t_j')$.

Following this definition and from Figure 4.17(a) we see that each collision according to (1°) and (2°) causes at least two "delays" (either one 2-delay on one of the two time-paths or one 1-delay on each of the two time-paths in collision). Later we shall show that the fact $x_1 = y_1$ causes at least 3 delays in x_1 for some special time-paths.

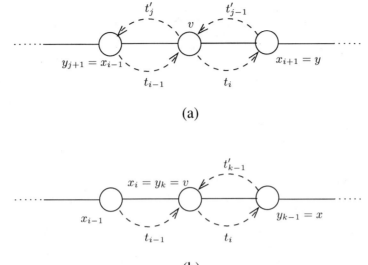

Fig. 4.17. Collisions between two time-paths

Let $X[T] = x_1, \ldots, x_m[t_1, \ldots, t_{m-1}]$ and $Y[T'] = y_1, \ldots, y_k[t'_1, \ldots, t'_{k-1}]$ be two time-paths of A going in opposite directions. We say that $\boldsymbol{X[T]}$ **and** $\boldsymbol{Y[T']}$ **have a relative collision in a node** $\boldsymbol{v}$ iff

(i) $x_m = y_k = v$ or

(ii) $\exists i \in \{1, \ldots, m-1\}$ such that $x_i = y_k = v$, and $t'_{k-1} < t_i$ (Figure 4.17(b)) or

(iii) $\exists j \in \{1, \ldots, k-1\}$ such that $x_m = y_j = v$ and $t_{m-1} < t'_j$.

We note that the relative collision defined above is termed relative because one time-path has already finished its work in the node v when the second time-path still flows via the node v. On the other hand we are also speaking about a collision in this case because the information cumulated in v cannot be distributed in the rounds in which the second time-path reaches and leaves the node v.

In what follows we shall say that two time-paths $\boldsymbol{X[T]}$ **and** $\boldsymbol{Y[T']}$ **have a conflict** if there is either a collision or a relative collision between $X[T]$ and $Y[T']$.

Let $X[T] = x_1, \ldots, x_m[t_1, \ldots, t_{m-1}]$ and $Y[T'] = y_1, \ldots, y_k[t'_1, \ldots, t'_{k-1}]$ be two time-paths of A going in the same direction. We say that $\boldsymbol{X[T]}$ **and** $\boldsymbol{Y[T']}$ **overlap in the node** $\boldsymbol{u}$ when one of the following conditions holds.

(i) There exists a node $u = x_i = y_j$ for some integers $i \in \{1, \ldots, m-1\}, j \in \{1, \ldots, k-1\}$ such that $\max\{t_{i-1}, t'_{j-1}\} < \min\{t_i, t'_j\}$ (we set $t_0 = t'_0 = 0$).

(ii) $x_m = y_j = u$ for some $j \in \{1, 2, \ldots, k-1\}$ and $t'_{j-1} < t_{m-1} < t'_j$.

(iii) $y_k = x_i = u$ for some $i \in \{1, 2, \ldots, m-1\}$ and $t_{i-1} < t'_{k-1} < t_i$.

We still need one more notation. Let X and Y be two simple paths in the cycle C_n, and let Y be contained in X. We shall denote this fact by $Y \subseteq X$.

For each node v in C_A we shall fix some time-paths bringing the cumulative message to v. Let $v \in C_A$, and let (l_v, r_v) be an edge satisfying (a), (b), (c) of Observation 4.3.16.

Let $L_v = L(l_v, v)$ $[R_v = R(r_v, v)]$ denote the path from l_v $[r_v]$ to v which does not go through the edge (l_v, r_v). We shall call the edge (l_v, r_v) **suitable for** v when the following property (d) holds.

(d) There exists two time-paths $L_v[z_1, \ldots, z_m]$ and $R_v[k_1, \ldots, k_b]$ such that $z_m \leq t_v^A$ and $k_b \leq t_v^A$.

Now, we fix one suitable edge (l_v, r_v) for each node v in C_A. The time-path $L_v[T_A]$ for $T_A = t_1^A, \ldots, t_m^A$ will be called the **left time-path of** A **for** v if, for any time-path $L_v[z_1, \ldots, z_m]$ of A, $t_i^A \leq z_i$ for each $i \in \{1, \ldots, m\}$. Analogously, $R_v[T_A]$ is called the **right time-path of** A **for** v if, for any time-path $R_v[a_1, \ldots, a_m]$ of A, $t_i^A \leq a_i$ for each $i \in \{1, \ldots, m\}$.

Let $L_v[T_A]$ $[R_v[T_A]]$ be the left [right] time-path of A for some $v \in C_A$. The number $t_v^A - t(L_v[T_A]) = k$ $[t_v^A - t(R_v[T_A])] = k$ is called the **k-relative delay of** L_v $[R_v]$ and denoted by $rd(L_v[T_A])$ $[rd(R_v[T_A])]$.

Clearly, the relative delays can be caused by relative collisions.

Now, we formulate an important property of nodes in C_A.

Lemma 4.3.17. *For each two nodes x and y in C_A the left (right) time-paths $L_x[T_A]$ and $L_y[T_A]$ ($R_x[T_A]$ and $R_y[T_A]$) do not* **overlap**.

Proof. The proof is done by contradiction. Let $L_x[T_A]$ and $L_y[T_A]$ overlap in a node v (Figure 4.18(a) and Figure 4.19). Without loss of generality we can assume that x is laid on the path L_y (if x does not lay on the path L_y then y is laid on L_x and the proof can be realized in the same way by exchanging the roles of x and y). Obviously, in the node v each of the two time-paths $L_x[T_A]$ and $L_y[T_A]$ has learned all pieces of information distributed in the nodes on the paths from l_x to v and from l_y to v. Thus a node on the path between x and y which has learned the cumulative message at least in the round $\min\{t_x^A, t_y^A\}$ must exist. It can be shown that the time-path $L_y[T_A]$ $[R_x[T_A]]$ disseminates the cumulative message to y $[x]$ exactly in the round t_y^A $[t_x^A]$. Clearly, this is a contradiction with the fact that $y \in C_A$ $[x \in C_A]$. If $R_x[T_A]$ and $R_y[T_A]$ overlap the contradiction can be achieved in the same way. □

Now, we give a lemma which enables us to add the number of delays caused by collisions among the left and right time-paths of nodes in C_A.

Lemma 4.3.18. *For each two nodes $x, y \in C_A$ there are two conflicts (between $L_x[T_A]$ and $L_y[T_A]$ on one side and $R_y[T_A]$ and $R_x[T_A]$ on the other side) causing at least 4 delays and relative delays.*

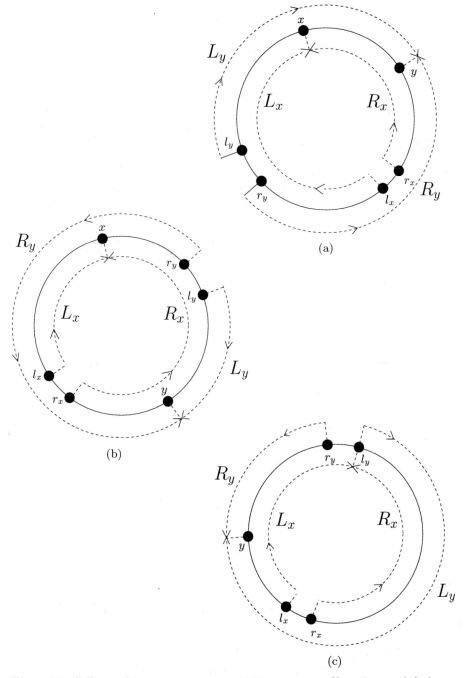

Fig. 4.18. Collisions between two communication processes (four time-paths) that accumulate the entire information to two different nodes x and y from C_A

Proof. We shall consider two possibilities according to the positions of $x, y, l_x,$ l_y, r_x, r_y.

I $(L_x \not\subseteq L_y) \wedge (L_y \not\subseteq L_x) \wedge (R_x \not\subseteq R_y) \wedge (R_y \not\subseteq R_x)$ (Figure 4.18(a) and Figure 4.18(b)).

II $(L_x \subseteq L_y) \vee (L_y \subseteq L_x) \vee (R_x \subseteq R_y) \vee (R_y \subseteq R_x)$ (Figure 4.19).

I. Case I is further divided into the following two subcases I.1 (Figure 4.18(a)) and I.2 (Figure 4.18(b)) according to the positions of x, y, l_x, l_y, r_x, r_y.

    I.1 $(I) \wedge (L_x \not\subseteq R_y) \wedge (R_x \not\subseteq L_y) \wedge (R_y \not\subseteq L_x) \wedge (L_y \not\subseteq R_x)$,

    I.2 $(I) \wedge ((L_x \subseteq R_y) \vee (R_x \subseteq L_y) \vee (R_y \subseteq L_x) \vee (L_y \subseteq R_x))$.

I.1 Let us first prove that there are two collisions in case (I.1). We give the proof only for the case described in Figure 4.18(a); all other cases of type (I.1) are symmetrical.

    First, we show that there is a collision between $L_x[T_A]$ and $R_y[T_A]$ which causes at least two delays. Clearly, if $r_y \neq l_x$ then there is a collision between $L_x[T_A]$ and $R_y[T_A]$ because $L_x[T_A]$ starts from l_x in the direction of r_y, $R_y[T_A]$ has to go through l_x in order to reach y. We shall handle the special case when $r_y = l_x$. According to the definition of a collision this is also a collision. We shall prove that this special collision also causes at least two delays. We claim that none of $R_y[T_A]$ and $L_x[T_A]$ can start in the first round. Let us assume that $R_y[T_A]$ does (the case for $L_x[T_A]$ is analogous). Then $R_x[T_A]$ cannot start in the first round because r_x receives the message from r_y in this round. But this is a contradiction to Lemma 4.3.17 because the time-paths $R_y[T_A]$ and $R_x[T_A]$ overlap at the node r_x.

    Now, we have two possibilities. If there is a collision between $L_y[T_A]$ and $R_x[T_A]$ then the proof for case (I.1) is done. If there is no collision between $L_y[T_A]$ and $R_x[T_A]$, then either $L_y[T_A]$ reaches y before $R_x[T_A]$ goes through y or $R_x[T_A]$ reaches x before $L_y[T_A]$ goes through x. Let us deal only with the case that $L_y[T_A] = L_y[t_1, \ldots, t_m]$ reaches y before $R_x[T_A]$ goes through y because the second case can be solved analogously. We see that the node y is reached by $R_x[T_A]$ after the round t_m. Since $R_x[T_A]$ and $R_y[T_A]$ do not overlap according to Lemma 4.3.17 we conclude that $R_y[T_A]$ cannot reach y before $R_x[T_A]$ leaves y. Thus, this relative collision of $L_y[T_A]$ and $R_x[T_A]$ causes a 2-relative delay for $L_y[T_A]$.

I.2 Let us now prove Lemma 4.3.18 for the case $L_x \subseteq R_y$ and $L_y \subseteq R_x$ described in Figure 4.18(b). All other cases of the type (I.2) are symmetrical, i.e., $R_x \subseteq L_y$ and $R_y \subseteq L_x$ (Figure 4.18(c) for the border case with $x = l_y$).

    We have $L_x \subseteq R_y$ and $L_y \subseteq R_x$. We show that there is a conflict between $L_x[T_A]$ and $R_y[T_A]$. If $R_y[T_A]$ leaves x before $L_x[T_A]$ reaches x then $R_y[T_A]$ and $L_x[T_A]$ have a collision in a node leading on the way L_x. If $L_x[T_A]$ reaches x before $R_y[T_A]$ leaves x then we have a relative collision in the node x which causes either two relative delays ($L_x[T_A]$ reaches x before

$R_y[T_A]$ does) or one relative delay and one delay ($R_y[T_A]$ reaches x first). To show that there is a conflict between $L_y[T_A]$ and $R_x[T_A]$ one can use the same consideration described above for the conflict between $L_x[T_A]$ and $R_y[T_A]$.

II We show that this case (Figure 4.19) cannot occur. We shall distinguish the following three possibilities:
(i) $R_x[T_A]$ reaches x before $L_y[T_A]$ does.
(ii) $L_y[T_A]$ reaches y before $R_x[T_A]$ does.
(iii) $R_x[T_A]$ and $L_y[T_A]$ have a conflict in a node between x and y.
We consider case (i); cases (ii) and (iii) can be solved analogously. Following Lemma 4.3.17 and Figure 4.19 we see that $L_x[T_A]$ reaches x before $L_y[T_A]$ reaches it. Thus, x has learned the cumulative message before $L_y[T_A]$ has reached x. This implies that $L_y[T_A]$ brings the cumulative message from x to y which is a contradiction to the fact that $y \in C_A$.

□

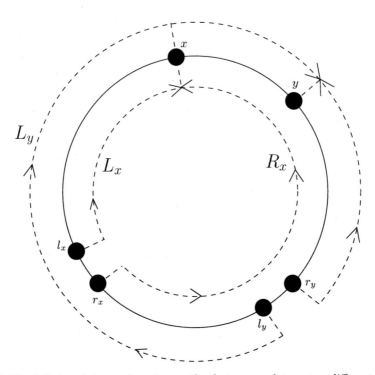

Fig. 4.19. Collisions between four time-paths that accumulate to two different nodes x and y from C_A

Now, we are ready to formulate our lower bound for $\sum_{i=1}^{l_A} t_{u_i}^A$.

Lemma 4.3.19. *The following inequality (4.4) holds*

$$\sum_{i=1}^{l_A} t_{u_i}^A \geq l_A \frac{(n-1)}{2} + l_A(l_A - 1) + \frac{l_A}{2}. \qquad (4.4)$$

Proof.

$$(*) \quad \sum_{i=1}^{l_A} t_{u_i}^A = \sum_{i=1}^{l_A} \frac{t(L_{u_i}[T_A]) + rd(L_{u_i}[T_A]) + t(R_{u_i}[T_A]) + rd(R_{u_i}[T_A])}{2}$$

$$= \sum_{i=1}^{l_A} \frac{d_l(l_{u_i}, u_i) + d_r(r_{u_i}, u_i)}{2}$$

$$+ \sum_{i=1}^{l_A} \frac{d(L_{u_i}[T_A]) + rd(L_{u_i}[T_A]) + d(R_{u_i}[T_A]) + rd(R_{u_i}[T_A])}{2}$$

$$\geq l_A \cdot \frac{(n-1)}{2} + \frac{1}{2} \left(\frac{4 \cdot l_A(l_A - 1)}{2} + l_A \right)$$

$$= l_A \cdot \frac{(n-1)}{2} + l_A(l_A - 1) + \frac{l_A}{2}.$$

The inequality (*) follows from the following facts:

(i) The paths $l_v, \ldots, v$ and $v, \ldots, r_v$ include all edges of C_n except (l_v, r_v), i.e., $d_l(l_v, v) + d_r(r_v, v) = n - 1$.
(ii) Following Lemma 4.3.18 we see that any pair of nodes x, y from C_A brings at least 4 delays to $d(L_x[T_A]) + rd(L_x[T_A]) + d(R_x[T_A]) + rd(R_x[T_A]) + d(L_y[T_A]) + rd(L_y[T_A]) + d(R_y[T_A]) + rd(R_y[T_A])$, and Lemma 4.3.17 assures that these delays are disjoint from any delays caused by another pair of nodes from C_A.
The number of distinct pairs of nodes in C_A is $l_A \frac{l_A - 1}{2}$.
(iii) For each node $x \in C_A$, there is one relative delay resulting from the relative collision between $L_x[T_A]$ and $R_x[T_A]$.

$\square$

Applying the inequality (4.4) of Lemma 4.3.19 in the inequality (4.3) we obtain

$$n \leq 2l_A(\lceil \frac{n}{2} \rceil + f(n)) - 2[l_A \frac{n-1}{2} + l_A(l_A - 1) + \frac{l_A}{2}] \qquad (4.5)$$

which directly implies

$$f(n) \geq \frac{n}{2l_A} - \lceil \frac{n}{2} \rceil + \frac{1}{l_A}(l_A \cdot \frac{n-1}{2} + l_A(l_A - 1) + \frac{l_A}{2})$$

$$\geq \frac{n}{2l_A} + l_A + \frac{n-1}{2} - \lceil \frac{n}{2} \rceil - \frac{1}{2}.$$

Thus, we obtain

$$f(n) \geq \frac{n}{2l_A} + l_A - 1 \text{ for } n \text{ even, and}$$

$$\geq \frac{n}{2l_A} + l_A - \frac{3}{2} \text{ for } n \text{ odd.} \tag{4.6}$$

Since (4.6) holds for any l_A (note that (4.6) holds for any gossip algorithm A) we obtain

$$f(n) \geq \min_{1 \leq l_A \leq n} \left\{ \frac{n}{2l_A} + l_A - a \;\middle|\; a = 1 \text{ for } n \text{ even, } a = \frac{3}{2} \text{ for } n \text{ odd} \right\}$$

$$= \min_{l_A} \{ g_n(l_A) | 1 \leq l_A \leq n \}.$$

Since the function $g_n(l_A)$ has the global minimum $l_A = \sqrt{\frac{n}{2}}$, on the real interval $[1, n]$ and f is the function into the set of positive integers, we obtain

$$f(n) \geq \lceil 2 \cdot \sqrt{\frac{n}{2}} \rceil - 1 \text{ for } n \text{ even, and}$$

$$f(n) \geq \lceil 2 \cdot \sqrt{\frac{n}{2}} - \frac{3}{2} \rceil \text{ for } n \text{ odd.}$$

Since $r(C_n) \geq \lceil \frac{n}{2} \rceil + f(n)$ the proof of Theorem 4.3.15 is completed.

$\square$

Corollary 4.3.20. *If A is an optimal gossip algorithm for C_n that meets the lower bound of Theorem 4.3.15, then A must have the number l_A of concentrators between $\lfloor \sqrt{\frac{n}{2}} \rfloor$ and $\lceil \sqrt{\frac{n}{2}} \rceil$.*

The next section shows that the designed optimal gossip algorithms in cycles are useful instruments for designing efficient gossip algorithms in several fundamental interconnection networks.

4.4 Gossiping in Hypercube-like Networks of Constant Degree

In this section we show how to construct effective algorithms for gossiping in CCC_k-networks and BF_k-networks using the optimal algorithm for gossiping in C_k.

Remember that both CCC_k and BF_k consists of 2^k cycles of k nodes denoted by $C_\alpha(k)$ for $\alpha \in \{0, 1\}^k$. Remember also that (i, α) denotes a node in the i-th level of CCC_k or BF_k for some $\alpha \in \{0, 1\}^k$. First, we define set-to-set broadcasting as a useful method for designing communication algorithms. Let A and B be two sets of nodes. The **set-to-set broadcasting from A to B** is a communication process in which each node in B has learned all pieces of information distributed in A. In particular, the execution of a set-to-set broadcast from a set of nodes A to A assures that every node in A has learned the cumulative message of A.

In what follows we present algorithms for set-to-set broadcasting from the i-th level to the i-th level in BF_k, and for set-to-set broadcasting from the i-th level to the $((i-1) \bmod k)$-th level in CCC_k.

SET $CCC_k(i)$

 for $j = 0$ to $k - 1$ do
  for all $\alpha \in \{0, 1\}^k$ do in parallel
   begin
    exchange information between
    $((i + j) \bmod k, \alpha)$ and $((i + j) \bmod k, \alpha((i + j) \bmod k))$
    { * needs two rounds * };
    if $j < k - 1$ then
    $((i + j) \bmod k, \alpha)$ sends to $((i + j + 1) \bmod k, \alpha)$
    { * needs 1 round * }
   end;

Exercise 4.4.1. Prove that for every positive integer k the communication algorithm SET CCC_k executes the set-to-set broadcasting from the i-th level to the $((i - 1) \bmod k)$-th level in CCC_k.

SET $BF_k(i)$

 for $j = 0$ to $k - 1$ do
  for all $\alpha \in \{0, 1\}^k$ do in parallel
   begin
    $((i+j) \bmod k, \alpha)$ sends to $((i+j+1) \bmod k, \alpha((i+j) \bmod k))$
    $((i+j) \bmod k, \alpha)$ sends to $((i+j+1) \bmod k, \alpha)$
   end;

Exercise 4.4.2. Prove that for every positive integer k, the communication algorithm SET $BF_k(i)$ executes the set-to-set broadcasting from the i-th level to the i-th level in BF_k.

Now, gossiping in CCC_k and BF_k can be done as follows.

Algorithm GOSSIP-BF_k

1. Use in parallel for all α the 1st phase of the optimal algorithm for gossiping in C_k concentrating the cumulative message of C_k in $l = \lfloor \sqrt{\lceil \frac{k}{2} \rceil} \rfloor$ "regularly distributed" nodes in C_k to concentrate the cumulative message of $C_\alpha(k)$ of BF_k in l nodes (v_i, α) for $1 \le i \le l$.
2. For all $i \in \{v_j \mid 1 \le j \le l\}$ do in parallel set-to-set broadcasting from the i-th level to the i-th level on BF_k.
   After the execution of the second step of the one-way communication algorithm GOSSIP-BF_k, every node from $\{(v_i, \alpha) \mid 1 \le i \le l, \alpha \in \{0, 1\}^k\}$ knows the cumulative message $I(BF_k)$.
3. Use in parallel for all α the 2nd phase of the optimal algorithm for gossiping in the cycle to broadcast the cumulative message of BF_k contained in the nodes $(v_i, \alpha), 1 \le i \le l$, to the other nodes in the cycle $C_\alpha(k)$.

Algorithm GOSSIP-CCC_k

1. Use in parallel for all $\alpha \in \{0,1\}^{2^k}$ the 1st phase of the optimal algorithm for gossiping in C_k to concentrate the cumulative message of $C_\alpha(k)$ of CCC_k in $l = \lfloor \sqrt{\lceil \frac{k}{2} \rceil} \rfloor$ nodes (v_i, α) for $1 \leq i \leq l$.

2. For all $i \in \{v_j \mid 1 \leq j \leq l\}$ do in parallel set-to-set broadcasting from the i-th level of CCC_k to the $((i-1) \bmod k)$-th level of CCC_k.
   After the execution of the second step of the one-way communication algorithm GOSSIP-CCC_k every node from $\{((v_i - 1) \bmod k, \alpha) \mid 1 \leq i \leq l, \alpha \in \{0,1\}^k\}$ knows the cumulative message $I(CCC_k)$.

3. Use in parallel for all α the 2. phase of the optimal algorithm for gossiping in the cycle to broadcast the cumulative message of CCC_k contained in the nodes $((v_i - 1) \bmod k, \alpha)$ of $CCC_k, 1 \leq i \leq l$, to the other nodes in the cycle $C_\alpha(k)$.

Analyzing the complexity of the above-stated procedures we obtain:

Theorem 4.4.3. *For every $k \geq 3$,*

$$\mathrm{r}(CCC_k) \leq \mathrm{r}(C_k) + 3k - 1 \leq \lceil \frac{7k}{2} \rceil + \lceil 2\sqrt{\lceil \frac{k}{2} \rceil} \rceil - 2.$$

Theorem 4.4.4. *For every $k \geq 3$,*

$$\mathrm{r}(BF_k) \leq \mathrm{r}(C_k) + 2k \leq \lceil \frac{5k}{2} \rceil + \lceil 2\sqrt{\lceil \frac{k}{2} \rceil} \rceil - 1.$$

To conclude this section we show that this technique can also be used for gossiping in the two-way mode.

Theorem 4.4.5. *For every positive even integer $k \geq 3$*

$$\mathrm{r}_2(CCC_k) \leq \frac{k}{2} + 2k = 5 \cdot \lceil \frac{k}{2} \rceil,$$

and for every positive odd integer $k \geq 3$

$$\mathrm{r}_2(CCC_k) \leq \lceil \frac{k}{2} \rceil + 2k + 2 = 5 \cdot \lceil \frac{k}{2} \rceil.$$

Proof. To do gossiping in CCC_k the following algorithm of three phases can be used.

1. Use the optimal algorithm for gossiping in C_k in two-way mode to do gossiping in parallel on all cycles $C_\alpha(k)$ of CCC_k.

2. For all odd $i \leq k - 1$ do in parallel set-to-set broadcasting from the i-th level to the $((i-1) \bmod k)$-th level on CCC_k.

3. For all odd $j \leq k - 1$ do in parallel:
   The j-th level learns in parallel from the $(j-1)$-th level [the $(k-1)$-th level learns in parallel in one special round when k is odd].

The result of Theorem 4.4.5 follows directly from the fact $r_2(C_k) = \frac{k}{2}$ for k even and $r_2(C_k) = \lceil \frac{k}{2} \rceil + 1$ for k odd proved in Theorem 4.3.10 and from the fact that the information exchange in the algorithm $SET\ CCC_k$ performed in the two-way mode runs in one round.

$\square$

As CCC_k is a subgraph of BF_k, we have the following corollary for two-way gossiping in BF_k:

Corollary 4.4.6. *Let $k \geq 3$ be an integer. Then*

$$r_2(BF_k) \leq \frac{k}{2} + 2k = 5 \cdot \lceil \frac{k}{2} \rceil \text{ for } k \text{ even, and}$$

$$r_2(BF_k) \leq \lceil \frac{k}{2} \rceil + 2k + 2 = 5 \cdot \lceil \frac{k}{2} \rceil \text{ for } k \text{ odd.}$$

4.5 Gossiping in Complete Graphs

The aim of this section is to present optimal one-way and two-way gossip algorithms in K_n. While the gossip problem in K_n is relatively readily solvable in the two-way mode, the design of the optimal one-way gossip algorithm in K_n requires a method a little more elaborated. The presentation of this method of counting precisely the necessary and sufficient growth of the amount of information disseminated in any gossip algorithm in K_n is the main methodological contribution of this section.

We consider first the two-way communication mode and show that for every $n \in I\!N$ the complete graph K_n is a minimal gossip graph.

Theorem 4.5.1.

$r_2(K_n) = \lceil \log_2 n \rceil$ *for every positive, even integer n, and*
$r_2(K_n) = \lceil \log_2 n \rceil + 1$ *for every positive, odd integer n.*

Proof. The lower bound was already presented in Lemma 4.2.1. We start with the upper bound for even n. Then we reduce the case for odd n to this case[1].

First, we have to show that $r_2(K_n) \leq \lceil \log_2 n \rceil$ holds for even n. Let $n = 2m$. We partition the set of processors into two sets Q, R of size m. Let us denote the processors by $Q[i], R[i], 0 \leq i \leq m - 1$. The following algorithm doubles the information at each node in every step.

Algorithm 2-WAY-GOSSIP-K_n

for all $i \in \{0, \ldots, m - 1\}$ do in parallel
    exchange information between $Q[i]$ and $R[i]$;
    for $t = 1$ to $\lceil \log_2 m \rceil$ do

---

[1] Note that the proof for $n = 2^k$, $k \in I\!N$, was already given in Lemma 4.2.3 because the hypercube H_k is a subgraph of K_n.

<u>for</u> all $i \in \{0, \ldots, m-1\}$ <u>do</u> in <u>parallel</u>
 exchange information between $Q[i]$ and $R[(i + 2^{t-1}) \bmod m]$;

Now we prove that the communication algorithm 2-WAY-GOSSIP-K_n is a two-way gossip algorithm for K_{2m}. Let $q[i], r[i], 0 \le i \le m-1$, denote the pieces of information stored by processors $Q[i], R[i]$ before starting the algorithm. Set $\alpha[i] = \{q[i], r[i]\}$ for $0 \le i \le m-1$.

After the execution of the first instruction, processors $Q[i]$ and $R[i]$ both store $\alpha[i]$. It is easy to verify by induction on t that after round $t, 1 \le t \le \lceil \log_2 m \rceil$, processors $Q[i]$ and $R[(i + 2^{t-1}) \bmod m]$ both store the set of information

$$\bigcup_{0 \le j \le 2^t - 1} \alpha[(i + j) \bmod m].$$

Therefore, after $1 + \lceil \log_2 m \rceil = \lceil \log_2 n \rceil$ rounds, all nodes have received the complete information.

Now, let $n = 2m + 1$. Number the nodes of K_n from 1 to n. The following algorithm performs gossiping in K_n:

1. Send the information of the node $i + m$ to the node i for all $2 \le i \le m+1$.
   {After this step, the cumulative message is distributed in the nodes $1, 2, \ldots, m+1$.}
2. If $m+1$ is even, gossip in $1, 2, \ldots, m+1$. If not, gossip in $1, 2, \ldots, m+2$.
   {After this step, each of the nodes $1, 2, \ldots, m+1$ knows the cumulative message.}
3. Send the information of the node i to the node $i + m$ for all $2 \le i \le m+1$.
   {After this step, each node knows the cumulative message.}

If $m + 1$ is even, the above algorithm takes

$$
\begin{aligned}
r_2(K_{m+1}) + 2 &= \lceil \log_2(m+1) \rceil + 2 \\
&= \lceil \log_2\left(\tfrac{n+1}{2}\right) \rceil + 2 \\
&= \lceil \log_2(n+1) \rceil + 1 \\
&= \lceil \log_2 n \rceil + 1
\end{aligned}
$$

rounds. If $m + 1$ is odd, then $n + 1$ is not a power of two, and hence the algorithm takes

$$
\begin{aligned}
r_2(K_{m+2}) + 2 &= \lceil \log_2(m+2) \rceil + 2 \\
&= \lceil \log_2\left(\tfrac{n+3}{2}\right) \rceil + 2 \\
&= \lceil \log_2(n+3) \rceil + 1 \\
&= \lceil \log_2 n \rceil + 1
\end{aligned}
$$

rounds.   □

The algorithm described in Theorem 4.5.1 does not use all the edges of the complete graph. In fact, since the algorithm uses only $\lceil \log_2 n \rceil$ rounds (we consider here only the case where n is even), for every node at most $\lceil \log_2 n \rceil$ of its edges are used. Thus the algorithm defines a graph of degree at most $\lceil \log_2 n \rceil$. This graph is a minimal gossip graph and we denote it by $\boldsymbol{Gos_n}$.

Gos_n is defined for even $n, n = 2m$, and has n nodes which are denoted by $Q[i]$ and $R[i]$, $0 \le i \le m - 1$. The edges connect $Q[i]$ and $R[i]$ for every i, $0 \le i \le m - 1$. Furthermore for every i, $0 \le i \le m - 1$, and for every t, $1 \le t \le \lceil \log_2 m \rceil$, there are edges connecting $Q[i]$ with $R[(i + 2^{t-1}) \mod m]$. The graph Gos_{12} is shown in Figure 4.20.

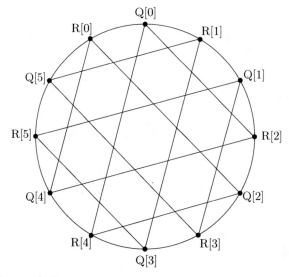

Fig. 4.20. The graph Gos_{12}

Because of the construction of Gos_n the following result follows from Theorem 4.5.1.

Corollary 4.5.2. *For every even positive integer n:*

$$\mathrm{r}_2(Gos_n) = \lceil \log_2 n \rceil.$$

It is not difficult to see that if n is a power of 2, i.e., $n = 2^k$ for some k, then the graph Gos_n is isomorphic to H_k, the hypercube of dimension k. Thus, a special consequence of Corollary 4.5.2 is that the hypercube is a minimal gossip graph (which was already shown in Lemma 4.2.3). We have already seen that the hypercube is very well-suited for information dissemination in Lemma 3.2.2, where we showed that the hypercube is a minimal broadcast graph. Of course, for even n every minimal gossip graph is also a minimal broadcast graph and Corollary 4.5.2 can be viewed as a generalization of Lemma 3.2.2.

Finding an optimal gossip strategy in the one-way mode is more sophisticated. The number of rounds in this case is determined by the growth of the Fibonacci numbers defined in Section 2.3.

Let $\Phi = \frac{1}{2}(1 + \sqrt{5})$. Using $\Phi^2 = \Phi + 1$, Exercise 2.3.30 provides $\Phi^{i-2} \leq F(i) \leq \Phi^{i-1}$ for all $i \geq 2$.

We will consider the gossip problem in the one-way mode only for an even number of nodes. Results for an odd number of nodes are left as an advanced exercise to the reader.

Theorem 4.5.3. *For every positive, even integer n, and for every positive integer k with $F(k) \geq \frac{n}{2}$:*

$$\mathrm{r}(K_n) \leq k + 1.$$

Proof. The algorithm is somewhat similar to the algorithm presented in the proof of Theorem 4.5.1. Again the set of processors is partitioned into two equal-sized subsets Q and R. In each round either all processors from Q send their information to all processors from R or vice versa, i.e., in each round either all processors from Q are writing and all processors from R are reading, or all processors from Q are reading and all processors from R are writing. Let $n = 2m$, and let us denote the processors by $Q[i], R[i], 0 \leq i \leq m - 1$.

Algorithm 1-WAY-GOSSIP-K_n

> $t := 0$;
> <u>for</u> all $i \in \{0, \ldots, m - 1\}$ <u>do in parallel</u> $R[i]$ sends to $Q[i]$;
> <u>for</u> all $i \in \{0, \ldots, m - 1\}$ <u>do in parallel</u> $Q[i]$ sends to $R[i]$;
> <u>while</u> $F(2t + 1) < m$ <u>do</u> <u>begin</u>
> $t := t + 1$;
>   <u>for</u> all $i \in \{0, \ldots, m - 1\}$ <u>do in parallel</u>
> $R[(i + F(2t - 1)) \bmod m]$ sends to $Q[i]$;
>   <u>if</u> $F(2t) < m$ <u>then</u>
>   <u>for</u> all $i \in \{0, \ldots, m - 1\}$ <u>do in parallel</u>
> $Q[(i + F(2t)) \bmod m]$ sends to $R[i]$
> <u>end</u>;

When this algorithm stops (since $F(2t + 1) \geq m$ or $F(2t) \geq m$ holds), then it has performed $2t$ rounds or $2(t - 1) + 1$ rounds respectively, within the while loop. Therefore, the algorithm performs $(k - 1) + 2$ rounds, where k is the smallest integer such that $F(k) \geq m$.

In order to prove the correctness of the algorithm, let again $q[i], r[i], 0 \leq i \leq m - 1$, denote the pieces of information stored by processors $Q[i], R[i]$ before starting the algorithm. Set $a[i] = \{q[i], r[i]\}, 0 \leq i \leq m - 1$.

After the execution of the first two instructions, processors $Q[i]$ and $R[i]$ both store $\alpha[i]$. It is not difficult to verify by induction on t that after t runs through the while loop of the above algorithm

$$Q[i] \text{ stores} \bigcup_{0 \leq j \leq F(2t+1)-1} \alpha[(i+j) \bmod m],$$

$$R[i] \text{ stores} \bigcup_{0 \leq j \leq F(2t+2)-1} \alpha[(i+j) \bmod m].$$

If k is an odd number, $k = 2t + 1$, then after t runs of the while loop all processors store the whole information. If k is an even number, $k = 2t+2$, then the first instruction in run $t + 1$ also has to be executed before all processors store the whole information.

□

Exercise 4.5.4. Prove, for every positive, odd integer n, and for every positive integer k with $F(k) \geq \frac{n}{2}$,

$$\mathrm{r}(K_n) \leq k + 2.$$

Now, we give a method for proving the optimality of the upper bounds presented above. For this purpose, we introduce a new problem called the Network Counting Problem (NCP).

The information stored by each of the n processors will be an integer. At the beginning all the integers will be equal to one. The processors are communicating in the one-way mode, i.e., in each round either a processor sends its integer or it receives an integer. If it receives an integer, then it adds this integer to its own integer. Again we are interested in the number of rounds needed until all processors store an integer which is greater or equal to n.

It is clear that any algorithm for solving the gossip problem also solves NCP and that a lower bound for NCP is also a lower bound for the gossip problem. There exists a straightforward algorithm for solving NCP. The set of processors is partitioned into groups, each of two processors. Within such a group the processors alternately send their information to each other, i.e., after t rounds one of them stores $F(t + 1)$ and the other one $F(t + 2)$.

Therefore the above algorithm needs $k - 1$ rounds, where k is the smallest integer such that $n \leq F(k)$.

We shall prove now the lower bound on $\mathrm{r}(K_n)$ and we shall do this by proving a lower bound for the Network Counting Problem (NCP). The idea is to represent the "state" of K_n after a number of rounds as a **state vector** $(x_1, x_2, \ldots, x_n)$ where x_i is the number assigned to the node i. We will show that one round of the NCP can be described by the multiplication of the current state vector by some associated matrix and we shall use methods from matrix theory to prove the lower bound in the following way. We shall show that matrices that correspond to communication rounds have a bounded norm. Since the vector $(1, 1, \ldots, 1)$ has a small norm and the vector $(i_1, i_2, \ldots, i_n)$ with $i_j \geq n$ for $i = 1, \ldots, n$ has a large norm, one needs to multiply $(1, 1, \ldots, 1)$ by many matrices in order to get a vector whose norm is at least the norm of $(n, n, \ldots, n)$. Hence, one obtains a lower bound on the number of rounds to solve the NCP problem. The main instrument of this approach is the term "vector norm". Different vector norms may lead to different lower bounds and so we have to choose an appropriate norm in order to get the optimal lower bound.

Before fixing the norm, we present some fundamental facts about norms. Let $||..||$ be any vector norm over $\mathbb{R}^n$, i.e.,

$$||x|| = 0 \Leftrightarrow x = 0^n,$$

$$||\alpha \cdot x|| = |\alpha| \cdot ||x||,$$

$$||x + y|| \leq ||x|| + ||y||$$

for all $\alpha \in \mathbb{R}, x, y \in \mathbb{R}^n$.

The matrix norm associated with a vector norm $||..||$ is defined by

$$||A|| = sup_{x \neq 0} \frac{||Ax||}{||x||}.$$

This matrix norm fulfills the following properties:

$$||A|| = 0 \Leftrightarrow A = 0$$
$$||A + B|| \leq ||A|| + ||B||$$
$$||\alpha A|| = \alpha \cdot ||A||$$
$$||A \cdot B|| \leq ||A|| \cdot ||B||$$
$$||A \cdot x|| \leq ||A|| \cdot ||x||$$

for all $A, B \in \mathbb{R}^{n^2}, x \in \mathbb{R}^n, \alpha \in \mathbb{R}, \alpha \geq 0$.

It turns out that for proving our lower bounds the Euclidean vector norm, defined by

$$||x|| = \sqrt{\Sigma_{i=1}^n |x_i|^2} \text{ for any } x = (x_1, .., x_n),$$

is appropriate. It is well-known that the spectral norm is associated with this vector norm as the following matrix norm

$$||A|| = \text{ spectral norm}(A) = \sqrt{|\lambda_{max}(A^T \cdot A)|},$$

where A^T is the transposed matrix of A and λ_{max} denotes the eigenvalue of the maximal absolute value.

Let us now consider one round in an algorithm solving the NCP. Let $x, y \in \mathbb{N}^n$ be the state vectors of numbers stored by the n processors before and after that round. We associate an $n \times n$ matrix A with entries $a_{i,j} \in \{0, 1\}$ with this round, whereby

(i) $a_{ii} = 1$ for all $i = 1, \ldots, n$,

(ii) $a_{ij} = 1$ for $i \neq j$ $\Leftrightarrow$ processor j sends its number to processor i.

Observe that $Ax = y$ holds, and since we are working in the one-way mode A fulfills the following properties:

(i) $a_{ii} = 1$ for all $i = 1, \ldots, n$

(ii) $a_{ij} = 1$ for some $i \neq j$ implies $a_{iv} = a_{vj} = 0$ for all $v \neq i, j$, $a_{vi} = 0$ for all $v \neq i$, and $a_{jv} = 0$ for all $v \neq j$.

We will denote the class of matrices with the properties (i) and (ii) by $P(1, 1, n)$. We are interested in determining the largest spectral norm $\|A\|$ among all matrices $A \in P(1, 1, n)$.

Note that every matrix $A \in P(1, 1, n)$ can be transformed by using coordinate transformations into a matrix with "blocks" of the form $B = \begin{pmatrix} 1 & 1 \\ 0 & 1 \end{pmatrix}$ along the main diagonal, i.e.,

$$TAT^{-1} = \begin{pmatrix} B & & & & & 0 \\ & B & & & & \\ & & \cdot & & & \\ & & & \cdot & & \\ & & & B & & \\ & & & 1 & & \\ & & & & \cdot & \\ 0 & & & & & 1 \end{pmatrix}$$

The spectral norms of A, TAT^{-1} and B coincide and the spectral norm of B is easily computable.

$$B^T \cdot B = \begin{pmatrix} 1 & 0 \\ 1 & 1 \end{pmatrix} \begin{pmatrix} 1 & 1 \\ 0 & 1 \end{pmatrix} = \begin{pmatrix} 1 & 1 \\ 1 & 2 \end{pmatrix}$$

$$\Rightarrow (2 - \lambda)(1 - \lambda) - 1 = 0$$

$$\Rightarrow \lambda^2 - 3\lambda + 1 = 0$$

$$\Rightarrow \lambda_{max}(B^T B) = \frac{3}{2} + \sqrt{\frac{5}{4}}.$$

This implies the following equality for every matrix A associated with a round in the one-way mode.

$$\|A\| = \sqrt{\lambda_{max}(A^T A)} = \frac{1}{2}(1 + \sqrt{5}). \tag{4.7}$$

Theorem 4.5.5. *Let n be an even positive integer. Every algorithm that solves the NCP in the telegraph communication mode needs at least $2 + \lceil \log_\phi \frac{n}{2} \rceil$ rounds.*

Proof. Assume that a communication algorithm that solves the NCP in r rounds for a positive integer r exists. Let $A_j, 1 \leq j \leq r$, be the matrix associated with round j of the algorithm. Let $\alpha_i, 1 \leq i \leq n$, be the number of pieces of information gathered by processor i during the first $r - 2$ rounds of the algorithm, i.e.,

$$
\alpha := \begin{pmatrix} \alpha_1 \\ \cdot \\ \cdot \\ \cdot \\ \alpha_n \end{pmatrix} = A_{r-2} \cdot \ldots \cdot A_2 \cdot A_1 \cdot \begin{pmatrix} 1 \\ \cdot \\ \cdot \\ \cdot \\ 1 \end{pmatrix}.
$$

Applying equality (4.7) we obtain the following upper bound on $||\alpha||$

$$
||\alpha|| \leq \Big(\prod_{i=1}^{r-2} ||A_i|| \Big) \cdot ||(1, ..., 1)|| \leq \Phi^{r-2} \cdot \sqrt{n}. \tag{4.8}
$$

Our next aim is to derive a lower bound on $||\alpha||$.

We denote by $inf(i, t)$ the number of pieces of information gathered by node i in the first t rounds of the algorithm. Since this algorithm solves the NCP in r rounds, $inf(i, r) \geq n$ for all $i = 1, \ldots, n$.

In the final round, at most $\frac{n}{2}$ processors can gather more information, i.e.,

$$
inf(i, r - 1) \geq n \text{ for at least } \frac{n}{2} \text{ processors } i.
$$

Some indices i may already exist, where $\alpha_i = inf(i, r - 2) \geq n$. However, if $\alpha_i < n$ and $inf(i, r - 1) \geq n$, then there exists some processor j with $\alpha_i + \alpha_j \geq n$ sending its information in round $r-1$ to processor i. We distinguish between three cases:

(1) $\alpha_i \geq n$,
(2) $\alpha_i < n$ and $\alpha_j \geq n$,
(3) $\alpha_i < n, \alpha_j < n$ and $\alpha_i + \alpha_j \geq n$.

Let c_k be the number of indices for which $(k), 1 \leq k \leq 3$, holds. Then $c_1 \geq c_2$ and $c_1 + c_2 + c_3 \geq \frac{n}{2}$, and therefore

$$
2c_1 + c_3 \geq \frac{n}{2}
$$

holds.

Furthermore we use the fact that, for arbitrary numbers $\beta, \gamma \in \mathbb{R}$ with $\beta + \gamma = n$, the expression $\beta^2 + \gamma^2$ has the minimal value for $\beta = \gamma = \frac{n}{2}$. Putting all this together we obtain the following lower bound on $||\alpha||$:

$$
||\alpha|| = \sqrt{\Sigma_{i=1}^n \alpha_i^2} \geq \sqrt{c_1 n^2 + c_3 \cdot 2 \cdot \frac{n^2}{4}} \geq n \cdot \sqrt{\frac{1}{2}(2c_1 + c_3)} \geq \frac{n}{2}\sqrt{n} \tag{4.9}
$$

The upper bound (4.8) and the lower bound (4.9) on $||\alpha||$ provide

$$\frac{n}{2} \cdot \sqrt{n} \leq ||\alpha|| \leq \Phi^{r-2} \cdot \sqrt{n},$$

which implies

$$r \geq 2 + \lceil \log_\Phi \lceil \frac{n}{2} \rceil \rceil.$$

□

The upper bound from Theorem 4.5.3 and the lower bound from Theorem 4.5.5 are very close. The following lemma shows that their difference is at most 1 and that they are equal for infinitely many n.

Lemma 4.5.6. *Let $n = 2m$ be some even integer, and let $t_1 := 1 + k$ [where k is the smallest integer such that $m \leq F(k)$] be the upper bound from Theorem 4.5.3 and $t_2 := 2 + \lceil \log_\Phi m \rceil$ be the lower bound from Theorem 4.5.5. Then $t_1 = t_2$ holds for infinitely many m and $t_1 \leq t_2 + 1$ holds for all m.*

Proof. Φ fulfills $\Phi^2 = \Phi + 1$ and we have already mentioned that this implies $\Phi^{i-2} \leq F(i) \leq \Phi^{i-1}$ for all $i \geq 2$.

Consider $n \in \mathbb{N}$ such that $n = 2 \cdot F(k)$ for some k. Then $t_1 = k + 1$ and $t_2 = 2 + \lceil \log_\Phi F(k) \rceil = 2 + k - 1 = k + 1$. Therefore $t_1 = t_2$ holds for such n.

Let $n = 2m$ be an arbitrary, positive integer. If i is determined by $\Phi^{i-1} < m \leq \Phi^i$, then $t_2 = 2 + i$.

Let k be the smallest positive integer such that $F(k) \geq m$. Since $\Phi^{k-2} \leq F(k) \leq \Phi^{k-1}$, we obtain either $i = k - 1$ or $i = k - 2$. This implies $t_1 = k + 1 \leq i + 3$.

□

Now we know that the difference between the upper bound and the lower bound is at most one and this makes us more ambitious. We would like to know the exact value. The following Table 4.1 shows the upper bound for the gossip problem and the upper and lower bound for the network counting problem for values of n up to 22.

Table 4.1. Upper and lower bounds for gossip and NCP

n	2	4	6	8	10	12	14	16	18	20	22
upper bound gossip	2	4	5	6	6	7	7	7	8	8	8
upper bound NCP	2	4	5	5	6	6	7	7	7	7	8
lower bound NCP	2	4	5	5	6	6	7	7	7	7	7

4.6 Overview

As a summary of this chapter, Tables 4.2 and 4.3 contain overviews of the best currently known time bounds for gossiping in the one-way and two-way modes for common interconnection networks and the according references in this paper and in the literature.

In the tables, $even(n) = 1$ if n is even and 0 otherwise, and $odd(n) = 1$ if n is odd and 0 otherwise. Most of the lower bounds are derived from the lower bounds for broadcasting.

Table 4.2. Gossip times for common networks in the two-way mode

graph	no. nodes	diameter	lower bound	upper bound
K_n	n	1	$\lceil \log_2 n \rceil + odd(n)$ [Kn75]	$\lceil \log_2 n \rceil + odd(n)$ [Kn75]
H_k	2^k	k	k Lemma 4.2.3	k Lemma 4.2.3
P_n	n	$n-1$	$n - even(n)$ Theorem 4.3.1	$n - even(n)$ Theorem 4.3.1
C_n	n	$\lfloor n/2 \rfloor$	$\lceil n/2 \rceil + odd(n)$ Theorem 4.3.10, [FP80]	$\lceil n/2 \rceil + odd(n)$ Theorem 4.3.10, [FP80]
CCC_k	$k \cdot 2^k$	$\lfloor 5k/2 \rfloor - 2$	$\lceil 5k/2 \rceil - 2$ Theorem 3.3.2, [LP88]	$\lceil 5k/2 \rceil - 2, k$ even $\lceil 5k/2 \rceil + 1, k$ odd Theorem 4.4.5, [SS03]
SE_k	2^k	$2k - 1$	$2k - 1$ Theorem 3.3.5, [HJM93]	$2k + 5$ Theorem 3.3.5, [Pe98]
BF_k	$k \cdot 2^k$	$\lfloor 3k/2 \rfloor$	$1.9770k$ Theorem 3.4.6, [FP99b]	$2.25 \cdot k + o(k)$ Corollary 4.4.6, [Si01]
DB_k	2^k	k	$1.5965k$ Theorem 3.4.10, [FP99b]	$2k + 5$ Theorem 3.3.7, [Pe98]

In [FP01], families of bounded-degree graphs have been constructed in which the gossip time in the two-way mode matches the lower bound from Theorem 3.3.2 in [LP88] for degrees 3 and 4 and for the generalized lower bound of [BHLP92] for degree at least 5, thus showing that these bounds are optimal. Improved general lower bounds on the gossiping time of d-bounded degree networks in the one-way mode are derived in [FP99b]. Thus, the main open problems left are to find improved bounds for gossiping in specific networks of bounded degree, and to improve the general upper and lower bounds on the gossiping time of d-bounded degree networks in the one-way mode.

Table 4.3. Gossip times for common networks in the one-way mode

graph	no. nodes	diameter	lower bound	upper bound
K_n	n	1	$1.44 \log_2 n$ [EM89]	$1.44 \log_2 n$ [ES79, EM89]
H_k	2^k	k	$1.44k$ [EM89]	$1.88k$ [Kr92]
P_n	n	$n-1$	$n + odd(n)$ Theorem 4.3.1	$n + odd(n)$ Theorem 4.3.1
C_n	n even n odd	$\lfloor n/2 \rfloor$ $\lfloor n/2 \rfloor$	$n/2 + \lceil \sqrt{2n} \rceil - 1$ $\lceil n/2 \rceil + \lceil \sqrt{2n - 1/2} \rceil - 1$ Theorem 4.3.15, [HJM93]	$n/2 + \lceil \sqrt{2n} \rceil - 1$ $\lceil n/2 \rceil + \lceil 2\sqrt{\lceil n/2 \rceil} \rceil - 1$ Theorem 4.3.13, [HJM93]
CCC_k	$k \cdot 2^k$	$\lfloor 5k/2 \rfloor - 2$	$\lceil 5k/2 \rceil - 2$ Theorem 3.3.2, [LP88]	$\lceil 7k/2 \rceil + \lceil 2\sqrt{\lceil k/2 \rceil} \rceil - 2$ Theorem 4.4.3, [HJM90]
SE_k	2^k	$2k - 1$	$2k - 1$ Theorem 3.3.5, [HJM93]	$3k + 3$ Theorem 3.3.5, [Pe98]
BF_k	$k \cdot 2^k$	$\lfloor 3k/2 \rfloor$	$1.9770k$ Theorem 3.4.6, [FP99b]	$\lceil 5k/2 \rceil + \lceil 2\sqrt{\lceil k/2 \rceil} \rceil - 1$ Theorem 4.4.3, [HJM90]
DB_k	2^k	k	$1.5965k$ Theorem 3.4.10, [FP99b]	$3k + 3$ Theorem 3.3.7, [Pe98]

5

Systolic Communication

*We demand that life should have a sense,
but it has only as much sense
as we ourselves are able to give it.*

Hermann Hesse

5.1 Introduction

One of the most intensively investigated areas of computation theory is the study and comparison of the computational power of distinct interconnection networks as candidates for use as parallel architectures for existing parallel computers. There are several approaches that enable us to compare the efficiency and the "suitability" of different parallel architectures from distinct points of view. One extensively used approach deals with the possibility of simulating one network using another without any essential increase of computational complexity (parallel time, number of processors). Such an effective simulation of a network A using a network B surely exists if the network A can be embedded into B (more details and an overview about this research direction can be found in [MS90]).

Another approach to measuring the computational power of interconnection networks is to investigate which class of computing problems can be computed by a given class of networks. Obviously, this question is reasonable only with additional restrictions on the networks because each class of networks with an unbounded number of processors (like paths, grids, complete binary trees, hypercubes, etc.) can recognize all recursive sets. These additional restrictions mostly restrict the time of computations (for example, to $\log_2 n$ using complete binary trees or to real time using paths) and/or the kind of computation assuring a regular flow of data in the given network. A nice concept for the study of the power of networks from this point of view was introduced by Culik II et al. [CGS84], and investigated in [IK84, CC84, IPK85, CSW84, CGS83, IKM85]. This concept considers classes of languages recognized only by systolic computations on the given parallel architecture (network) in the shortest possible time for a given network. The notion "systolic computation" has been introduced by Kung [Ku79], and it means that the computation consists only of the repetition of simple computation and communication steps in a periodic way. The reason we prefer systolic computations is based on the fact that each processor of a network executing a systolic algorithm works very regularly repeating only a short sequence of simple instructions during the whole computation. Thus, the hardware and/or software execution of systolic algorithms is essentially cheaper than the execution of parallel algorithms containing many irregularities in the data flow or in the behavior of the processors.

The last of the approaches mentioned here helping in the search for the best (most effective) structures of interconnection networks is the study of the complexity of information dissemination in networks (for an overview see [HHL88, HKMP95]). This approach is based on the observation that the execution of the communication (data flow between the processes) of several parallel algorithms on networks requires at least as much (or sometimes even more) time as the computation time of the processors. This means that the time spent on communication is an important parameter of the quality of interconnection networks. To get a comparison of networks from the communication point of view, the complexity of the execution of some basic communication tasks like broadcast (one processor wants to tell all other processors something), accumulation (one processor wants to get some pieces of information from all other processors) and gossip (each processor wants to tell each other something) is investigated for different networks.

The aim of this chapter is to combine the ideas of the last two approaches mentioned above to get a concept of systolic communication algorithms enabling us to study the communicational power of networks when a very regular behavior of each processor of the network is required.

The first step in this direction was made by Liestman and Richards [LR93b] who introduced a very regular form of communication based on graph coloring. (This kind of communication has later been called "periodic gossiping" by Labahn et al. [LHL94].) This "periodic" communication was introduced in order to solve a special gossip problem introduced by Liestman and Richards [LR93a] and was called "perpetual gossiping", where each processor may get a new piece of information at any time from outside of the network and the never halting communication algorithm has to broadcast it to all other processors as soon as possible.

The concept of Liestman and Richards [LR93b] includes some restrictions which bound the possibility of systolic communication in an unnecessary way (more about this in the next section). Another drawback is that the complexity considered in [LR93b, LHL94] is the number of systolic periods and not the number of communication rounds, i.e., only rough approximations on the precise number of rounds executed are achieved in [LR93b, LHL94]. Here, we introduce a more general concept of systolic (periodic) dissemination of information in order to evaluate the quality of interconnection networks from this point of view. Our main aim in the investigation of systolic communication is not only to establish the complexity of systolic execution of basic communication tasks in distinct networks, but also to learn how much must be paid for the change from an arbitrary "irregular" communication to a nice, regular systolic one.

This chapter is organized as follows. In Section 5.2 we give the formal definitions, and in Section 5.3 broadcasting is investigated. In Section 5.4 some simple observations about systolic gossiping are presented. Different possibilities to define systolic communication modes are discussed in Section 5.5. Systolic gossip on different networks such as paths, cycles, trees, grids, butterflies, and cube-connected-cycles is designed in Section 5.6. A survey of systolic communication is given in Section 5.7.

5.2 Definitions

We start this section by giving the formal definition of four types of systolic algorithms and then discussing the differences between them.

Definition 5.2.1. Let $A = E_1, E_2, \ldots, E_m$ be a two-way communication algorithm[1] for $G = (V, E)$. A is called a **two-way p-systolic communication algorithm** if $E_i = E_{i+p}$ for all $1 \leq i \leq m - p$. We will call $E_1, E_2, \ldots, E_p$ the **period/cycle** of algorithm A and p the **period-length** of A.

Thus a p-systolic algorithm consists of a sequence $P = E_1, E_2, \ldots, E_p$ which is repeated $\lfloor m/p \rfloor$ times followed by the first $m - p \cdot \lfloor m/p \rfloor$ matchings. Furthermore the structure of A is

$$(E_1, E_2, \ldots, E_p)^{\lfloor m/p \rfloor}, E_1, E_2, \ldots, E_{m-p \cdot \lfloor m/p \rfloor}.$$

Obviously, each communication algorithm is p-systolic for some sufficiently large p. We will mainly consider p-systolic communication algorithms for fixed p for some classes of networks. In this approach, p is a constant independent of the sizes of the networks of the class. This means that our p-systolic algorithms are simply performed through the repetition of a cycle of p simple instructions by any processor of the network.

An example of a two-way 6-systolic gossip algorithm is given in Figure 5.1. Instead of labeling each edge e with all its active rounds (i.e., rounds in which e is used), we just mark e with its first p active rounds. Thus e is labeled with $\{i \mid e \in E_i, 1 \leq i \leq p\}$. Furthermore, if an edge e is marked with i, then e is also active in rounds $i + j \cdot p \leq m$ for all $j \in \mathbb{N}$.

The above example is the graph D_{20} (see Figure 2.13). The algorithm described in Figure 5.1 takes 19 rounds to complete a gossip, which is one round more than $r_2(D_{20})$.

Exercise 5.2.2. Show that any two-way p-systolic gossip algorithm for D_n ($n \geq 8$) with $p < n/3$ uses strictly more than $r_2(D_n)$ rounds.

---

[1] Remember that a two-way communication algorithm (see Section 2.4) for a graph $G = (V, E)$ is a sequence $E_1, E_2, \ldots, E_m$ of some sets (matchings) $E_i \subseteq E$, where for each $i \in \{1, \ldots, m\}, \forall (x_1, y_1), (x_2, y_2) \in E_i : |\{x_1, y_1\} \cap \{x_2, y_2\}| \in \{0, 2\}$.

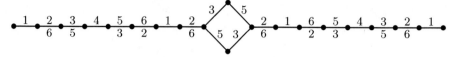

Fig. 5.1. Example of a two-way 6-systolic gossip algorithm

We will now introduce a new restriction on p-systolic algorithms. So far during each period in a p-systolic algorithm an edge may be active more than once. In the following definition this will be excluded.

Definition 5.2.3. *Let A be a two-way p-systolic communication algorithm for $G = (V, E)$ with period $E_1, E_2, \ldots, E_p$. A is called* **two-way p-periodic** *if each edge $e \in E$ is in at most one E_i, i.e., if $e \in E_i \cap E_j$ implies $i = j$.*

Thus a p-periodic communication algorithm for $G = (V, E)$ is given by an edge-coloring with p colors of a spanning subgraph $G' = (V, E')$ of G. As an example we give a p-periodic gossip algorithm for G_{18} in Figure 5.2.

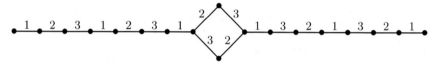

Fig. 5.2. Example of a two-way 3-periodic gossip algorithm

It is easy to see that this 3-periodic gossip algorithm takes 22 rounds. Thus this algorithm takes four rounds more than an optimal non-p-periodic one. It is an easy exercise to show that any p-periodic gossip algorithm uses 22 rounds on D_{18}. Furthermore one may construct examples where the difference is even larger.

Exercise 5.2.4. Show that 22 rounds are necessary for a two-way p-periodic gossip algorithm for D_{18}.

Exercise 5.2.5. Construct a graph $G = (V, E)$ with degree k and n nodes such that any two-way p-periodic gossip algorithm for G takes a time of at least $r_2(G) + (n - k) \cdot (k - 2)$.

We will now give the definition of a one-way p-systolic communication algorithm. Let us recall a part of the definition of a one-way communication algorithm for a graph $G = (V, E)$ from Section 2.4. A one–way communication algorithm is a sequence $E_1, E_2, \ldots, E_m$ of sets (matchings) $E_i \subseteq \overline{E}$, where $\overline{E} = \{(v \to u), (u \to v) \mid (u, v) \in E\}$, and if $(x_1 \to y_1), (x_2 \to y_2) \in E_i$ and $(x_1, y_1) \neq (x_2, y_2)$ for some $i \in \{1, \ldots, m\}$ then $x_1 \neq x_2 \land x_1 \neq y_2 \land y_1 \neq x_2 \land y_1 \neq y_2$.

Definition 5.2.6. *Let A be a one-way communication algorithm for $G = (V, E)$ with the sequence $E_1, E_2, \ldots, E_m$ of matchings. A is called a* **one-way p-systolic communication algorithm** *if $E_i = E_{i+p}$ for all $1 \leq i \leq m - p$.*

Fig. 5.3. Example of a one-way 6-systolic gossip algorithm

An example of a one-way 6-systolic communication algorithm is given in Figure 5.3. In the one-way mode we will also introduce the restriction that an edge may be active at most once during the period. An example is given in Figure 5.4

Definition 5.2.7. *Let A be a one-way p-systolic communication algorithm for $G = (V, E)$ with period $E_1, E_2, \ldots, E_p$. A is called* **one-way p-periodic** *if each edge $e \in E$ is in at most one E_i, i.e., if $e \in \overline{E_i} \cap \overline{E_j}$ implies $i = j$.*

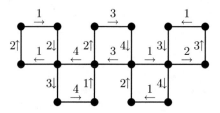

Fig. 5.4. Example of a one-way 4-periodic gossip algorithm

Exercise 5.2.8. Give examples where a p-systolic or p-periodic concept is used in reality, e.g., traffic control, cars, distribution of parcels, and television.

It remains to define the complexity measures for the above modes. Let us recall that $com(A)$ denotes the uniform complexity of a communication algorithm A (see Definition 2.4.4). Similarly to that definition we define the complexities here.

Definition 5.2.9. *Let $G = (V, E)$ be a connected graph and let p be an integer, $p \geq 2$. The* **two-way p-systolic/p-periodic and one-way p-systolic/p-periodic gossip complexity** *of G is defined by:*

$$[p]\text{-}sr_2(G) = \min(\{\text{com}(A) \mid A \text{ is a two-way p-systolic gossip algorithm for } G \}, \infty).$$

$$[p]\text{-}pr_2(G) = \min(\{\text{com}(A) \mid A \text{ is a two-way p-periodic gossip algorithm for } G \}, \infty).$$

$$[p]\text{-}sr(G) = \min(\{\text{com}(A) \mid A \text{ is a one-way p-systolic gossip algorithm for } G \}, \infty).$$

$$[p]\text{-}pr(G) = \min(\{\text{com}(A) \mid A \text{ is a one-way p-periodic gossip algorithm for } G \}, \infty).$$

Exercise 5.2.10. Let $p \in I\!N$. Describe for each of the gossip complexities of Definition 5.2.9 the class of graphs for which the gossip complexity is not ∞.

Directly from the definition of these modes we get the following straightforward observation.

Observation 5.2.11. *Let $G = (V, E)$ be a graph and $p \in I\!N$.*

$$r_2(G) \leq [p]\text{-}sr_2(G) \leq [p]\text{-}pr_2(G)$$
$$\quad \mathsf{I\land} \qquad\qquad \mathsf{I\land}$$
$$r(G) \leq [p]\text{-}sr(G) \leq [p]\text{-}pr(G)$$

Definition 5.2.12. *Let $G = (V, E)$ be a graph, $v \in V$, and let A be a one–way [two–way] communication algorithm. The* **p-systolic/p-periodic broadcast/accumulation complexity of G [from/towards v]** *is defined by:*

$$[p]\text{-}sb(v, G) = \min(\{\text{com}(A) \mid A \text{ is a one-way p-systolic broadcast algorithm for } G \text{ and } v\}, \infty),$$

$$[p]\text{-}sb(G) = \max\{[p]\text{-}sb(v, G) \mid v \in V\},$$

$$[p]\text{-}sb_{min}(G) = \min\{[p]\text{-}sb(v, G) \mid v \in V\},$$

$$[p]\text{-}pb(v, G) = \min(\{\text{com}(A) \mid A \text{ is a one-way p-periodic broadcast algorithm for } G \text{ and } v\}, \infty),$$

$$[p]\text{-}pb(G) = \max\{[p]\text{-}pb(v, G) \mid v \in V\},$$

$$[p]\text{-}pb_{min}(G) = \min\{[p]\text{-}pb(v, G) \mid v \in V\},$$

$$[p]\text{-}sa(v, G) = \min(\{\text{com}(A) \mid A \text{ is a one-way p-systolic accumulation algorithm for } G \text{ and } v\}, \infty),$$

$$[p]\text{-}sa(G) = \max\{[p]\text{-}sa(v, G) \mid v \in V\},$$

$$[p]\text{-}sa_{min}(G) = \min\{[p]\text{-}sa(v, G) \mid v \in V\},$$

$$[p]\text{-}pa(v, G) = \min(\{\text{com}(A) \mid A \text{ is a one-way p-periodic accumulation algorithm for } G \text{ and } v\}, \infty),$$

$$[p]\text{-}pa(G) = \max\{[p]\text{-}pa(v, G) \mid v \in V\},$$

$$[p]\text{-}pa_{min}(G) = \min\{[p]\text{-}pa(v, G) \mid v \in V\}.$$

Observation 5.2.13. *Let $g = (V, E)$ be a graph, $v \in V$ and $p \in I\!N$.*

$$b(v, G) \leq [p]\text{-sb}(v, G) \leq [p]\text{-pb}(v, G)$$
$$\|\qquad\qquad\|$$
$$a(v, G) \leq [p]\text{-pb}(v, G) \leq [p]\text{-pa}(v, G)$$

$$b(G) \;\leq\; [p]\text{-sb}(G) \;\leq\; [p]\text{-pb}(G)$$
$$\|\qquad\qquad\|$$
$$a(G) \;\leq\; [p]\text{-pb}(G) \;\leq\; [p]\text{-pa}(G).$$

Exercise 5.2.14. Prove Observation 5.2.13.

Exercise 5.2.15. Define a new systolic (resp. periodic) mode where the direction of passing a message on an edge may change in each period. Compare these new modes to the presented ones.

5.3 Systolic Broadcast

The main result of this section shows that, in some sense, the broadcast complexity is the same as the d-periodic broadcast complexity for any network of constant degree d. We will explain the idea of this proof using Figure 5.5.

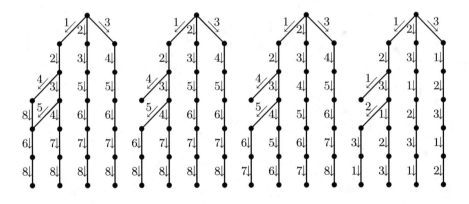

Fig. 5.5. A broadcast algorithm transformed into a periodic algorithm

The following steps are done from left to right in Figure 5.5:

1. All communication actions, in which the broadcast message is sent to a node which has already received it, are deleted.
2. The situations where a node delays the message, i.e., it has received the message and does not forward it to any uninformed neighbor in the next step, are eliminated.
3. In the last step the period is determined. If an edge carries the message in the round j, then it will be activated in each round that is a multiple of $(j - 1) \bmod d + 1$.

The technical details in the proof of Lemma 5.3.1 complete our consideration.

Lemma 5.3.1. *For every class of networks $\{G_i\}_{i=1}^{\infty}$ and for every positive integer d such that G_i has the degree at most d for any $i \in \mathbb{N}$,*

$$[d]\text{-pb}(v, G_i) = [d]\text{-sb}(v, G_i) = b(v, G_i)$$

for any $i \in \mathbb{N}$ and any node v of G_i.

Proof. Let $B = B_1, B_2, \ldots, B_m$ be a broadcast algorithm from v in $G_i = (V, E)$ for some positive integer i. W.l.o.g. we may assume $B_i \cap B_j = \emptyset$ for $i \neq j$. One can reconstruct B to get another broadcast algorithm $A = A_1, A_2, \ldots, A_m$ for v and G_i with the following three properties:

(i) $\displaystyle\bigcup_{i=1}^{m} A_i \subseteq \bigcup_{j=1}^{m} B_j$, $com(B) = com(A)$,

(ii) $T = (V, \displaystyle\bigcup_{i=1}^{m} A_i)$ is a directed tree with the root v (all edges are directed from the root to the leaves), and

(iii) for every node $w \in V$, if w gets $I(v)$ in time t and the degree (the indegree plus the outdegree) of w in T is k, then w submits $I(v)$ in the rounds $t+1, t+2, \ldots, t+k-1$ to all its $k-1$ descendants in T.

The construction of the broadcast algorithm

$$A' = A'_1, A'_2, \ldots, A'_m$$

from B, with the property

$$T = (V, \bigcup_{i=1}^{m} A'_i)$$

is a directed tree with the root v, is due to the proof of Lemma 2.5.5.

Assume now that there is an inner node v' with $d' \leq d$ descendants $v'_1, \ldots, v'_{d'}$ which receives message $I(v)$ in round r. W.l.o.g. the message $I(v)$ is forwarded to v'_i in round $r + r_i$ ($1 \leq i \leq d'$) and $1 \leq r_1 < r_2 < \ldots < r_{d'}$. Note that by the construction of Lemma 2.5.5 each edge is active only once in A'. Thus the descendants of v'_i are not active before round $r + r_i$ and we can reschedule the communication from v' to v'_i in round $r + i$. We apply this construction with $r = 0$ on v and in a top-down fashion to all its descendants and subtrees. Finally we get algorithm A with the above properties.

To define a d-periodic broadcast algorithm

$$C = (C_1, C_2, \ldots, C_d)^r, C_1, C_2, \ldots, C_j$$

for some $r \in \mathbb{N}$, $j \in \{0, \ldots, d-1\}$, $m = r \cdot d + j$, it is sufficient to specify C_i for $i = 1, \ldots, d$. For $1 \leq s \leq d$,

$C_s = \{e \mid e \in A_{s+j\cdot d}, j \in I\!N, s + j \cdot d \le m\}$ for $s \in \{1, \ldots, d\}$.

Since $A_i \subseteq C_{(i-1) \bmod d+1}$, it is obvious that C is a broadcast algorithm for G and v. With Observation 5.2.13 and the just-presented fact $[d]$-$\mathrm{pb}(v, G_i) \le b(v, G_i)$ the lemma is proven.

□

Clearly, the same consideration as in the proof of Lemma 5.3.1 leads to the following result.

Lemma 5.3.2. *Let d be a positive integer. Let $\{G_i\}_{i=1}^{\infty}$ be a class of networks, where for every $i \in I\!N$, G_i has degree bounded by d. Then, for every $i \in I\!N$ and every node v of G_i*

$$[d]\text{-}\mathrm{sa}(v, G_i) = [d]\text{-}\mathrm{pa}(v, G_i) = a(v, G_i).$$

Thus for networks of constant degree we have the main result of this section.

Theorem 5.3.3. *Let d be a positive integer. Let $\{G_i\}_{i=1}^{\infty}$ be a class of networks, where for every $i \in I\!N$, G_i has degree bounded by d. Then, for every $i \in I\!N$*

$$[d]\text{-}\mathrm{sb}(G_i) = [d]\text{-}\mathrm{sp}(G_i) = [d]\text{-}\mathrm{sa}(G_i) = [d]\text{-}\mathrm{pa}(G_i) = a(G_i) = b(G_i).$$

If the given degree and period differ, then the situation changes.

Exercise 5.3.4. Show that the following problem is NP-complete:
Decide for given p, b and a graph $G = (V, E)$ whether $[p]$-$\mathrm{sb}(G) \le b$ holds.

Change your construction to prove the following:

Exercise 5.3.5. Show that the following problem is NP-complete:
Decide for given p, b and a graph $G = (V, E)$ whether $[p]$-$\mathrm{sr}(G) \le b$ holds.

Exercise 5.3.6. Can you modify your solutions for Exercises 5.3.4 and 5.3.5 in order to prove NP-completeness of these decision problems for $[p]$-$\mathrm{pr}(G)$ and $[p]$-$\mathrm{sr}_2(G)$, too?

5.4 General Observations for Systolic Gossip

Since there are no differences in the complexities of general broadcast algorithms and systolic broadcast algorithms the main questions we pose are the following two:

1. What is the k-systolic gossip complexity of fundamental interconnection networks? What is the most appropriate k (if it exists) for a given network?

2. How much must be paid for the systolization of communication algorithms in concrete interconnection networks (i.e., what is the difference between the complexity of general gossip and the complexity of systolic gossip)?

The next important question is what relation holds between general gossip complexity and systolic gossip complexity. The following sections show that, as opposed to broadcast (accumulation), there are already essential differences between the complexities of general gossip and systolic gossip.

Here we shall still deal with the relation between gossip and broadcast. We have shown in Chapter 2 and Chapter 4 that $r(G) \leq$ min a$(G)+$ min b$(G) = 2$·min b(G) and $r_2(G) \leq 2$·min b$(G) - 1$ for any graph G, and that for trees and some cyclic graphs with "weak connectivity" the equalities $r(G) = 2$·min b(G) and $r_2(G) = 2$·min b$(G) - 1$ hold. The idea of the proof of $r(G) \leq$ min a$(G)+$ min b(G) is very simple: One node of G first accumulates $I(G)$, and then it broadcasts $I(G)$ to all other nodes. Unfortunately, we cannot use this scheme to get systolic gossip from systolic broadcast and systolic accumulation, because we have to use every edge of an optimal broadcast (accumulation) scheme in both directions in each repetition of the cycle of a systolic gossip algorithm which increases the time for the broadcast phase twice. Thus, using this straightforward idea we only obtain the following:

Theorem 5.4.1. *Let G be a communication network of degree bounded by some positive integer k. Then*

$$[2k]\text{-}sr(G) \leq 4 \cdot [k]\text{-min sb}(G) + 2k.$$

Proof. Let $B = B_1, B_2, \ldots, B_m$, $m = $ min b$(G) = [k]$-min sb(G), be an optimal broadcast algorithm for G and some node v of G with the properties (i), (ii), (iii) as in Lemma 5.3.1. Let $T_B = (V, \bigcup_{i=1}^{m} B_i)$. Obviously, $A = A_1, A_2, \ldots, A_m$, where $A_i = B_{m-i+1} = \{(x \to y)|(y \to x) \in B_{m-i+1}\}$ is an optimal accumulation algorithm for G and v. Moreover, $T_A = (V, \bigcup_{i=1}^{m} A_i)$ is the same scheme as T_B, only the edges are directed in opposite directions. Building

$$C_s = \bigcup_{j \in In(s)} B_j \text{ for } s \in \{1, \ldots, k\},$$

$$D_s = \bigcup_{j \in In(s)} A_j \text{ for } s \in \{1, \ldots, k\},$$

one obtains optimal k-systolic broadcast and accumulation algorithms

$$C = (C_1, C_2, \ldots, C_k)^r, C_1, C_2, \ldots, C_j \text{ and } D = (D_1, \ldots, D_k)^r, D_1, D_2, \ldots, D_j$$

respectively for some positive integer r, and $j \in \{0, \ldots, k - 1\}$. Now, we consider the $2k$-systolic communication algorithm

$$F = (D_1, \ldots, D_k, C_1, \ldots, C_k)^{2r+1}.$$

The initial part of F, $(D_1, \ldots, D_k, C_1, \ldots, C_k)^r D_1, \ldots, D_k$, is a one-way $2k$-systolic accumulation algorithm for G and the node v. The remaining part, $C_1, \ldots, C_k(D_1, \ldots, D_k, C_1, \ldots, C_k)^r$, is a one-way $2k$-systolic broadcast algorithm for G and the node v. Thus, F is a one-way $2k$-systolic gossip algorithm with $\mathrm{com}(F) = 2k \cdot (2r + 1) = 4r \cdot k + 2k \leq 4 \cdot [k]\text{-minsb}(G) + 2k$.

$\square$

Exercise 5.4.2. Does a similar result hold for periodic gossip?

5.5 Special Systolic Modes

This way of defining a one-way p-systolic communication algorithm is a natural one, but there is also another possibility for defining one. In the above definition the time and the communication direction of a transmission are fixed by the period. One may also consider fixing only the time of transmission by the period and opening up the possibility of determining the direction of transmission independently of the period.

Definition 5.5.1. *Let* $A = E_1, E_2, \ldots, E_m$ *be a one-way communication algorithm for* $G = (V, E)$. *A is called a* **free one-way p-systolic communication algorithm** *if* $U(E_i) = U(E_{i+p})$ *for all* $1 \leq i \leq m - p$, *where* $U(E') = \{(u, v) \mid (u \to v) \in E'\}$ *for* $E' \subset \overline{E}$.

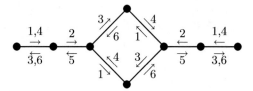

Fig. 5.6. A free one-way 3-periodic gossip algorithm

An example of a free one-way 3-systolic gossip algorithm is given in Figure 5.6. Each edge is marked with the rounds in which it is transmitting information. Note that in the example the gossiping time is $r(D_{16}) + 1$. Thus we are losing again one round in this mode. Using the idea from Exercise 5.2.2 one can prove the following assertion.

Exercise 5.5.2. Show that any free one-way p-systolic gossip algorithm for D_n ($n \geq 8$, see Figure 2.13) with $p < n/3$ uses strictly more than $r_2(D_n)$ rounds.

Definition 5.5.3. *Let $A = E_1, E_2, \ldots, E_m$ be a free one-way p-systolic communication algorithm for $G = (V, E)$.*

A is called **free one-way p-periodic** *if each edge $e \in E$ is active at most once during the period, i.e., $e \in \overline{E_i} \cap \overline{E_j} \Rightarrow i - j \equiv 0 \bmod p$.*

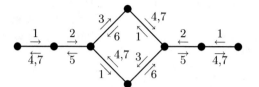

Fig. 5.7. A one-way free 3-systolic gossip algorithm

Note that it is not necessary to define free p-systolic/p-periodic complexity measures for broadcasting or accumulation, because in those algorithms the direction of information flow is fixed, see Corollary 2.5.6.

Definition 5.5.4. *Let $G = (V, E)$ be a graph. The* **free one-way p-systolic/ p-periodic gossip complexity of G** *is defined by:*

$$[p]\text{-}\mathbf{fsr}(G) = \min(\{\mathrm{com}(A) \mid A \text{ is a one-way p-systolic gossip} \\ \text{algorithm for } G \}, \infty),$$

$$[p]\text{-}\mathbf{fpr}(G) = \min(\{\mathrm{com}(A) \mid A \text{ is a one-way p-periodic gossip} \\ \text{algorithm for } G \}, \infty).$$

5.6 Systolic Gossip in Concrete Networks

Obviously, each communication algorithm is k-systolic for some sufficiently large k. But we want to consider k-systolic communication algorithms for fixed k for some classes of networks. In this approach, k is a constant independent of the sizes of the networks of the class. This means that our k-systolic algorithms are simply performed by the repetition of a cycle of k simple instructions by any processor of the network.

We observe that each k-systolic algorithm uses (activates) at most k adjacent edges of every node of the network during the whole work of the algorithm. Thus, there is no reason to consider classes of networks like hypercubes and complete graphs, because a k-systolic algorithm can use only a subgraph of these graphs with the degree bounded by k. For this reason, we shall investigate the systolic complexity of broadcast and gossip for constant-degree bounded classes of networks only. Our aim is not only to get some lower and upper bounds on the systolic broadcast and gossip complexity of some concrete networks, but also to compare the general complexities of unrestricted communication algorithms with the systolic ones. In this way, we can determine the price of our systolization, i.e., how many additional rounds are sufficient and necessary to go from an optimal broadcast (gossip) algorithm to an optimal systolic one.

The subsections that follow deal with the systolic gossip problem in concrete networks.

5.6.1 Systolic Gossip on a Path

In this section we consider systolic gossiping in the path P_n of n nodes.

For the two-way mode, we can find a systolic gossip algorithm for P_n with an optimal period length that works as efficiently as the algorithm in the general gossip mode.

Theorem 5.6.1.

(i) $[2]$-$sr_2(P_n) = n - 1 = r_2(P_n)$ *for even* $n \geq 2$,
(ii) $[2]$-$sr_2(P_n) = n = r_2(P_n)$ *for odd* $n \geq 3$.

Proof. The lower bounds for the general mode were given in Chapter 2 and Chapter 4. The upper bounds on the systolic gossip complexity can be obtained by modifying the gossip algorithms in the unrestricted two-way mode.

Consider the algorithm A for P_n (where $V(P_n) = \{x_1, \ldots, x_n\}$, $E(P_n) = \{(x_1, x_2), \ldots, (x_{n-1}, x_n)\}$) that has the following two systolic periods:

$$A_1 = \{(x_1, x_2), (x_3, x_4), (x_5, x_6), \ldots\},$$
$$A_2 = \{(x_2, x_3), (x_4, x_5), (x_6, x_7), \ldots\}.$$

A simple analysis shows that algorithm A takes $n - 1$ rounds if n is even and n rounds if n is odd.

$\square$

Note that for an optimal gossip in P_n, the period lengths in Theorem 5.6.1 are the best possible. Any systolic algorithm for P_n must have at least period length 2.

Let us turn to the one-way mode of communication now. For the complexity of systolic gossiping in the path P_n of n nodes, we obtain upper and lower bounds which are tight up to a constant. An important observation contrasting with the two-way case is that there is no constant d such that $[d]$-$sr(P_n) = r(P_n)$ for every $n \in \mathbb{N}$. Instead, the next theorem shows that one can essentially gossip faster in P_n with a longer period k. For growing k, $r(P_n)$ can be approached more and more but never reached.

Theorem 5.6.2. *For any* $n \geq 2$, $k \geq 4$:

(i) $\frac{k}{k-2} \cdot (n - 2) \leq [k]$-$sr(P_n) \leq \frac{k}{k-2} \cdot n - 1$ *for k even,*
(ii) $\frac{k}{k-2} \cdot (n - 2) \leq [k]$-$sr(P_n) \leq \frac{k}{k-2} \cdot (n - 1) + 1$ *for k odd.*

Proof. Let us first describe the upper bounds.

(i) *The upper bound for even k:*

Let us first assume that n is a multiple of $k-2$. Then the path P_n is divided into subpaths $B_1, B_2, \ldots, B_{n/(k-2)}$ of $k-2$ nodes as follows:

$$B_i := \{(k-2)\cdot(i-1)+1, (k-2)\cdot(i-1)+2, \ldots, (k-2)\cdot i\}$$

for $1 \leq i \leq n/(k-2)$.

In each period, the systolic one-way gossip algorithm A does the following:

1. Gossip in B_i for all $1 \leq i \leq n/(k-2)$.
2. Exchange the information between the endnodes of adjacent blocks.

As $k-2$ is even, Step 1. takes $k-2$ rounds by using the optimal gossip algorithm in each block B_i, $1 \leq i \leq n/(k-2)$. Step 2. can be achieved in 2 rounds. Thus, the period length of the systolic algorithm is k.

For a complete gossip, it is enough to ensure that the message I_1 from the left end of the path reaches the right end of the path, and that the message I_2 from the right end of the path reaches the left end. With each period (of length k), I_1 moves one block to the right, and I_2 moves one block to the left. Hence, after $n/(k-2)-1$ periods, i.e., after $k \cdot (n/(k-2)-1)$ rounds, I_1 has moved to block $B_{\lceil n/(k-2)\rceil}$, and I_2 has moved to block B_1. Now, the gossip in the first $k-2$ rounds of the next period suffices to get I_1 and I_2 to the endpoints. Hence, the overall time is at most

$$k \cdot \left(\frac{n}{k-2}\right) - 2.$$

If n is not a multiple of $k-2$, we consider the gossip scheme A for a path $P_{n'}$ of $n' := (k-2) \cdot \lceil \frac{n}{k-2} \rceil$ nodes (where $V(P_{n'}) = \{x_1, x_2, \ldots, x_{n'}\}$). Consider the subpath $P_n = \{x_{\ell+1}, \ldots, x_{\ell+n}\}$ of n nodes where

$$\ell := \left\lfloor \frac{(k-2)\cdot\lceil \frac{n}{k-2}\rceil - n}{2} \right\rfloor \geq \frac{k-2}{2}.$$

Then the scheme A restricted from $P_{n'}$ to P_n achieves gossiping on P_n in

$$\left(k \cdot \left\lceil \frac{n}{k-2}\right\rceil - 2\right) - 2 \cdot \left(\left\lfloor \frac{(k-2)\cdot\lceil \frac{n}{k-2}\rceil - n}{2}\right\rfloor + 1\right)$$

$$\leq \left(k \cdot \left\lceil \frac{n}{k-2}\right\rceil - 2\right) - \left((k-2)\cdot\left\lceil \frac{n}{k-2}\right\rceil - n\right) - 1$$

$$= 2 \cdot \left\lceil \frac{n}{k-2}\right\rceil + n - 3 \leq \frac{k}{k-2}\cdot n - 1$$

rounds (starting from round ℓ). To make A start with round 1 from the node x_ℓ, the old systolic period $A_1, \ldots, A_k$ has to be rotated by $\ell - 1$ positions to $A_\ell, A_{\ell+1}, \ldots, A_k, A_1, \ldots, A_{\ell-1}$.

(ii) *The upper bound for odd k:*

Let us first assume that $n - 1$ is a multiple of $k - 2$. Then the path P_n is divided into subpaths $B_1, B_2, \ldots, B_{(n-1)/(k-2)}$ of $k - 1$ nodes as follows:

$$B_i := \{(k - 2) \cdot (i - 1) + 1, (k - 2) \cdot (i - 1) + 2, \ldots, (k - 2) \cdot i + 1\}$$

for $1 \leq i \leq (n - 1)/(k - 2)$.

Note that two adjacent blocks overlap by one node. In each period, the systolic one-way gossip algorithm performs a complete gossip in each block B_i for all $1 \leq i \leq (n - 1)/(k - 2)$. For doing this, the optimal gossip algorithm is used in each block. As the number of nodes, $k - 1$, is even in each block, the gossip takes $k - 1$ rounds. The only problem is how communication conflicts between two adjacent blocks can be avoided. For this purpose, note that the optimal gossip algorithm (Section 4.3) in the path $P_m = \{x_1, \ldots, x_m\}$ of m nodes uses the edges (x_1, x_2) and (x_{m-1}, x_m) only in the first and the last round. Hence, if we start the gossip in blocks B_i, i odd, in the first round of each systolic period, and in blocks B_i, i even, in the second round of the period, blocks B_i, i odd, only use their leftmost and rightmost edges in rounds 1 and $k - 1$ of the period, and blocks B_i, i even, only use their leftmost and rightmost edges in rounds 2 and k of each period. This way, no conflict will occur.

For the analysis of the complexity of the algorithm, let us again consider the messages I_1 and I_2 from the left and the right ends of the path. With each period (of length k), I_1 moves one block to the right, and I_2 moves one block to the left. Hence, after $(n - 1)/(k - 2)$ periods, i.e., after

$$k \cdot \left(\frac{n - 1}{k - 2}\right)$$

rounds, I_1 has reached the right end of the path and I_2 has reached the left end.

If $n - 1$ is not a multiple of $k - 2$, we consider the gossip scheme A for a path $P_{n'}$ of $n' := (k - 2) \cdot \lceil \frac{n-1}{k-2} \rceil + 1$ nodes (where $V(P_{n'}) = \{x_1, x_2, \ldots, x_{n'}\}$). Consider the subpath $P_n = \{x_{\ell+1}, \ldots, x_{\ell+n}\}$ of n nodes where

$$\ell := \left\lfloor \frac{(k - 2) \cdot \lceil \frac{n-1}{k-2} \rceil - (n - 1)}{2} \right\rfloor \geq \left\lceil \frac{k - 2}{2} \right\rceil = \frac{k - 1}{2}.$$

Then the scheme A restricted from $P_{n'}$ to P_n achieves gossiping on P_n in

$$k \cdot \left\lceil \frac{n - 1}{k - 2} \right\rceil - 2 \cdot \left(\left\lfloor \frac{(k - 2) \cdot \lceil \frac{n-1}{k-2} \rceil - (n - 1)}{2} \right\rfloor + 1 \right)$$

$$\leq \frac{k}{k - 2} \cdot (n - 1) + 1$$

rounds (starting from round ℓ). To make A start with round 1 from the node x_ℓ, the old systolic period $A_1, \ldots, A_k$ has to be rotated by $\ell - 1$ positions to $A_\ell, A_{\ell+1}, \ldots, A_k, A_1, \ldots, A_{\ell-1}$.

This completes the proof of the upper bounds of Theorem 5.6.2. To derive the lower bounds, we start by introducing the concept of viewing a gossip algorithm as a set of time-paths. This concept was successfully used in Chapter 4 to get an optimal gossip algorithm for cycles.

Definition 5.6.3. *Let $G = (V, E)$ be a graph, and let $X = x_1, \ldots, x_m$ be a simple path (i.e., $x_i \neq x_j$ for $i \neq j$) in G. Let $A = A_1, \ldots, A_s$ be a communication algorithm in one-way mode. Let $T = t_1, \ldots, t_{m-1}$ be an increasing sequence of positive integers such that $(x_i \rightarrow x_{i+1}) \in A_{t_i}$ for $i = 1, \ldots, m-1$. We say that $X[t_1, \ldots, t_{m-1}]$ is a* **time-path of A** *because it provides the information flow from x_1 to x_m in A. If $t_{i+1} - t_i - 1 = k_i \geq 0$ for some $i \in \{1, \ldots, m-2\}$ then we say that*

$$X[t_1, \ldots, t_{m-1}] \text{ has a } k_i\text{-delay at the node } x_{i+1}.$$

The **global delay of** $X[t_1, \ldots, t_{m-1}]$ *is*

$$\mathbf{d}(\mathbf{X}[\mathbf{t_1}, \ldots, \mathbf{t_{m-1}}]) = t_1 - 1 + \sum_{i=1}^{m-2} k_i.$$

The **global time of** $X[t_1, \ldots, t_{m-1}]$ *is*

$$m - 1 + d(X[t_1, \ldots, t_{m-1}]).$$

Obviously, the necessary and sufficient condition for a communication algorithm to be a gossip algorithm in a graph G is the existence of time-paths between all pairs of nodes in G.

So, one can view the gossip algorithm for a graph G as a set of time-paths between any ordered pair of nodes.

The complexity (the number of rounds) of a communication algorithm can be measured as

$$\max\{\text{global time of } X[T] \mid X[T] \text{ is a time-path of } A\}.$$

To see a gossip algorithm as a set of time-paths is mainly helpful for proving lower bounds. A conflict of two time-paths (the meeting of two time-paths going in "opposite directions" at the same node and at the same time of the systolic period) causes some delays in these time-paths (because of the restriction given by the communication modes). Too many unavoidable conflicts mean too many delays, and so one can get much better lower bounds for gossiping in some graphs G than the trivial diameter lower bound. A combinatorial analysis providing lower bounds by analyzing the number of conflicts and delays requires a precise definition and use of these two notions. Thus, we define these notions for the one-way communication mode and the path P_n as follows. Note that the next definition essentially differs from the definition of conflicts in [HJM93] because that definition allows at most one conflict between two time-paths going in opposite directions on the same physical path of the network. The essential point here is that the time-paths from one endpoint to another are performed in a systolic manner and that this systolic realization causes conflicts and delays in nodes where the crucial information pieces flowing between the two end-points do not meet in a physical time.

Definition 5.6.4. *Let* $P_n = (\{x_1, x_2, \ldots, x_n\}, \{(x_i, x_{i+1}) | i = 1, \ldots, n\})$ *be a path of n nodes. Let $A = (A_1, A_2, \ldots, A_k)^r A_1, A_2, \ldots, A_s$ be a k-systolic one-way gossip algorithm for P_n for some $k, r \in \mathbb{N}$, $0 \le s < k$. Let $X = x_1, x_2, \ldots, x_n$ and $Y = x_n, x_{n-1}, \ldots, x_1$. An **X-direction of A** is $X[A] = (S_1, S_2, \ldots, S_{n-1})$, where $S_i = \{1 \le j \le k | (x_i \to x_{i+1}) \in A_j\}$ for $i = 1, \ldots, n-1$. A **Y-direction of A** is $Y[A] = (Q_1, Q_2, \ldots, Q_{n-1})$, where $Q_i = \{1 \le j \le k | (x_{i+1} \to x_i) \in A_j\}$ for $i = 1, \ldots, n-1$.*

For any $p \in \{1, \ldots, n-1\}$ and $q \in \{1, \ldots, k\}$, let

$$\text{next}_q(x_p \to x_{p+1}) = \min\{a, b+k \mid a, b \in S_p \text{ and } q < a \le k, \ 1 \le b \le q\},$$

and

$$\text{next}_q(x_{p+1} \to x_p) = \min\{a, b+k \mid a, b \in Q_p \text{ and } q < a \le k, \ 1 \le b \le q\}.$$

*Let i be a positive integer, $1 \le i \le k$. We say that **the directions of A $X[A]$ and $Y[A]$ have an i-collision in a node x_m** for some $m \in \{2, \ldots, n-1\}$ if one of the following four conditions hold (see Figure 5.8):*

(i) $i \in S_{m-1}$, $\text{next}_i(x_{m+1} \to x_m) < \text{next}_i(x_m \to x_{m+1}) < \text{next}_i(x_m \to x_{m-1})$,

(ii) $i \in Q_m$, $\text{next}_i(x_{m-1} \to x_m) < \text{next}_i(x_m \to x_{m-1}) < \text{next}_i(x_m \to x_{m+1})$,

(iii) $i \in S_{m-1}$, $\text{next}_i(x_{m+1} \to x_m) < \text{next}_i(x_m \to x_{m-1}) < \text{next}_i(x_m \to x_{m+1})$,

(iv) $i \in Q_m$, $\text{next}_i(x_{m-1} \to x_m) < \text{next}_i(x_m \to x_{m+1}) < \text{next}_i(x_m \to x_{m-1})$.

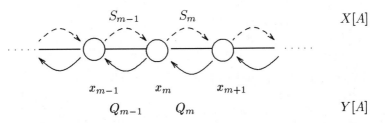

Fig. 5.8. An i-collision in a node x_m

If one of the cases (i) or (ii) happens, then we say that the i-collision in x_m causes at least a 1-delay on $X[A]$ and at least a 1-delay on $Y[A]$. If (iii) happens, then we say that the i-collision in x_m causes at least a 2-delay on $X[A]$. If (iv) happens, then we say that the i-collision causes at least a 2-delay on $Y[A]$. For every $i \in \{1, \ldots, k\}$, every i-collision in x_m is called a **collision** *in x_m.*

Let $X[T]$ and $Y[T']$ be two time-paths of A for some $T = t_1, t_2, \ldots, t_{n-1}$ and $T' = t'_1, t'_2, \ldots, t'_{n-1} = r_{n-1}, \ldots, r_1$. Let $i \in \{1, \ldots, k\}$ and $m \in \{2, \ldots, n-1\}$. We say that **the time-path $X[T]$ meets the X-direction $X[A]$ in the node x_m in the i-th periodical moment** *if*

$(t_{m-1} = kj + d,$ *where* $1 \le d \le i)$ *and*

$((t_m = kj + d_1,$ *where* $i < d_1 \le k)$ *or*

$(t_m = k(j+1) + d_2,$ *where* $1 \le d_2 < i))$.

We say that **$Y[T']$ meets $Y[A]$ in the node x_m in the i-th periodical moment** *if*

$(r_m = kj + d,$ *where* $1 \le d \le i)$ *and*

$((r_{m-1} = kj + d_1,$ *where* $i < d_1 \le k)$ *or*

$(r_{m-1} = k(j+1) + d_2,$ *where* $1 \le d_2 < i))$.

Finally, we say that there is a (systolic) **conflict** *between $X[T]$ and $Y[T']$ in a node x_m if there exists $i \in \{1, \ldots, k\}$ such that $X[A]$ and $Y[A]$ have an i-collision in the node x_m and*

(i) $X[T]$ meets $X[A]$ in the node x_m in the $((i+1) \bmod k + 1)$-th periodical moment, and

(ii) $Y[T']$ meets $Y[A]$ in the node x_m in the $((i+1) \bmod k + 1)$-th periodical moment.

□

We see that the conflict between the time-paths $X[T]$ and $Y[T']$ in a node x_m means that there exist $i \in \{1, \ldots, k\}$, $p, q \in I\!N$ such that

(i) the piece of information $I(x_1)$ flowing to x_n according to $X[T]$ is visiting x_m (still not in x_{m+1}) after the execution of $pk + i + 1$ rounds,
(ii) the piece of information $I(x_n)$ flowing to x_1 according to $Y[T']$ is visiting x_m after the execution of $qk + i + 1$ rounds,
(iii) there is an i-collision between $X[A]$ and $Y[A]$ in the node x_m.

Despite the fact that $I(x_1)$ and $I(x_n)$ possibly do not meet at x_m at the same time (p may differ from q), the systolic realization defined by the directions $X[A]$ and $Y[A]$ causes that the delay on $X[A]$ ($Y[A]$) caused by the the i-collision in x_m is also a delay on the time-path $X[T]$ ($Y[T']$).

Overall, it can be stated that the sum of the delays on $X[T]$ and $Y[T']$ caused by any conflict is at least 2.

Now, we are prepared to prove the lower bounds of Theorem 5.6.2. Let $V(P_n) = \{x_1, \ldots, x_n\}$, $E(P_n) = \{(x_1, x_2), \ldots, (x_{n-1}, x_n)\}$, and consider the paths $X = x_1, \ldots, x_n$ and $Y = x_n, \ldots, x_1$. Each gossip algorithm A for P_n must contain two time-paths $X[T]$ and $Y[T']$ for some $T = t_1, \ldots, t_{n-1}$ and $T' = t'_1, \ldots, t'_{n-1}$.

The aim will be to bound the number of conflicts between $X[T]$ and $Y[T']$ from below. The fact that each conflict causes a delay of at least 2 on $X[T]$ and $Y[T']$ will then give rise to a lower bound on the number of rounds of A.

To make the lower bound proof more transparent, we first show a lower bound which is weaker than that stated in Theorem 5.6.2, but which is less technical to prove.

Lemma 5.6.5. *For any $n \geq 2$:*

(i) $[k]\text{-}sr(P_n) \geq \frac{k+2}{k} \cdot (n-1) - 1$ *for k even,*
(ii) $[k]\text{-}sr(P_n) \geq \frac{k+1}{k-1} \cdot (n-1) - 1$ *for k odd.*

Proof of Lemma 5.6.5. The core of the proof is to show that there are at least $\left\lceil \frac{n-1}{\lfloor k/2 \rfloor} \right\rceil - 1$ conflicts in the inner nodes of P_n. This is done by first establishing that the distance between two neighboring conflicts is not too "large". (Two conflicts c_1 in node x_i and c_2 in x_j, $i < j$, are called *neighboring* if there is no conflict in $x_{i+1}, \ldots, x_{j-1}$ between $X[T]$ and $Y[T']$.)

Claim 1. The distance between two neighboring conflicts c_1 in x_i and c_2 in x_j, $i < j$, is at most $\lfloor k/2 \rfloor$.

Proof of Claim 1. Consider the time-paths $X[T] = X[t_1, t_2, \ldots, t_{n-1}]$ and $Y[T'] = Y[t'_1, t'_2, \ldots, t'_{n-1}]$. For any $\ell \in \{1, \ldots, n-1\}$, let $\mathbf{diff}(\ell) := (t_\ell - t'_\ell) \bmod k$ measure the time-difference in the systolic period between the communication $x_\ell \to x_{\ell+1}$ and $x_{\ell+1} \to x_\ell$, and let

$$\mathrm{reldiff}(i, \ell) := (t_\ell - t'_\ell) - (t_i - t'_i - (t_i - t'_i) \bmod k)$$

denote the time-difference relative to the time-difference in (x_i, x_{i+1}). As there is a conflict c_1 in x_i, we have

$$\mathrm{reldiff}(i, i) \geq 1.$$

Following the definition of the time-paths $X[T]$ and $Y[T']$, we have $t_1 < t_2 < \ldots < t_{n-1}$ and $t'_1 > t'_2 > \ldots > t'_{n-1}$. Hence,

$$\mathrm{reldiff}(i, i+1) \geq 3,$$
$$\mathrm{reldiff}(i, i+2) \geq 5,$$

$$\vdots$$

$$\mathrm{reldiff}(i, i+s) \geq 2s + 1 \quad \text{for } s \geq 1.$$

As soon as

$\mathrm{reldiff}(i, i + s) \geq k$ (i.e., $\mathrm{reldiff}(i, i + s - 1) < k$ and $\mathrm{reldiff}(i, i + s) \geq k$),

there is a conflict in x_{i+s}. It follows that there is a conflict in at least one of the nodes $x_{i+1}, x_{i+2}, \ldots, x_{i+s}$ if $2s + 1 \geq k$ or $s \geq \lfloor k/2 \rfloor$ respectively. This completes the proof of Claim 1. $\square$

The proof of Claim 1 shows that the largest distance between two neighboring conflicts c_1 in x_i and c_2 in x_j, $i < j$, can only be achieved if $\mathrm{reldiff}(i, i+s)$ is as small as possible for any s, i.e.,

$\mathrm{reldiff}(i, i) = 1,$

$\mathrm{reldiff}(i, i + 1) = 3,$

$\vdots$

$\mathrm{reldiff}(i, i + s) = 2s + 1 \quad \text{for } s \geq 1.$

For the corresponding time-paths $X[T]$ and $Y[T']$, this means that

$(t_i - t_i') \bmod k = 1,$

$(t_{i+1} - t_{i+1}') \bmod k = 3,$

$\vdots$

$(t_{i+s} - t_{i+s}') \bmod k = 2s + 1 \quad \text{for } s \geq 1,$

and for the directions

$$X[A] = (S_1, S_2, \ldots, S_{n-1}) \text{ and } Y[A] = (Q_1, Q_2, \ldots, Q_{n-1})$$

it follows that for $r = \mathrm{Mod}(t_i)$,

$\mathrm{Mod}(r - 1) \in Q_i \qquad r \in S_i$

$\mathrm{Mod}(r - 2) \in Q_{i+1} \qquad \mathrm{Mod}(r + 1) \in S_{i+1}$

$\mathrm{Mod}(r - 3) \in Q_{i+2} \qquad \mathrm{Mod}(r + 2) \in S_{i+2}$

$\vdots \qquad\qquad\qquad \vdots$

(where $\mathrm{Mod}(m) := (m - 1) \bmod k + 1$ for any $m \in I\!N$). This optimal pattern of length $\lfloor k/2 \rfloor$ on the X- and Y-direction between the two neighboring conflicts c_1 and c_2 will be referred to as pattern P_{opt}^1.

Proof of Lemma 5.6.5 continued. According to Claim 1, there is a conflict in each $\lfloor k/2 \rfloor$ steps. Hence, the number ℓ of conflicts in the inner nodes of P_n is at least $\left\lceil \frac{n-1}{\lfloor k/2 \rfloor} \right\rceil - 1$. As each conflict causes an overall delay of at least 2 on $X[T]$ and $Y[T']$, one of the time-paths incurs a delay of at least ℓ. Therefore, the message on this time-path needs at least $(n - 1) + \ell$ rounds. Applying $\ell \geq \left\lceil \frac{n-1}{\lfloor k/2 \rfloor} \right\rceil - 1$ leads to a lower bound on the number of rounds of at least

$$(n-1) + \left(\frac{n-1}{\lfloor k/2 \rfloor} - 1 \right) = \begin{cases} \frac{k+2}{k} \cdot (n-1) - 1 & \text{if } k \text{ is even,} \\ \frac{k+1}{k-1} \cdot (n-1) - 1 & \text{if } k \text{ is odd.} \end{cases}$$

This completes the proof of Lemma 5.6.5.   □

Proof of Theorem 5.6.2 continued. A technically more involved consideration than that of Lemma 5.6.5 provides the precise lower bound of Theorem 5.6.2.

Lemma 5.6.6. *For any* $n, k \geq 3$:

$$[k]\text{-}sr(P_n) \geq \frac{k}{k-2} \cdot (n-2) .$$

Proof of Lemma 5.6.6. The core of the proof is to show an improved lower bound on the number of conflicts in the inner nodes of P_n. To obtain this improved bound, it is not enough to bound the distance between two neighboring conflicts from above. Instead, we will argue about the distance between s successive conflicts. The improvement in the argument derives from the fact that the average distance between two neighboring conflicts is less than the maximum distance. Technically, we prove the following fact.

Claim 2. Let $c_1, c_2, \ldots, c_s$ be s successive conflicts. Then the distance between c_1 and c_s is at most

$$(s-1) \cdot (k/2 - 1) + 1.$$

If Claim 2 is true, one can easily complete the proof of Lemma 5.6.6 in the following way. Using Claim 2, we see that the number s of conflicts in the inner nodes of P_n must satisfy

$$((s+2) - 1) \cdot (k/2 - 1) + 1 \geq n - 1$$

(if this inequality is not true, then the inner nodes contain at least $s + 1$ conflicts). This implies

$$s \geq 2 \cdot \frac{n-2}{k-2} - 1.$$

As each conflict causes an overall delay of at least 2 on $X[T]$ and $Y[T']$, one of these time-paths has a delay of at least s. Therefore, the number of executed rounds of any k-systolic gossip algorithm on P_n is at least

$$(n-1) + s \geq \frac{k}{k-2} \cdot (n-2).$$

Thus, to complete the proofs of Theorem 5.6.2 and Lemma 5.6.6 it is sufficient to prove Claim 2.

Claim 2 will be proved separately for even k and odd k. The proof itself will be an induction on the number s of conflicts. For the inductive step, an additional property about the structure of the conflicts is needed. Hence, Claim 2 is reformulated in an appropriate way.

For this purpose, let us first specify some further notation. For two conflicts c_1 in x_i and c_2 in x_j, $i < j$, let $\text{dist}(c_1, c_s)$ denote the distance between c_1 and c_2 on the path, i.e., the number of edges between x_i and x_j. Consider the time-paths $X[T] = X[t_1, t_2, \ldots, t_{n-1}]$ and $Y[T'] = Y[t'_1, t'_2, \ldots, t'_{n-1}]$. Let c be a conflict in some inner node x_i of P_n. Then

$$\text{rdiff}(c) := (t_i - t'_i) \bmod k \text{ and } \text{ldiff}(c) := (t'_{i-1} - t_{i-1}) \bmod k.$$

$\text{rdiff}(c)$ [$\text{ldiff}(c)$] measures the time-difference in the systolic period between the communication $x_i \rightarrow x_{i+1}$ and $x_{i+1} \rightarrow x_i$ [$x_{i-1} \rightarrow x_i$ and $x_i \rightarrow x_{i-1}$]. For the conflict situations described in (i)-(iv) of Definition 5.6.4, we have the following time-differences:

(i) $\text{ldiff}(c) \geq 3$, $\text{rdiff}(c) \geq 1$,
(ii) $\text{ldiff}(c) \geq 1$, $\text{rdiff}(c) \geq 3$,
(iii) $\text{ldiff}(c) \geq 2$, $\text{rdiff}(c) \geq 2$,
(iv) $\text{ldiff}(c) \geq 2$, $\text{rdiff}(c) \geq 2$.

Note that the pattern P^1_{opt} achieving optimal length $\lfloor k/2 \rfloor$ between two neighboring conflicts c_1 and c_2 fulfills $\text{rdiff}(c_1) = 1$ and

$$\text{ldiff}(c_2) = \begin{cases} 1 & \text{if } k \text{ is even,} \\ 2 & \text{if } k \text{ is odd.} \end{cases}$$

Now, we are able to reformulate Claim 2 as follows.

Claim 2a. Let $k \in \mathbb{N}$, k even. Let $c_1, c_2, \ldots, c_s$ be s successive conflicts, $s \geq 2$. Then the following statements hold:

(1) $\text{dist}(c_1, c_s) \leq (s-1) \cdot (k/2 - 1) + 1$,

(2) $\text{dist}(c_1, c_s) = (s-1) \cdot (k/2 - 1) + 1$
$\implies (\text{rdiff}(c_1) = 1 \wedge \text{ldiff}(c_s) = 1)$.

Claim 2b. Let $k \in \mathbb{N}$, k odd. Let $c_1, c_2, \ldots, c_s$ be s successive conflicts, $s \geq 2$. Then the following statements hold:

(1) If s is odd:
    (1a) $\text{dist}(c_1, c_s) \leq \left(\frac{s-1}{2}\right) \cdot (k-2) + 1$,

    (1b) $\text{dist}(c_1, c_s) = \left(\frac{s-1}{2}\right) \cdot (k-2) + 1$
    $\implies (\text{rdiff}(c_1) = 1 \wedge \text{ldiff}(c_s) = 1)$.

(2) If s is even:

$$\text{dist}(c_1, c_s) \leq \left(\frac{s-2}{2}\right) \cdot (k-2) + \lfloor k/2 \rfloor .$$

Clearly, Claim 2 follows immediately from Claims 2a and 2b. So, the only thing left to show for the proof of Lemma 5.6.6 is the validity of Claims 2a and 2b.

Proof of Claim 2a. (Induction on s)
The induction base for $s = 2$ is clear from the remark about P^1_{opt} above. Assume that (1) and (2) hold for $s \in \mathbb{N}$, $s \geq 2$. We try to extend $X[A]$ and $Y[A]$ from c_s to c_{s+1}.

(i) If $\text{dist}(c_1, c_s) \leq (s - 1) \cdot (k/2 - 1)$, then

$$\text{dist}(c_1, c_{s+1}) \leq \text{dist}(c_1, c_s) + k/2 \leq ((s + 1) - 1) \cdot (k/2 - 1) + 1$$

according to Claim 1 in the proof of Lemma 5.6.5. Hence, (1) holds for $s + 1$.
It remains to show that (2) holds for $s + 1$. To obtain $\text{dist}(c_1, c_{s+1}) = ((s+1) - 1) \cdot (k/2 - 1) + 1$, $\text{dist}(c_1, c_s) = (s - 1) \cdot (k/2 - 1)$ must hold, and the extension from c_s to c_{s+1} must be of length $k/2$. The only pattern for achieving this is P^1_{opt} from the proof of Claim 1 in Lemma 5.6.5. For this pattern, $\text{rdiff}(c_s) = \text{ldiff}(c_{s+1}) = 1$ holds. Hence, $\text{ldiff}(c_s) \geq 3$. Consider the conflicts $c_1, c_2, \ldots, c_s$. As $\text{ldiff}(c_s) \geq 3$, the distance between c_1 and c_s can be increased by one by moving c_s one node to the right on the path (and by extending $X[A]$ and $Y[A]$). Then, the distance between c_1 and c_s is $(s - 1) \cdot (k/2 - 1) + 1$, and the induction hypothesis yields $\text{rdiff}(c_1) = 1$. Hence, (2) holds for $s + 1$.
(ii) If $\text{dist}(c_1, c_s) = (s - 1) \cdot (k/2 - 1) + 1$, then $\text{rdiff}(c_1) = \text{ldiff}(c_s) = 1$ according to the induction hypothesis. Hence, the extension from c_s to c_{s+1} must start with $\text{rdiff}(c_s) = 3$. The same argumentation as in the proof of Claim 1 in Lemma 5.6.5 shows that $\text{dist}(c_s, c_{s+1}) \leq k/2 - 1$ must hold, and if $\text{dist}(c_s, c_{s+1}) = k/2 - 1$ then $\text{ldiff}(c_{s+1}) = 1$. Hence, (1) and (2) hold for $s + 1$.

This completes the proof of Claim 2a.   □

Proof of Claim 2b (1). (Induction on s)
Let $s = 3$. Then

$$\text{dist}(c_1, c_s) = \text{dist}(c_1, c_3) \leq 2 \cdot \lfloor k/2 \rfloor = k - 1 = \left(\frac{s-1}{2}\right) \cdot (k - 2) + 1$$

follows from Claim 1 in the proof of Lemma 5.6.5. The only way to achieve $\text{dist}(c_1, c_3) = k - 1$ is to construct $X[A]$ and $Y[A]$ by using the pattern P^1_{opt} from the proof of Claim 1 in Lemma 5.6.5 between c_1 and c_2, which leads to $\text{rdiff}(c_1) = 1$, $\text{ldiff}(c_2) = 2$. Hence, an optimal extension from c_2 to c_3 has to start with $\text{rdiff}(c_2) = 2$, have length $\lfloor k/2 \rfloor$ and lead to $\text{ldiff}(c_3) = 1$ (by using the same arguments as in the proof of Claim 1 of Lemma 5.6.5). This whole optimal pattern between c_1 and c_3 is referred to as pattern P^2_{opt}.
Assume that (1a) and (1b) hold for $s \in \mathbb{N}$, $s \geq 3$ odd. We try to extend $X[A]$ and $Y[A]$ from c_s to c_{s+2}.

(i) If $\text{dist}(c_1, c_s) \leq \left(\frac{s-1}{2}\right) \cdot (k-2)$, then

$$\text{dist}(c_1, c_{s+2}) \leq \text{dist}(c_1, c_s) + (k-1) \leq \left(\frac{(s+2)-1}{2}\right) \cdot (k-2) + 1.$$

Hence, (1a) holds for $s + 2$.

It remains to show that (1b) holds for $s + 2$. To obtain $\text{dist}(c_1, c_{s+2}) = \left(\frac{(s+2)-1}{2}\right) \cdot (k-2) + 1$, $\text{dist}(c_1, c_s) = \left(\frac{s-1}{2}\right) \cdot (k-2)$ must hold, and the extension from c_s to c_{s+2} must be of length $k-1$. The only pattern for achieving this is P_{opt}^2. For this pattern, $\text{rdiff}(c_s) = \text{ldiff}(c_{s+2}) = 1$ holds. Hence, $\text{ldiff}(c_s) \geq 3$. Consider the conflicts $c_1, c_2, \ldots, c_s$. As $\text{ldiff}(c_s) \geq 3$, the distance between c_1 and c_s can be increased by one by moving c_s one node to the right on the path (and by extending $X[A]$ and $Y[A]$). Then, the distance between c_1 and c_s is $\left(\frac{s-1}{2}\right) \cdot (k-2) + 1$, and the induction hypothesis yields $\text{rdiff}(c_1) = 1$. Hence, (1b) holds for $s + 2$.

(ii) If $\text{dist}(c_1, c_s) = \left(\frac{s-1}{2}\right) \cdot (k-2) + 1$, then $\text{rdiff}(c_1) = \text{ldiff}(c_s) = 1$ according to the induction hypothesis. Hence, the extension from c_s to c_{s+2} must start with $\text{rdiff}(c_s) = 3$. The same argumentation as for P_{opt}^2 shows that $\text{dist}(c_s, c_{s+2}) \leq k-2$ must hold, and if $\text{dist}(c_s, c_{s+2}) = k-2$ then $\text{ldiff}(c_{s+2}) = 1$. Hence, (1a) and (1b) hold for $s + 2$. □

Proof of Claim 2b (2). According to Claim 2b (1), $\text{dist}(c_1, c_{s-1}) \leq \left(\frac{s-2}{2}\right) \cdot (k-2) + 1$. If $\text{dist}(c_1, c_{s-1}) \leq \left(\frac{s-2}{2}\right) \cdot (k-2)$, then

$$\text{dist}(c_1, c_s) \leq \text{dist}(c_1, c_{s-1}) + \text{dist}(c_{s-1}, c_s) \leq \left(\frac{s-2}{2}\right) \cdot (k-2) + \lfloor k/2 \rfloor$$

according to Claim 1 of Lemma 5.6.5. If $\text{dist}(c_1, c_{s-1}) = \left(\frac{s-2}{2}\right) \cdot (k-2) + 1$, then $\text{rdiff}(c_1) = \text{ldiff}(c_{s-1}) = 1$ according to Claim 2b (1). Hence, $\text{rdiff}(c_s) \geq 3$, $\text{dist}(c_{s-1}, c_s) \leq \lfloor k/2 \rfloor - 1$, and it follows that

$$\text{dist}(c_1, c_s) \leq \text{dist}(c_1, c_{s-1}) + \text{dist}(c_{s-1}, c_s) \leq \left(\frac{s-2}{2}\right) \cdot (k-2) + \lfloor k/2 \rfloor$$

This completes the proof of Claim 2b.   □

This completes the proofs of Lemma 5.6.6 and Theorem 5.6.2, too.   □

Thus, Theorem 5.6.2 provides upper and lower bounds on $[k]\text{-}sr(P_n)$ which differ only in a small constant independent of n and k.

Corollary 5.6.7. *For any $n \geq 2$, $k \geq 4$:*

$$[k]\text{-}sr(P_n) = \frac{k}{k-2} \cdot (n-2) + c_{n,k} \quad \text{for some constant } 0 \leq c_{n,k} \leq 3.$$

5.6.2 Systolic Gossip on a Cycle

The aim of this section is to investigate the complexity of systolic gossip in cycles in one-way and two-way modes. Similarly as in the non-systolic case (Chapter 4), it is much easier to find optimal systolic two-way gossip algorithms than to get very closed lower and upper bounds on the complexity of one-way systolic gossip in cycles.

We start with the simple two-way case. The next assertion is based on the simple fact that the optimal two-way gossip algorithm in cycles can be seen as a systolic one for any even period length.

Theorem 5.6.8. *For any even $n \geq 4$, and any even $k \geq 2$:*

$$[k]\text{-}sr_2(C_n) = n/2.$$

If n is odd, the situation is a little more complicated. But, also in this case, already a period of length 4 is enough to get an optimal systolic two-way gossip algorithm for C_n with odd n. We omit the proof of the following assertion and leave it as an exercise for the reader.

Theorem 5.6.9. *For any odd $n \geq 3$:*

$$[2]\text{-}sr_2(C_n) = n = r_2(P_n),$$
$$[3]\text{-}sr_2(C_n) \leq \lceil (2n)/3 \rceil + 2,$$
$$[4]\text{-}sr_2(C_n) = \lceil n/2 \rceil + 1 = r_2(C_n).$$

Exercise 5.6.10. Prove Theorem 5.6.9.

We call attention to the fact that odd period lengths are not appropriate for systolic gossip in cycles. One can even prove for odd k that $[k]\text{-}sr_2(C_n)$ does not match the trivial lower bound $\lceil n/2 \rceil$ given by the diameter, as in the case of even k.

Now, we give the results for the one-way case which presents the main contribution of this section. The proof is technical and we omit it here.

Theorem 5.6.11. *For any $k \geq 6$ and any $n \geq 2$:*

$$\frac{n}{2} \cdot \frac{k+2}{k} - 2 \leq [k]\text{-}sr(C_n) \leq \frac{n}{2} \cdot \frac{k+2}{k} + 2k - 2 \quad \text{for } k \text{ even,}$$

$$\frac{n}{2} \cdot \frac{k+2}{k} - 2 \leq [k]\text{-}sr(C_n) \leq \frac{n}{2} \cdot \frac{k+1}{k-1} + 2k - 2 \quad \text{for } k \text{ odd.}$$

Idea of the Proof. The upper bounds are based on some analysis showing that the best strategy for one-way systolic gossiping in cycles is to totally prefer the information flow in one direction over the flow in the opposite direction. The lower bound is based on a careful analysis of the number of "collisions" between messages flowing in opposite directions in the cycle.   □

Finally, we observe that there are some similarities in the results for paths and in the results for cycles stated above. For both path and cycle we do not need to pay for the systolization in the two-way case, but we have to pay with $const \cdot n$ additional rounds for any period length k in the one-way case.

5.6.3 Systolic Gossip on Complete Trees

In this section we investigate the systolic gossip complexity of complete, balanced k-ary trees. The main result of this section is that there exist systolic gossip algorithms with constant period whose complexity matches the lower bound for general (non-systolic) algorithms.

Let us first remember the lower bound for gossiping in complete, balanced k-ary trees. It is shown in Chapter 4 that the gossip complexity in two-way mode $r_2(T)$ for any tree T is exactly $2 \cdot \min b(T) - 1$, and that for one-way mode $r_1(T) = 2 \cdot \min b(T)$ holds. For a complete, balanced k-ary tree T_k^h of height h it is not hard to see that $\min b(T_k^h)$ is given by $k \cdot h$. This implies the following proposition:

Theorem 5.6.12. *For a complete, balanced k-ary tree T_k^h of height h and any period p*

i) $[p]\text{-}sr(T_k^h) \geq r(T_k^h) = 2 \cdot k \cdot h$
ii) $[p]\text{-}sr_2(T_k^h) \geq r(T_k^h) = 2 \cdot k \cdot h - 1.$

To describe our algorithms we introduce the following notations. In a systolic algorithm with period p each vertex has to repeat a communication pattern of length p. For the two-way mode of communication we specify such a pattern using a string of length p over the alphabet $C_1, C_2, \ldots, C_k, P, N$. The semantics of this specification is that any vertex v performs a communication with its i-th child (parent, resp.) in round j, iff the pattern of v contains C_i (P, resp.) at position $j \bmod p$. The letter N indicates that no communication is performed. In one-way mode we use the alphabet $C_1^\uparrow, C_1^\downarrow, \ldots, C_k^\uparrow, C_k^\downarrow, P^\uparrow, P^\downarrow, N$, where $\uparrow$ ($\downarrow$, resp.) indicates that the flow of information is directed towards the root (towards the leaves, resp.). A gossip algorithm can now be given by specifying a communication pattern for each vertex. Note that the patterns of incident vertices have to be compatible in the sense that whenever the pattern of some vertex v being the i-th child of its parent $p(v)$ indicates a parent communication ($P, P^\uparrow$ or $P^\downarrow$), the pattern of $p(v)$ has to contain the matching communication ($C_i, C_i^\uparrow$, or $C_i^\downarrow$) at the corresponding position.

Another point of view emphasizing this compatibility constraint is to specify a round of communication by a (directed) matching in the tree, where vertices communicate in the given round, iff an edge from the matching connects the vertices. Thus a sequence of p matchings can be used alternatively to specify a systolic algorithm with period p.

Note that there exists no systolic algorithm of period $\leq k$, if $h > 1$, because in this case there are vertices of degree $k + 1$. Any algorithm with period $\leq k$ would ignore some edge and no information between the components of the tree connected by this edge can be exchanged.

Now we are able to state our first result, namely a nearly optimal gossiping scheme with minimal period in the two-mode-way of communication.

Theorem 5.6.13. *For $k \geq 2$ and $h \geq 0$*

$$[k+1]\text{-}sr_2(T_k^h) \leq 2 \cdot k \cdot h.$$

Proof. We present a gossiping scheme of period $k+1$ by specifying the communication pattern of every vertex. All occurring patterns are cyclic shifts of $S = (P, C_1, C_2, \ldots, C_k)$, provided we substitute the parent communication for the root and the child communications for all leaves by N. In the following we will assume that these obvious substitutions are applied where appropriate, without explicit mention. Let $S_i = (C_i, C_{i+1}, \ldots, P, \ldots C_{i-1})$ be the pattern obtained by cyclically shifting the string S i positions to the left. Thus $S = S_0 = S_{k+1}$ holds. The actual patterns for the gossiping scheme are now obtained recursively as follows:

i) the root uses pattern $S_{h \bmod (k+1)}$,
ii) if v is the i-th child of $p(v)$ and $p(v)$ uses S_j, then v uses $S_{(j-i) \bmod (k+1)}$.

Four simple observations are in order:

1. The patterns are chosen in such a way that the parent communication of each vertex, being the i-th child of its parent, aligns with letter C_i in the pattern of its parent. Thus the given patterns obey the compatibility constraints.
2. The subtree of the first child of the root performs the given gossiping scheme for T_k^{h-1}. And all vertices of the i-th subtree perform the pattern of the corresponding vertex of the first tree shifted $(i-1)$ positions to the right.
3. In round $k(h-1) + i \equiv i + 1 - h \equiv h + i \pmod{(k+1)}$ the root performs a communication with its i-th child according to $S_{h \bmod (k+1)}$.
4. The leftmost leaf has pattern $S_0 = S$, and therefore starts with a parent communication.

We now show, by induction on h, that this communication scheme performs simultaneously a fast accumulation and a fast broadcasting with respect to the root. From these results we then can conclude our claim. First, we show that after $k \cdot h$ rounds the cumulative message of T_k^h is known to the root.

For $h = 0$ this statement is true. Assume that it holds for all trees T_k^{h-1}, $h > 0$. In T_k^h we consider now the i-th child r_i of the root. By induction hypothesis and observation 2.) we can conclude that the cumulative message of r_i's subtree is known to r_i after round $k \cdot (h-1) + i - 1$, for $1 \leq i \leq k$. Observation 3.) now states that in the next round $k \cdot (h-1) + i$ the cumulative message of the i-th subtree is given to the root. Thus after round $k \cdot (h-1) + k = k \cdot h$ all messages have arrived in the root. At this point it is also worthwhile to mention that not only the root holds the cumulative message after $k \cdot h$ rounds, but also its k-th child. This is a consequence of the two-way mode of communication. Before round $k \cdot h$ the root knows at least all messages not contained in the k-th subtree and its k-th child knows the complementary information. Since the information in two-way mode is exchanged, both vertices learn the cumulative message in this last round.

Next we consider the broadcast capabilities of our scheme. By induction on h it follows that any information known to the root of T_k^h before round t is broadcasted to all vertices after round $t+kh-1$, if in round t a communication with its first child is performed, and after round $t + kh$ otherwise. For the induction step we observe that all children of the root obtain the broadcast information before round $t + k$, if in round t the root communicates with its first child, and before round $t+k+1$ otherwise. Since for all vertices each parent communication is directly followed by a communication with the first child, we can inductively assume that the broadcast in the subtrees is finished after round $(t+k)+k(h-1)-1 = t+kh-1$, or round $(t+k+1)+k(h-1)-1 = t+kh$, respectively.

Concerning the gossip complexity of the communication scheme we now can argue as follows. After kh rounds the cumulative message is known to the root and its k-th child. According to the communication pattern of the root in round $kh+i+1$ the i-th child is informed. The broadcasting of the cumulative message in the i-th subtree is therefore finished after round $kh + i + 2 + k(h - 1) - 1$, for $1 \leq i \leq k - 1$, and in the k-th subtree of the root after round $kh + 1 + k(h - 1) - 1$. Thus the time-critical subtree is the $(k-1)$-st subtree. The broadcast in this tree and the entire gossip is finished after round $2kh$.

$\square$

The above algorithm is not time-optimal. When the root has received the cumulative message for the first time – after kh rounds – this message is delayed by one round because of the N in the communication pattern of the root. To overcome this delay the root should perform a pattern like $(C_1, C_2, \ldots, C_k, C_1, C_2, \ldots, C_{k-1}, \ldots)$. But such a pattern does not fit within $k + 1$ rounds, thus we have to increase the period to prove the following theorem.

This new time-optimal algorithm will consist of two parts. Most of the nodes of T_k^h will perform exactly the same pattern as specified in Theorem 5.6.13. To be precise, the period is $2 \cdot (k + 1)$ and the pattern is $C_1, C_2, \ldots, C_k, P, C_1, C_2, \ldots, C_k, P$. The nodes of the three top levels of the tree will follow some special patterns. Note that by using a period of $2 \cdot (k+1)$ the optimal algorithm for a T_k^1 is already systolic. We use now the optimal algorithm for the communication within the top two levels. Level three produces the correct interaction between the top part and level four, where the algorithm for subtrees from Theorem 5.6.13 is implemented. Before presenting the new algorithm in Theorem 5.6.14 we take a closer look at the algorithm from Theorem 5.6.13. The subtree rooted at the first son of the root is the only one which has to start the communication in the first round. As a consequence all other subtrees may start their communications one round earlier. The subtree rooted at the $(k-1)$-th son of the root is the only one which has to communicate in round $2 \cdot k \cdot h$. Thus we have to shift this subtree by one round. Due to our first observation this is possible iff the $(k-1)$-th son of the root is different to the first son of the root. Thus the next Theorem 5.6.14 deals with the case $k \geq 3$, and the case $k = 2$ is solved in Theorem 5.6.15.

Theorem 5.6.14. *For $k \geq 3$ and $h \geq 0$*

$$[2 \cdot (k + 1)]\text{-}sr_2(T_k^h) = 2 \cdot k \cdot h - 1 = r_2(T_k^h).$$

Proof. Within this new algorithm are several patterns:

$$S^r = (N, C_1, C_2, \ldots, C_k, C_1, C_2, \ldots, C_{k-1}, N, N)$$
$$S^s = (P, \underbrace{N, \ldots, N}_{k-1 \text{ times}}, P, N, C_1, C_2, \ldots, C_k)$$
$$S^{s'} = (P, \underbrace{N, \ldots, N}_{k-1 \text{ times}}, P, C_1, C_2, \ldots, C_k, N)$$
$$S^t = (N, C_1, C_2, \ldots, C_k, P, C_1, C_2, \ldots, C_k)$$
$$S^u = (P, C_1, C_2, \ldots, C_k, P, C_1, C_2, \ldots, C_k).$$

Let S_i^r ($S_i^s, S_i^{s'}, S_i^t, S_i^u$) be the pattern obtained from S^r ($S^s, S^{s'}, S^t, S^u$) by cyclically shifting the string i positions to the left. The gossiping scheme for T_k^h is defined in the following recursive way:

i) the root uses pattern $S_{h \bmod (2 \cdot (k+1))}^r$,

ii) the i-th child of the root ($1 \leq i \leq k - 2$) uses pattern $S_{(h-i) \bmod (2 \cdot (k+1))}^s$,

iii) the $k - 1$-th child of the root uses pattern $S_{(h-k+1) \bmod (2 \cdot (k+1))}^{s'}$,

iv) the k-th child of the root uses pattern $S_{(h+1) \bmod (2 \cdot (k+1))}^t$,

v) if v is the i-th child of $p(v)$ and $p(v)$ uses S_j^s, then v uses $S_{(j-i) \bmod (2 \cdot (k+1))}^t$,

vi) if v is the i-th child of $p(v)$ and $p(v)$ uses $S_j^{s'}$,
    then v uses $S_{(j-i+1) \bmod (2 \cdot (k+1))}^t$,

vii) if v is the i-th child of $p(v)$ and $p(v)$ uses S_j^t or S_j^u, then v uses $S_{(j-i) \bmod (2 \cdot (k+1))}^u$.

Several simple observations are in order:

1. The patterns are chosen in such a way that the parent communication of each vertex, being the i-th child of its parent, aligns with letter C_i in the pattern of its parent. Thus the given patterns obey the compatibility constraints. This is true because of the following:

    If the root uses pattern S_j^r, then the i-th son ($1 \leq j \leq k-1$) uses pattern S_{j-i}^s or $S_{j-i}^{s'}$ and the last son uses pattern S_{j+1}^t. Note that S^s and $S^{s'}$ have the P communication at the same position.

    If a node uses pattern S_j^s, then the i-th son ($1 \leq j \leq k$) uses pattern S_{j-i}^t.

    If a node uses pattern $S_j^{s'}$, then the i-th son ($1 \leq j \leq k$) uses pattern S_{j-i+1}^t. Note that the communications with the children in $S^{s'}$ are shifted one position to the left compared with S^s.

If a node uses pattern S_j^t or S_j^u, then the i-th son ($1 \leq j \leq k$) uses pattern S_{j-i}^u.

It is easy to see the compatibility constraints are valid in all cases.

2. If a node v uses pattern S^t, then the subtree rooted at v performs the communication pattern from Theorem 5.6.13. This is true because all descendants will use the pattern S^u.

3. The leftmost leaf has pattern $S_0^v = S$, and therefore it starts with a parent communication.

4. Let f_i be the i-th son of the root. The subtree of f_k performs the given gossiping scheme for T_k^{h-1} from Theorem 5.6.13. Note that f_k uses pattern S^t.

5. Let f_{11} be the first son of f_1. The subtree of f_{11} perform the given gossiping scheme for T_k^{h-2} from Theorem 5.6.13. Note that f_{11} use pattern S^t.

6. The node f_{11} sends the cumulative message to f_1 without any delay. Due to Theorem 5.6.13 at time $k \cdot (h-2)$ the cumulative message of the subtree rooted at f_{11} has arrived in f_{11}. The node f_{11} uses pattern $S_{(h-2) \bmod 2 \cdot (k+1)}^t$ which has a parent communication at time $k \cdot (h-2)+1$.

7. Any node v at level two sends the cumulative message to its parent $p(v)$ without any delay.

8. Within the top three levels of T_k^h any cumulative message is passed to the parent node without any delay. Thus the root receives the cumulative message at time kh. A more detailed proof from Observation 1 will also produce this result.

9. The root sends the cumulative message one step prior to the algorithm from Theorem 5.6.13.

10. A node f_i ($1 \leq i \leq k-2$) delays the cumulative message by one step before sending it to its sons. Thus all nodes within the subtree rooted at f_i receive the cumulative message within $2 \cdot k \cdot h - 1$ steps. Note that within this part of the tree the new algorithm behaves like the algorithm from Theorem 5.6.13. But in that algorithm there is only one leaf which receives the cumulative message at time $2 \cdot k \cdot h$. This leaf is a descendent of f_{k-1}.

11. All nodes within the subtree rooted at f_{k-1} receive the cumulative message within $2 \cdot k \cdot h - 1$ steps. Note that by using pattern $S^{s'}$ the sending down of the cumulative message is not delayed.

12. All nodes within the subtree rooted at f_k receive the cumulative message within $2 \cdot k \cdot h - k$ steps.

From all the above remarks we conclude the validity of this algorithm.

□

Note that the above algorithm does not work in the case $k = 2$. But using the same technical considerations from Theorem 5.6.14 we get the following theorem:

Theorem 5.6.15. *For binary trees of height $h \geq 0$,*

$$[9]\text{-}sr_2(T_2^h) = 4 \cdot h - 1 = r_2(T_2^h).$$

Proof. We just define the algorithm. Within this algorithm are several patterns:

$$
\begin{aligned}
S^r &= (N, C_1, C_2, C_1, N, N, N, N, N) \\
S^s &= (P, N, P, C_1, C_2, N, N, C_1, C_2) \\
S^t &= (N, N, P, N, C_1, C_2, P, N, N, N) \\
S^{t'} &= (N, N, P, C_1, C_2, N, P, N, N, N) \\
S^u &= (N, C_1, C_2, P, C_1, C_2, N, C_1, C_2) \\
S^v &= (P, C_1, C_2, P, C_1, C_2, P, C_1, C_2).
\end{aligned}
$$

Let S_i^r $(S_i^s, S_i^t, S_i^{t'}, S_i^u, S_i^v)$ be the pattern obtained from

$$S^r(S^s, S^t, S^{t'}, S^u, S^v)$$

by cyclically shifting the string i positions to the left. The gossiping scheme for T_k^h is defined in the following recursive way:

i) the root uses pattern $S_{h \bmod 9}^r$,

ii) the i-th child of the root $(1 \leq i < k)$ uses pattern $S_{(h-i) \bmod 9}^s$,

iii) the k-th child of the root uses pattern $S_{(h-k) \bmod 9}^u$,

iv) if v is the i-th child of $p(v)$, and $p(v)$ is the first son of the root, and $p(v)$ uses S_j^s, then v uses $S_{(j-i) \bmod 9}^t$,

v) if v is the i-th child of $p(v)$, and $p(v)$ is not the first son of the root, and $p(v)$ uses S_j^s, then v uses $S_{(j-i) \bmod 9}^{t'}$,

vi) if v is the i-th child of $p(v)$, and $p(v)$ uses S_j^t, then v uses $S_{(j-i) \bmod 9}^u$,

vii) if v is the i-th child of $p(v)$, and $p(v)$ uses $S_j^{t'}$, then v uses $S_{(j-i+1) \bmod 9}^u$,

viii) if v is the i-th child of $p(v)$, and $p(v)$ uses S_j^u, then v uses $S_{(j-i) \bmod 9}^v$,

ix) if v is the i-th child of $p(v)$, and $p(v)$ uses S_j^v, then v uses $S_{(j-i) \bmod 9}^v$.

Using the arguments similar to those from Theorem 5.6.14 one can verify the correctness of this algorithm and prove that it runs in $4n - 1$ rounds. $\qquad\square$

Exercise 5.6.16. Complete the details of the proof of Theorem 5.6.15.

Next we turn to the one-way mode of communication. We will derive a systolic gossip algorithm that requires $2kh + 1$ rounds and has a period of $(3 + \lceil \frac{3}{k-1} \rceil)(k + 1)$. Thus for binary trees a period of length 18, for ternary trees a period of length 20, and for k-ary trees with $k \geq 4$ a period of length $4(k + 1)$ are sufficient.

For $i \geq 0$, we will recursively derive communication schemes $A_\uparrow^i$ and $A_\downarrow^i$ having the following properties when applied to T_k^h:

These communication schemes perform a parent communication at the root. Such a communication is interpreted as an output from the root to the environment (if $P^\uparrow$ is specified), or as an input to the root of the tree from the environment (if $P^\downarrow$ is specified).

P1 $A_\downarrow^i$ as well as $A_\uparrow^i$ perform gossiping in $2kh + 1 + i$ rounds.

P2 $A_\downarrow^i$ guarantees accumulation of all messages in the root after $kh+i$ rounds, and outputs the cumulative message to the environment in round $kh+i+k+2 = k(h+1)+i+2$. Moreover $A_\uparrow^i$ broadcasts any message received from the environment in round r using kh additional rounds, i.e., the message is distributed after round $r + kh$.

P3 $A_\uparrow^i$ guarantees accumulation of all messages in the root after $kh+i$ rounds also, but outputs the cumulative message to the environment in round $kh + i + 1$. Moreover $A_\uparrow^i$ broadcasts any message received from the environment in round r using $k(h+1)+1$ additional rounds, i.e., the message is distributed after round $r + k(h + 1) + 1$.

Surely, given communication schemes $A_\uparrow^0$ and $A_\downarrow^0$ it is trivial to obtain schemes for $A_\uparrow^i$ and $A_\downarrow^i$ just by shifting cyclically all communication patterns of $A_\uparrow^0$ and $A_\downarrow^0$ respectively, i positions to the right. Thus we concentrate on deriving schemes $A_\uparrow^0$ and $A_\downarrow^0$.

Up to cyclic shifts $h^* + 1$ different patterns of length $(h^* + 1)(k + 1)$ are used, for some appropriate constant h^* to be chosen later. All vertices of depth $d \equiv 0 (\bmod h^*)$ use (up to cyclic shifts) one of the following $[0 \bmod h^*]$-patterns

either $(C_1^\uparrow, \ldots, C_k^\uparrow, P^\downarrow, C_1^\downarrow, \ldots, C_k^\downarrow, P^\uparrow, N, \ldots, N)$

or $(C_1^\uparrow, \ldots, C_k^\uparrow, P^\uparrow, C_1^\downarrow, \ldots, C_k^\downarrow, N, \ldots, N, P^\downarrow)$.

All vertices of depth $d \equiv j (\bmod h^*), 1 \le j \le h^* - 1$ use (up to cyclic shifts) the $[j \bmod h^*]$-pattern

$$(C_1^\uparrow, \ldots, C_k^\uparrow, P^\uparrow, \underbrace{N, \ldots, N}_{(j(k+1)-1)\text{times}}, P^\downarrow, C_1^\downarrow, \ldots, C_k^\downarrow, N, \ldots, N).$$

Note that after an appropriate shift these patterns are compatible with each other. Each pattern involves two communications with the same child, say C_ℓ. More precisely, if $C_\ell^\uparrow$ occurs in a $[(j - 1) \bmod h^*]$-pattern at position t, then $C_\ell^\downarrow$ occurs at position $t + j(k+1)$. The same distance, namely $j(k+1)$ occurs in the $[j \bmod h^*]$-pattern for the next level between $P^\uparrow$ and $P^\downarrow$. Note especially that for both $[0 \bmod h^*]$-patterns the cyclic distance between $P^\uparrow$ and $P^\downarrow$ is exactly $h^*(k + 1)$.

Consider now $T_k^{h^*}$. By fixing the pattern for the root and using the pattern given above, we indeed fix the entire communication scheme, provided the compatibility constraints are obeyed. (Note that in the leaves we have to use one of the root patterns, but since all child communications are substituted by

N, both patterns become indistinguishable when cyclical shifts are allowed.)
In general the only choice in designing our communication scheme after fixing
the patterns is the choice of the pattern in vertices of depth $d \equiv 0 (\bmod h^*)$.
For scheme $A_\uparrow^0$ we use in the root the pattern

$$(C_1^\uparrow, \ldots, C_k^\uparrow, P^\uparrow, C_1^\downarrow, \ldots, C_k^\downarrow, N, \ldots, N, P^\downarrow)$$

shifted such that $C_k^\uparrow$ is performed in round kh^*. Similarly, for scheme $A_\downarrow^0$ we
use the pattern

$$(C_1^\uparrow, \ldots, C_k^\uparrow, P^\downarrow, C_1^\downarrow, \ldots, C_k^\downarrow, P^\uparrow, N, \ldots, N)$$

again shifted such that $(C_k^\uparrow)$ is performed in round kh^*. Actually, for $T_k^{h^*}$ both
choices lead to exactly the same scheme, except for the parent communications
of the root. It is instructive to have a closer look at the execution of this
scheme. For this purpose, it will be convenient to have the notion of signatures
of vertices.

Definition 5.6.17. *Assume that each edge e in T_k^h connecting some arbitrary
inner vertex v with its j-th child is labeled with $\ell(e) = j$, for $1 \le j \le k$. Let
$e_1, \ldots, e_s$ be the edges lying on the unique path from vertex w to the root. Then
the signature of w is defined as*

$$\mathrm{sig}(w) = \sum_{i=1}^{s} \ell(e_i).$$

Assume that in round r the root performs operation $C_k^\uparrow$ for some sufficiently large r. Since for any inner vertex v any information received from its
j-th child has to wait exactly $k - j$ rounds in v before it is delivered to the
next level or until round r is elapsed, in case v is the root, any information
given from leaf ℓ to its parent in round $r - k(h^* - 1) + \mathrm{sig}(\ell) + 1$ will arrive at
the root before or in round r. Note that the incurred overall delay time, i.e.,
the sum of delays incurred in the vertices on the path from ℓ to the root, for
a piece of information starting in leaf ℓ is just $kh^* - \mathrm{sig}(\ell)$, since the delay in
depth d and the contribution of the edge between depth d and $d + 1$ on the
root path of ℓ to $\mathrm{sig}(\ell)$ add up exactly to k, for $0 \le d < h$.

An obvious consequence is that accumulation in the root requires only kh^*
rounds, if the root pattern is adjusted such that operation $C_k^\uparrow$ is performed in
round kh^*. This is because leaf ℓ performs $P^\uparrow$ in round $\mathrm{sig}(\ell) - h^* \ge 1$, and
therefore all pieces of information reach the root before round kh^*.

With respect to the broadcast capabilities of this scheme we observe that at
any vertex v, except possibly at the root, a piece of information sent to the j-th
child incurs a delay of $j - 1$. For any message obtained from the environment,
the delay in the root is either $j - 1$ (when using $A_\downarrow^0$) or $(j - 1) + (k + 1)$ (when
using $A_\uparrow^0$). Thus the total delay incurred by a message received by the root

and being forwarded to ℓ is either $\mathrm{sig}(\ell) - h^*$ or $\mathrm{sig}(\ell) - h^* + (k+1)$. Since the forward path has length h^* the message arrives in ℓ at round $r' + \mathrm{sig}(\ell)$ or $r' + \mathrm{sig}(\ell) + (k+1)$ respectively, where r' is the round in which the root has received the input message. Since $h^* \leq \mathrm{sig}(\ell) \leq kh^*$ all leaves have obtained the message after kh^* additional rounds, if $A_\downarrow^0$ is used, and after $k(h^*+1)+1$ additional rounds, if $A_\uparrow^0$ is used.

The gossip in $T_k^{h^*}$ requires $2kh^* + 1$ rounds. Accumulation in the root is finished after kh^*, the next round is declared in the root as the parent communication and then the broadcast is started. Note that the broadcast requires kh^* rounds by either communication scheme, since in both schemes the root causes a delay of $j - 1$ for messages sent to the j-th child, if round $kh^* + 1$ is considered as the first round of the broadcast. Indeed, we can make a slightly more accurate statement. If each leaf ℓ delivers its message for the first time in round $\mathrm{sig}(\ell) - (h^* - 1)$, then ℓ receives the cumulative message in round $\mathrm{sig}(\ell) - (h^* - 1) + h^*(k+1)$.

Now it is easy to check that these schemes restricted to the first c levels of T_k^h achieve the accumulation, broadcast, and gossip capabilities postulated in P1, P2, and P3 for T_k^c with $c \leq h^*$.

The extension of these schemes for T_k^h of arbitrary height $h > h^*$ is quite easy. Assume that we have constructed appropriate schemes for trees of height $h - h^*$. We cut off the first $h^* + 1$ levels of T_k^h. This toptree of height h^* is handled exactly as $T_k^{h^*}$, where the root pattern is adjusted such that $C_k^\uparrow$ is performed in round kh.

Now consider the subtree T_ℓ of T_k^h rooted at an arbitrary leaf ℓ of the toptree. If $\mathrm{sig}(\ell) \leq h^* + k$ we use scheme $A_\uparrow^{\mathrm{sig}(\ell)-(h-1)}$ for this subtree, otherwise we use $A_\downarrow^{\mathrm{sig}(\ell)-(h-1)-(k+1)}$.

Note that according to P2 and P3, the root ℓ of T_ℓ delivers the cumulative message of T_ℓ to its parent in round $kh - k(h^* - 1) + \mathrm{sig}(\ell) + 1$, independent of the pattern used for T_ℓ. This guarantees that accumulation in the root is finished after round kh, as well as the compatibility of the pattern used in ℓ with the pattern in its parent. Moreover the cumulative message of T_k^h is given as output in round $kh + 1$ for $A_\uparrow^0$, and in round $k(h+1) + 2$ for $A_\downarrow^0$, as required.

To analyze the broadcast properties of the scheme, we consider scheme $A_\downarrow^0$ and note that any input message received by the root in round r arrives in leaf ℓ of the toptree in round $r + \mathrm{sig}(\ell)$ for $A_\downarrow^0$. Inductively we may assume that the broadcast in T_ℓ requires additional $k(h - h^*)$ rounds, if $\mathrm{sig}(\ell) > h^* + k$. In this case all vertices of T_ℓ have received the message after round $r + k(h - h^*) + \mathrm{sig}(\ell) \leq kh$, since $\mathrm{sig}(\ell) \leq kh^*$ holds for any leaf of the toptree. If $\mathrm{sig}(\ell) \leq h^* + k$, then scheme $A_\uparrow^{\mathrm{sig}(\ell)-h^*}$ is used for T_ℓ, requiring $k(h-h^*)+k+1$ additional rounds to broadcast in T_ℓ. In this case the broadcast in T_ℓ is finished after round $r + kh - (k-1)h^* + 2k + 2$. Note that $r + kh - (k-1)h^* + 2k + 2 < r + kh + 1$ holds, iff $h^* \geq 2 + \lceil \frac{3}{k-1} \rceil$ holds. In case scheme $A_\downarrow^0$ is used, obviously $k+1$ additional rounds, due to delays in the root, are required for the broadcast in T_ℓ. Thus the required broadcast capabilities are achieved only if $h^* \geq 2 + \lceil \frac{3}{k-1} \rceil$ holds. Recall that the required period is $h^* + 1$.

To see that the given scheme performs a gossip in $2kh + 1$ rounds, we note that accumulation in the root is achieved in kh rounds. Leaf ℓ of the toptree thus receives the cumulative message in round $kh + 1 + \mathrm{sig}(\ell)$. Performing the same analysis as in the case of broadcast (i.e., substituting r by $kh + 1$) yields the claimed result.

Summarizing the above discussion we have got the following result.

Theorem 5.6.18. *For $k \geq 2$ and $h \geq 0$,*

$$\left[\left(3 + \left\lceil \frac{3}{k-1} \right\rceil \right) (k+1) \right] \text{-}sr(T_k^h) \leq 2 \cdot k \cdot h + 1.$$

As in the case of the two-way-mode, the gossip scheme given above achieves a runtime that requires just one round of communication more than stated in the lower bound. In the rest of this section we will speed up the scheme by one round at the cost of increasing slightly the length of the period using an idea very similar to that applied in two-way-mode. Thus we obtain a systolic gossip scheme with an optimal number of communication rounds.

The improvement is based on the observation that only a few vertices receive the cumulative message in the last round. We first consider the one-way mode of communication.

Recall the gossip scheme applied to T_k^h. Let ℓ be any leaf of the top-tree of height h^*. The cumulative message was broadcasted successfully in T_ℓ after round $kh + 1 + k(h - h^*) + \mathrm{sig}(\ell)$, if $\mathrm{sig}(\ell) > h^* + k$. Thus, among all these subtrees, only the rightmost one requires $2kh + 1$ rounds to broadcast the cumulative message, because for all leaves ℓ of the top-tree, except the rightmost one, $\mathrm{sig}(\ell) \leq kh^* - 1$ holds. If $\mathrm{sig}(\ell) \leq h^* + k$, then the broadcast of the cumulative message is finished in T_ℓ after round $kh + 1 + k(h - h^*) + \mathrm{sig}(\ell) + k + 1 \leq 2kh + 2k + 2 - (k - 1)h^*$. If we now choose $h^* \geq 2 + \lceil \frac{4}{k-1} \rceil$, all these subtrees finish their broadcast in or before round $2kh$. Thus inductively we can conclude that whenever we apply the previous scheme with a period of $(3 + \lceil \frac{4}{k-1} \rceil)(k+1)$ only one vertex, namely the rightmost leaf, has not received the cumulative message after round $2kh$. We will now modify the scheme in such a way that all vertices receive the cumulative message either in the same round as before or one round earlier. Especially, all vertices in the rightmost subtree of the root will receive the cumulative message one round earlier, which guarantees a gossip complexity of $2kh$ rounds.

The modification is as follows:

1. The pattern of the root is changed from
   $(C_1^\uparrow, \ldots, C_k^\uparrow, P^\downarrow, C_1^\downarrow, \ldots, C_k^\downarrow, P^\uparrow, N, \ldots, N)$ to
   $(C_1^\uparrow, \ldots, C_k^\uparrow, C_1^\downarrow, \ldots, C_k^\downarrow, N, \ldots, N)$.

2. The pattern of the j-th child r_j of the root for $1 \leq j \leq k-1$ is changed from

$$(C_1^\uparrow, \ldots, C_k^\uparrow, P^\uparrow, \underbrace{N, \ldots, N}_{k \text{ times}}, P^\downarrow, C_1^\downarrow, \ldots, C_k^\downarrow, N, \ldots, N) \text{ to}$$

$$(C_1^\uparrow, \ldots, C_k^\uparrow, P^\uparrow, \underbrace{N, \ldots, N}_{(k-1) \text{ times}}, P^\downarrow, N, C_1^\downarrow, \ldots, C_k^\downarrow, N, \ldots, N).$$

3. The pattern of the k-th child r_k of the root is changed from

$$(C_1^\uparrow, \ldots, C_k^\uparrow, P^\uparrow, \underbrace{N, \ldots, N}_{k \text{ times}}, P^\downarrow, C_1^\downarrow, \ldots, C_k^\downarrow, N, \ldots, N) \text{ to}$$

$$(C_1^\uparrow, \ldots, C_k^\uparrow, N, P^\uparrow, \underbrace{N, \ldots, N}_{(k-1) \text{ times}}, P^\downarrow C_1^\downarrow, \ldots, C_k^\downarrow, N, \ldots, N).$$

4. All vertices in the subtree rooted at r_k, except r_k, obtain the pattern of the corresponding vertex in the subtree rooted at r_{k-1}.

The alignment of these patterns in time is such that the root pattern performs operation $C_k^\uparrow$ in round kh. All other patterns are then fixed by the compatibility constraints. This modification has the effect that the subtrees rooted at r_k and at r_{k-1} now work absolutely synchronously, except for the parent communication at r_k and at r_{k-1}. Moreover, all vertices with depth $d > 1$ not in the subtree rooted at r_k perform exactly the same communication as in the unmodified scheme. Especially, they deliver the cumulative message of their subtrees in the same round as before, and also expect the overall cumulative message in the same rounds as before. One can easily verify that $r_1, \ldots, r_{k-1}$ indeed perform their child communications exactly as before, especially that r_j, for $j \leq k-1$, holds the cumulative message of its subtree after round $k(h-1) + j$ and passes the overall cumulative message in round $kh + j + i + 1$ to its i-th child. It follows that the broadcast of the cumulative message in the subtree at r_j for $j \leq k-1$ is finished after round $2kh$ as before. Since the broadcast in the subtree at r_k is performed synchronously with the broadcast in the subtree of r_{k-1}, we can conclude that gossip is performed in $2kh$ rounds.

This yields the following result.

Theorem 5.6.19. *For $k \geq 2$ and $h \geq 0$,*

$$\left[\left(3 + \left\lceil \frac{4}{k-1} \right\rceil\right)(k+1)\right]\text{-}sr(T_k^h) \leq 2 \cdot k \cdot h = r(T_k^h).$$

Note that the lengths of the periods in the time-optimal and nearly time-optimal gossip schemes from Theorem 5.6.18 and Theorem 5.6.19 differ only for $k = 2$ and 4.

5.6.4 Systolic Gossip on Grids

The main result of this section is an optimal systolic algorithm for the gossip problem in two-dimensional grids. In [KP94], an algorithm was presented for the two-way mode of communication which completes gossiping on an $(m_1 \times m_2)$-grid[2] M (i.e., a grid consisting of $m_1 + 1$ columns and $m_2 + 1$ rows) in $d + O(1)$ rounds, where $d = m_1 + m_2$ is the diameter of the grid. We derive an algorithm for the one-way mode of communication that performs gossiping on M in d rounds, provided m_1 and m_2 are large enough. The period-length of the presented algorithm is 16. Since the diameter is the trivial lower bound on the number of required rounds for any gossip algorithm (even for a non-systolic one in the two-way mode of communication), the optimality of our algorithm is obvious. The fact itself that we do not need to pay for the systolization in two-dimensional grids is of special interest, because grids are typical parallel architectures.

Note that any solution in the one-way mode of communication is also a solution for the two-way mode. Thus, the possible advantages of using the two-way model solely lie in shorter periods and a weaker restriction on the size of the grids.

Exercise 5.6.20. Give a two-way gossip algorithm for a (4×6)-grid. Try to be as fast as possible.

Exercise 5.6.21. Give a one-way gossip algorithm for a (4×10)-grid. Try to be as fast as possible.

These exercises show already that the messages originating in the corners of the grid are the critical ones. These critical messages have to pass each other somewhere in the center of the grid. This passing needs some space and to generate this space the algorithm has three phases:

1. First we will collect all messages in a subset of nodes of the grid. These nodes form then a knowledge set.
2. A gossip is performed between the nodes of the knowledge set.
3. From the nodes of the knowledge set the other nodes are informed.

The first phase and the third phase will take constant time. To implement these three phases we split the grid into a center and a seam as shown in Figure 5.9. We will continue with the algorithm for the center of the grid.

Lemma 5.6.22. *For any $(m_1 \times m_2)$-grid M with $m_i \equiv 0 \pmod 4$ there exist a one-way 4-systolic communication algorithm A and a subset K of the nodes of M such that:*

1. *A performs a gossip between the nodes of K in time $m_1 + m_2$.*
2. *Let x be any node of the grid.*
   a) *After at most 8 rounds the information of x has reached some node of K.*
   b) *After at most 8 rounds x has been informed by some node of K.*

---

[2] Note that this differs from the notation in other chapters of the book, where $(m_1 \times m_2)$-grid denotes a grid consisting of m_1 columns and m_2 rows.

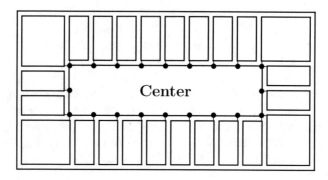

Fig. 5.9. Decomposition of the grid

3. $[4]$-$sr(M) \leq m_1 + m_2 + 16$.

Proof. The algorithm A is presented for a (16×8)-grid in Figure 5.10. The set of nodes from K is marked by square boxes. It is an easy exercise to show that:

1. The regular pattern of the algorithm may be repeated to fit any $(m_1 \times m_2)$-grid M with $m_i \equiv 0 \pmod 4$.
2. The algorithm A satisfies the above conditions.

$\square$

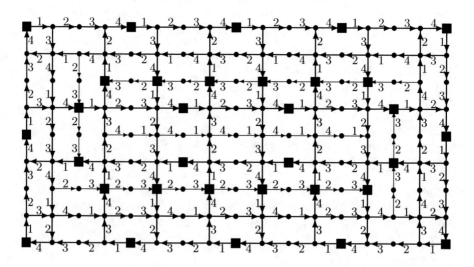

Fig. 5.10. One-way gossiping in a (16×8)-grid (near-optimal scheme)

Exercise 5.6.23. Show that the algorithm presented in Figure 5.10 may be repeated to fit any $(m_1 \times m_2)$-grid M with $m_i \equiv 0 \pmod 4$.

Theorem 5.6.24. *For any* $(m_1 \times m_2)$*-grid* M *with* m_i *even, and* $m_i \geq 28$,

$$[16]\text{-}sr(M) \; = \; m_1 + m_2.$$

Proof. The proof is performed as follows. First, one selects the center part for the near-optimal algorithm A described in Lemma 5.6.22. This central part will be a grid M' of size $(m'_1 \times m'_2)$ with $m'_i = \lfloor (m_i - 24)/4 \rfloor \cdot 4$. Thus there remains a seam of width at least 12 and at most 16. Here we count also the edges which connect the seam to the center. Each fourth node of the border of grid M' belongs to the knowledge set K from Lemma 5.6.22. The seam is now split into rectangular parts. For each of the border nodes of K, we will have one subgrid M_s of size $(a_1 \times a_2)$ with $11 \leq a \leq 14$, $b \in \{4, 12, 13, 14, 15\}$, and one additional edge connecting the subgrid to the node n_s from K (see Figure 5.9). For each of these subgrids it is easy to make the following construction of a systolic communication algorithm (see Figure 5.11 for an example) with a period of 16:

1. Accumulate the information of all nodes of M_s into n_s in $a_1 + a_2 + 1$ steps.
2. Transform the algorithm from Lemma 5.6.22 and change the period from 4 to 16 by repeating the algorithm four times.
3. Ensure that the messages from the corner nodes are forwarded at the corner nodes of M' without any delay. This may be done by changing the starting period.
4. Broadcast – without any delay – the accumulated message from n_s to all nodes of M_s.

The accumulation and broadcasting are carried out on two different communication trees. It is now an exercise to show that these communications trees may be embedded in a subgrid of any size without conflicts.

$\square$

Exercise 5.6.25. Let M_s be a subgrid in the corner of M. Show that the above algorithm may be constructed for any values a, b with $11 \leq a \leq 14$ and $b \in \{4, 12, 13, 14, 15\}$.

5.6.5 Systolic Gossip on Butterfly and CCC Networks

This section presents systolic algorithms on the butterfly network and the cube-connected-cycles. For these networks the situation is a little bit different to the previous ones. With these networks we get systolic algorithms which are similar to the non-systolic versions, because the way the information is distributed in the network is similar to the non-systolic algorithms.

We start by considering cube-connected-cycles.

Theorem 5.6.26. *Let* $m = 3 \cdot m'$ *for* $m' \in \mathbb{N}$. *Then*

$$[3]\text{-}sr_2(CCC_m) \leq \frac{8}{3} \cdot m + 1$$

holds.

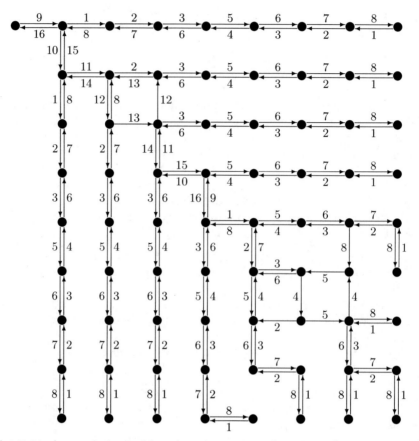

Fig. 5.11. Accumulation and broadcast in a corner of size 8×9 (optimal scheme)

Proof. The 3-systolic algorithm is defined as follows.

Any edge $\{(i, \alpha), ((i + 1) \bmod m, \alpha)\}$ is active in round $((i \cdot 2) \bmod 3) + 1$.
Any edge $\{(i, \alpha), (i, \alpha(i))\}$ is active in round $((i \cdot 2 + 1) \bmod 3) + 1$.

First we consider the broadcast of the information of node $(0, 0^m)$. This information reaches in the first step node $(0, 10^{m-1})$. In the next step the nodes $(1, 0^m)$ and $(1, 10^{m-1})$ are informed. After the third step also the nodes $(1, 010^{m-2})$ and $(1, 110^{m-2})$ are informed. This pattern is then repeated by the algorithm. Thus after step $2 \cdot i + 1$ the nodes $(i, \{0, 1\}^{i-1}0^{m-i})$ are informed. After $2 \cdot m - 1$ steps the information of node $(0, 0^m)$ has reached all nodes $(m - 1, \{0, 1\}^m)$.

Consider now a node $(m - 1, \alpha)$ for a fixed $\alpha \in \{0, 1\}^m$. We will check the transmission of the information from $(m - 1, \alpha)$ to the nodes (x, α). In the next step (the step $2 \cdot m$) the node $(0, \alpha)$ is informed and in step $2 \cdot m + 1$ the nodes $(1, \alpha)$ and $(m - 2, \alpha)$ are informed. This pattern is repeated, it takes two steps to inform 3 new nodes (x, α).

Thus it takes $2 \cdot m - 1 + 2 \cdot m'$ steps to broadcast the information of node $(0, 0^m)$ to all other nodes of CCC_m. The distribution of the information from nodes $(2, \{0, 1\}^m)$ starts in the third round.

$$
\begin{aligned}
[3]\text{-sr}_2(CCC_m) &\le 2 \cdot m - 1 + 2 \cdot m' + 2 \\
&= 2 \cdot m + 1 + 2 \cdot m/3 \\
&= 8 \cdot m/3 + 1.
\end{aligned}
$$

$\square$

The above algorithm works only for the case $m = 3 \cdot m'$. But it is easy to adapt the above idea for the other cases. The adaptation is done on one or two levels of the CCC_m. This adds a small constant to the gossiping time.

Theorem 5.6.27. *For every* $m \in I\!N$, $[3]\text{-sr}_2(CCC_m) \le \lfloor \frac{8}{3} \cdot m \rfloor + 5$.

Exercise 5.6.28. Give a formal proof for Theorem 5.6.27. Use and adapt the proof and the algorithm of Theorem 5.6.26.

Exercise 5.6.29. Give a valid one-way 2-systolic gossip algorithm for the cube-connected-cycles of dimension m.

Applying the idea of Theorem 5.6.26 one can obtain several other results.

Theorem 5.6.30. *For each* $m \in I\!N$, $[4]\text{-sr}(CCC_m) \le 4 \cdot m + 5$.

Exercise 5.6.31. Give a formal proof of Theorem 5.6.30. Do this by appropriately adapting the algorithm from Theorem 5.6.26.

Exercise 5.6.32. Check if Theorems 5.6.27 and 5.6.30 also hold for the periodic case.

Exercise 5.6.33. Show that there exists no 3-periodic algorithm for the cube-connected-cycles network of dimension m with running time in $O(m)$.

We are now ready to adapt the above algorithms to systolic gossiping in the butterfly network.

Exercise 5.6.34. Show for all $m \ge 3$: $[3]\text{-sr}_2(BF_m) \le [3]\text{-sr}_2(CCC_m)$.

Exercise 5.6.35. Show for all $m \ge 3$: $[3]\text{-sr}_2(BF_m) = [3]\text{-sr}_2(CCC_m)$.

Before we give a near-optimal systolic algorithm for the butterfly network in Theorem 5.6.37, we take a closer look at the algorithm based on the direct transformation from the non-systolic algorithms in Exercise 5.6.36.

Exercise 5.6.36. Let $m = 2 \cdot m'$ for an $m' \in I\!N$. Consider the following 4-systolic algorithm on BF_m:

An edge of the form $\{(i, \alpha), ((i+1) \bmod m, \alpha)\}$ is active in round $((i \cdot 2) \bmod 4) + 1$.

An edge of the form $\{(i, \alpha), ((i+1) \bmod m, \alpha(i))\}$ is active in round $((i \cdot 2 + 1) \bmod 4) + 1$.

Compute the running time of this algorithm.

Theorem 5.6.37. *For each* $m = 2 \cdot m'$ *for an* $m' \in \mathbb{N}$, $[4]\text{-sr}_2(BF_m) \le \frac{3}{2} \cdot m + 2$.

Proof. A 4-systolic algorithm can proceed as follows.

An edge of the form $\{(i, \alpha), ((i+1) \bmod m, \alpha)\}$ is active in round $(i \bmod 4) + 1$.

An edge of the form $\{(i, \alpha), (i, \alpha(i))\}$ is active in round $((i + 2) \bmod 4) + 1$.

The proof is done in the following steps:

1. We show that the information from node $(0, 0^m)$ is transported in $2 \cdot m$ steps to all nodes of the form $(0, \{0, 1\}^m)$.
2. If all nodes of the form $(0, \{0, 1\}^m)$ have some common information, then this is transported in $m/2$ step to all other nodes.

Step 1:
Consider any node $(0, a_0 a_1 a_2 \cdots a_{m-1})$. Let $b_i = a_i \otimes a_{(i+1) \bmod m}$ for all $0 \le i < m$.

We consider two paths between $(0, 0^m)$ and $(0, a_0 a_1 a_2 \cdots a_{m-1})$. Assume that the message is sent from $(0, 0^m)$ to $(0, a_0 a_1 a_2 \cdots a_{m-1})$ by the following path:

$(0, 0^m)$
$(1, a_0 0^{m-1})$
$(2, a_0 a_1 0^{m-2})$
$(3, a_0 a_1 a_2 0^{m-3})$
$\cdots$
$(i, a_0 a_1 a_2 \cdots a_{i-1} 0^{m-i})$
$\cdots$
$(0, a_0 a_1 a_2 \cdots a_{m-1})$

A close look at the above algorithm shows that it takes

$$\#_0(a_0 a_1 a_2 \cdots a_{m-1}) + 3 \cdot \#_1(a_0 a_1 a_2 \cdots a_{m-1})$$

steps.

If the message from $(0, 0^m)$ is sent to $(0, a_0 a_1 a_2 \cdots a_{m-1})$ by the following path

$(0, 0^m)$
$(m-1, 0^{m-1}a_{m-1})$
$(m-2, 0^{m-2}a_{m-2}a_{m-1})$
$\cdots$
$(i, 0^{m-i}a_{m-i}\cdots a_{m-2}a_{m-1})$
$\cdots$
$(1, 0a_1a_2\cdots a_{m-1})$
$(0, a_0a_1a_2\cdots a_{m-1}),$

then one can see that it takes

$$\#_1(a_0a_1a_2\cdots a_{m-1}) + 3\cdot\#_0(a_0a_1a_2\cdots a_{m-1})$$

steps for that path. By taking the fastest path we get a total running time of $2\cdot m$.

Step 2:
Assume that all nodes of the form $(0, \{0,1\}^m)$ have some common information. Any node of the form $(i, a_0a_1a_2\cdots a_{m-1})$ with $i \le m/2$ is informed by the node $(0, a_0a_1a_2\cdots a_{m-1})$ in $m/2$ steps by the following path:

$(0, a_0a_1a_2\cdots a_{m-1})$
$(1, a_0a_1a_2\cdots a_{m-1})$
$(2, a_0a_1a_2\cdots a_{m-1})$
$\cdots$
$(i, a_0a_1a_2\cdots a_{m-1})$

Any node of the form $(i, a_0a_1a_2\cdots a_{m-1})$ with $i > m/2$ is informed from node $(0, a_0a_1a_2\cdots \overline{a_{m-5}}a_{m-4}\overline{a_{m-3}}a_{m-2}\overline{a_{m-1}})$ in $m/2$ steps by the following path:

$(0, a_0a_1a_2\cdots \overline{a_{m-5}}a_{m-4}\overline{a_{m-3}}a_{m-2}\overline{a_{m-1}})$
$(m-1, a_0a_1a_2\cdots \overline{a_{m-5}}a_{m-4}\overline{a_{m-3}}a_{m-2}a_{m-1})$
$(m-2, a_0a_1a_2\cdots \overline{a_{m-5}}a_{m-4}\overline{a_{m-3}}a_{m-2}a_{m-1})$
$(m-3, a_0a_1a_2\cdots \overline{a_{m-5}}a_{m-4}a_{m-3}a_{m-2}a_{m-1})$
$(m-4, a_0a_1a_2\cdots \overline{a_{m-5}}a_{m-4}a_{m-3}a_{m-2}a_{m-1})$
$\cdots$
$(i, a_0a_1a_2\cdots a_{m-5}a_{m-4}a_{m-3}a_{m-2}a_{m-1})$

Combining both steps and considering that there are nodes which start transmitting their information after 2 steps we get the above result.

□

Theorem 5.6.38. *For each* $m \in I\!\!N$, $[4]$-$sr_2(BF_m) \le \lceil \frac{3}{2}\cdot m\rceil + 12$.

Exercise 5.6.39. Give a formal proof for Theorem 5.6.38.

For further details and lower bounds one can consult [NSV02], [FP99a] and [FP01].

5.7 Bibliographical Remarks

The main motivation behind this concept corresponds to the idea of Kung [Ku79] who introduced the so-called "systolic computations" as parallel computations with a cheap implementation due to a very regular, synchronized periodic behavior of all processors of the interconnection network during the whole execution of the computation (see also some of the first papers about systolic computations in trees and grids [CC84, CGS83, CGS84, CSW84]). Liestman and Richards [LR93a, LR93b] were the first who considered a very regular form of communication algorithms for broadcast and gossip. This form, later called "periodic" communication algorithms [LHL94, LR97, LR93b], was based on edge coloring of a given interconnection network and on periodic (cyclic) execution of communications via edges with the same color in one communication step. This "periodic" communication was introduced in order to solve a special gossip problem introduced by Liestman and Richards [LR93a] and called "perpetual gossiping", where each processor may get a new piece of information at any time from outside the network and the never-halting communication algorithm has to broadcast it to all other processors as soon as possible. The periodic gossiping based on graph coloring was implicitly used also in [XL92, XL94] in order to solve a completely practical problem – termination detection in synchronous interconnection networks. A little more general concept of regular communication was given by Hromkovič et al. [HKPU+94], who introduced k-systolic communication algorithms as a repetition of a given sequence of k communication steps (for some $k \in I\!N$, k independent of the number of processors of the network). This kind of communication was also considered in [KP94] for the problem of route scheduling under the two-way (telephone) model of communication.

Without going into the details of this concept, we note that the study of the power of networks from this point of view was introduced by Culik II et al. [CGS84], and investigated in [IK84, CC84, IPK85, CSW84, CGS83, IKM85]. This concept considers classes of languages recognized only by systolic computations on the given parallel architecture (network) in the shortest possible time for a given network.

Exercise 5.7.1. Give for the following list examples where a systolic concept is used: traffic control systems, cars, parcel distribution systems, and television.

Since there is no difference in the complexity of general broadcast algorithms and systolic broadcast algorithms (every broadcast algorithm can be systolized without any increase in the number of communication steps [HKPU+94]), the main research problems formulated in [HKPU+94] are the following two:

1. What is the k-systolic gossip complexity of fundamental interconnection networks? What is the most appropriate k (if it exists) for a given network?

2. How much must be paid for the systolization of communication algorithms in concrete interconnection networks (i.e., what is the difference between the complexity of general gossip and the complexity of systolic gossip)?

We presented solutions to these questions for some fundamental networks but there are several further communication structures of interest, for which there are no results about the systolic gossip complexity. Thus, there are many possibilities for continuing in the research on the topic presented in this chapter.

6

Fault-Tolerance

*For him who seeks the truth,
an error is nothing unknown.*

Johann Wolfgang von Goethe

6.1 Introduction

As interconnection networks grow in size and complexity, they become increasingly vulnerable to component failures. Links and/or nodes of the network may fail and these failures often result in delaying, blocking or even distorting of transmitted messages. It becomes important to design communication algorithms in such a way that the desired communication task be accomplished efficiently in spite of these faults, usually without knowing their location ahead of time. This chapter is devoted to the study of algorithms designed for this purpose. They are called fault-tolerant algorithms. We concentrate on the fundamental task of broadcasting messages. Since some nodes may now be faulty, the definition of broadcasting is reformulated by requiring that the source message be received by all fault-free nodes. Throughout the chapter we assume that the source is fault free: otherwise no broadcasting is possible. We work in the telephone model.

It is clear that no broadcasting algorithm can work properly for all fault types and configurations. For example, if all links of a network are faulty, and faults result in permanent blocking of all transmitted messages, there can be no hope of accomplishing any broadcasting task. On the other hand, such massive failures are extremely rare in practice, and they need not be of major concern in algorithm design. Much more frequent are faults that damage a limited number of components. Hence our goal will be to design broadcasting algorithms that work properly under some assumptions bounding the number of possible faults. Another thing that has to be specified is the nature of faults, i.e., the way in which they can damage transmitted messages. These two issues, the possible distributions and the description of different types of faults will be discussed in detail below, in two separate subsections.

There are two fundamental qualities that are demanded of a good fault-tolerant broadcasting algorithm. One of them is *robustness*, i.e., its capacity to work correctly in spite of the presence of faults of some assumed type which are distributed in unknown locations but according to some prescribed assumptions. The other one is *efficiency*: this requirement is similar to that in the fault-free situation. Similarly as in the rest of the book, we adopt broadcasting time as the efficiency criterion. Unfortunately, these two goals of robustness and efficiency are often contradictory. In fact, there is a well-known trade-off between them: the more robust a broadcasting algorithm is, the slower it works. The reason for this trade-off is simple. Robustness is acquired at the cost of redundancy which consists in repeating transmissions or routing messages via different paths in the interconnection network, in order to avoid faults. Redundancy almost always causes a slowdown in the broadcasting process. Hence the main challenge in the design of fault-tolerant broadcasting algorithms is the following: guarantee the required level of robustness incurring the smallest possible loss of broadcasting speed. This goal is both of primary practical importance and of utmost theoretical interest. This is also the main focus of the present chapter.

It should be stressed that all broadcasting algorithms discussed in this chapter are deterministic. This is the case also for algorithms designed to work under the probabilistic fault model: in this case, stochastic assumptions concern the distribution of faults, but broadcasting itself is deterministic. Also all lower bounds on running time concern deterministic broadcasting algorithms. Hence, throughout the chapter, we use the expression "broadcasting algorithm" as a shorthand for "deterministic broadcasting algorithm".

6.1.1 Fault Distribution

One of the crucial assumptions made about faults concerns their distribution. As mentioned before, some limitations on the number of possibly faulty components must be imposed, otherwise no communication is possible. Two commonly used fault models are the *bounded* model and the *probabilistic* model. In the bounded model, an upper bound k is imposed on the number of faulty components and their worst-case location is assumed. In the probabilistic model, faults are assumed to occur randomly and independently of each other, with specified probability. The choice between these two assumptions regarding fault occurrence influences the definition of the goal of broadcasting in the presence of faults.

In the bounded-fault model, we usually seek what is termed *k-tolerant* broadcasting. The source message must reach all fault-free nodes, provided that no more than k components (links or nodes or both, depending on the particular scenario) are faulty. In the probabilistic fault model, the communication goal cannot be achieved with certainty, since, with some small probability, all components may fail and preclude any message transmission. As a result, *almost safe* broadcasting is sought: all fault-free nodes must receive the source message with probability at least $1 - \frac{1}{n}$, if the number n of nodes is sufficiently large. This is an asymptotic-type requirement which not only implies that the reliability of broadcasting converges to 1 as n grows, but also prescribes the speed of this convergence. Usually, the lower bound $1 - \frac{1}{n}$ on reliability can be strengthened to $1 - \frac{1}{n^c}$, for any positive constant c, at the price of only a constant slowdown of broadcasting.

Designing efficient broadcasting algorithms whose reliability increases for networks of larger size is difficult for the following reason. In small networks, it is relatively easy to achieve fault-tolerant broadcasting using massive redundancy; due to the small scale of the network, resources used by such brute-force communication procedures (either time or number of messages) will not be excessive. In larger networks, however, the trade-off between reliability and efficiency becomes an important issue; for such networks, highly reliable and efficient algorithms are sought.

6.1.2 Fault Classification

In order to specify the fault model, we need to be precise about the way in which faults damage the affected components. The first concern is to specify which components are fault-prone: only links, only nodes, or both links and nodes. Furthermore, the nature of faults must be described. The two most commonly studied types are *crash* and *Byzantine* faults. If the fault is a crash, the faulty node does not send or receive messages, or the faulty link does not transmit messages; faulty components do not alter transmitted messages, they only fail to transmit them. Such faults are relatively benign. Although some information may be lost, at least the received messages can be trusted. Byzantine faults, on the other hand, are a worst-case fault scenario: faulty components can behave arbitrarily (even maliciously) as transmitters, by either blocking, rerouting, or altering transmitted messages in a way most detrimental to the broadcasting process. Byzantine failures that exhibit all these kinds of damaging behavior rarely occur in practice. They may be caused by a hostile agent whose aim is to destroy the communication process. On the other hand, the concept of Byzantine faults plays an important role in our study, representing a worst-case assumption. Broadcasting algorithms that work correctly in the presence of Byzantine faults can be used safely under any fault scenario.

Since Byzantine behavior includes creating messages at times not prescribed by a broadcasting algorithm, it has to be specified what we mean by an algorithm working in a Byzantine environment and satisfying the requirements of the telephone model. Of course, we cannot prevent a faulty node from sending many messages to different neighbors at the same time step. Neither can we prevent two faulty nodes from sending messages to a common neighbor simultaneously. Thus a natural requirement is that the algorithm be specified so that, at each time step, fault-free components obey the restrictions of the telephone model.

Another important characteristic of faults, which must be specified, is their duration. Faults can be either *permanent* (i.e., the status faulty/fault-free of a component does not change during the algorithm execution) or *transient* (i.e., the status of a component may change in each unit of time). Permanent faults usually correspond to hardware component failures, while transient faults may be caused, for example, by temporary magnetic interference. Transient link faults will be called *transmission faults*, as they affect a single transmission through a link in a single round. Another failure of this link in another round is considered to be another (transient) fault. Knowledge as to which types of faults are likely to occur is important in the design of a broadcasting algorithm. Repeated attempts to transmit the same message along the same link is useless in the case of permanent faults but may be essential if faults are of a transient nature.

6.1.3 Flexibility of Broadcasting Algorithms

Another issue which has to be decided in the design of broadcasting algorithms working in the presence of faults is their flexibility. This aspect does not occur when all components of the network are assumed to be functioning. In a fault-free environment, broadcasting algorithms have a simple form. All calls to be carried out in each time unit can be specified in advance, before the algorithm execution. In the presence of faults, however, a distinction must be made regarding this point, which significantly affects the efficiency and robustness of fault-tolerant communication. Broadcasting algorithms can be either *non-adaptive*, also called *oblivious*, where all calls must be scheduled in advance, or *adaptive*, where every node can schedule its next call based on information currently available to it. In adaptive algorithms, a node becomes aware of whether a call it attempted was successful (i.e., if the called node and the connecting link are fault-free); hence, different calls can be executed depending on the success or failure of previous ones. Even in this second case, however, nodes can only take advantage of *locally* available information; we do not assume the existence of any central monitor supervising the broadcasting process. Adaptive algorithms require more local control and memory at each node but are usually more efficient for the same fault-tolerance level. Non-adaptive broadcasting algorithms will be called *broadcasting schemes*.

6.2 Preliminary Results

As mentioned in the introduction, redundancy is the key ingredient in fault-tolerant broadcasting algorithms. When faults are permanent, this requirement translates into a condition of network connectivity: message transmission from node u to node v requires many pairwise disjoint paths in the network. This is captured by the following easy fact.

Proposition 6.2.1. *If nodes of the network are fault free and links are subject to permanent crash faults then k-tolerant broadcasting in the network is possible if and only if every node is connected to the source by at least $k + 1$ edge-disjoint paths.*

If crash faults can affect only nodes, or links and nodes, then k-tolerant broadcasting in the network is possible if and only if every node is connected to the source by at least $k + 1$ node-disjoint paths.

Proof. We prove the first part of the proposition. The second one is analogous. Suppose that every node is connected to the source by at least $k + 1$ edge-disjoint paths. Fix a node v and let $P_1, \ldots, P_{k+1}$ be edge-disjoint paths connecting v to the source. For any configuration of at most k faulty edges, at least one of these paths contains only fault-free edges. Hence the message can be transmitted from the source to v along this path.

Conversely, suppose that there is a node v for which the maximum number of edge-disjoint paths connecting v to the source does not exceed k. Then there exists a set of at most k edges whose removal disconnects the network, so that the source and v are in different connected components. If these edges are faulty, v cannot get the source message, hence k-tolerant broadcasting is impossible.

$\square$

A similar, although stronger, condition is necessary and sufficient for k-tolerant broadcasting in the presence of Byzantine faults. Now, since faults can corrupt messages, we have to guarantee a strict majority of fault-free paths, in order to recognize the correct message.

Proposition 6.2.2. *If nodes of the network are fault free and links are subject to permanent Byzantine faults then k-tolerant broadcasting in the network is possible if and only if every node is connected to the source by at least $2k + 1$ edge-disjoint paths.*

If Byzantine faults can affect only nodes, or links and nodes, then k-tolerant broadcasting in the network is possible if and only if every node is connected to the source by at least $2k + 1$ node-disjoint paths.

Proof. Again we prove only the first part of the proposition. The second one is analogous. Suppose that every node is connected to the source by at least $2k + 1$ edge-disjoint paths. Fix a node v and let $P_1, \ldots, P_{2k+1}$ be edge-disjoint paths connecting v to the source. Let m_i, for $1 = 1, \ldots, 2k + 1$, be the version of the source message relayed to v along path P_i. If no message is relayed along a given path, a default value is taken. Now v uses a voting mechanism considering the majority version as the original source message. Since there are at most k faulty edges, at least $k + 1$ of paths $P_1, \ldots, P_{2k+1}$ contain only fault-free edges, and consequently the version of the source message relayed along these paths is correct. It follows that the majority version of the source message exists and is indeed the correct message. Hence k-tolerant broadcasting is possible.

Conversely, suppose that there is a node v for which the maximum number of edge-disjoint paths connecting v to the source does not exceed $2k$. Then there exists a set of at most $2k$ edges whose removal disconnects the network, so that the source and v are in different connected components. Let $e_1, \ldots, e_t$, for $t \leq 2k$, be these edges. Suppose that $e_1, \ldots, e_k$ are faulty. If the source message is m and these faulty edges distort it to m', node v gets k copies of message m' and $t - k$ copies of the correct message m. However, v gets an identical set of relayed messages if the original source message is m', the faulty edges are $e_{k+1}, \ldots, e_t$ and they distort message m' to m. Hence v cannot decide if the correct message is m or m'. Since each of the two above fault configurations contains at most k faulty edges, it follows that k-tolerant broadcasting is impossible.

□

Pairwise disjoint paths also play an important role in constructing almost safe broadcasting algorithms under the probabilistic scenario. The events that such paths do not contain faulty components are independent, which facilitates the estimation of algorithm correctness. For a single path, the probability that it is free of faults is $(1-p)^x(1-q)^y$, where p is the link failure probability, q is the node failure probability, and x, y are the numbers of links and nodes on the path, respectively. The following easy fact is often used in the analysis of fault-tolerant broadcasting.

Proposition 6.2.3. *Assume that components of the network are subject to crash faults. Suppose that fault-free nodes u and v are joined by k node-disjoint paths, each of which is free of faulty components with probability at least α, for some positive constant α. Then it is possible to transmit a message from u to v with probability at least $1 - (1 - \alpha)^k$.*

Proof. Let $P_1, \ldots, P_k$ be node-disjoint paths joining u and v. Let E_i, for $i = 1, \ldots, k$, be the event that path P_i contains a faulty component. Thus $Prob(E_i) \leq 1 - \alpha$. Since paths P_i are node-disjoint, events E_i are independent. Hence the probability that all paths P_i contain a faulty component is

$$Prob(\bigcap_{i=1}^{k} E_i) = \prod_{i=1}^{k} Prob(E_i) \leq (1 - \alpha)^k.$$

Consequently, with probability at least $1 - (1 - \alpha)^k$, at least one of the paths is free of faulty components, and hence it is possible to transmit the message along this path.

□

Proposition 6.2.3 implies that logarithmic connectivity of a network is enough to guarantee almost safe broadcasting.

In our probabilistic considerations we will also use the following results. The first is a well-known lemma called the *Chernoff bound*.

Lemma 6.2.4. *Let X be the random variable denoting the number of successes in a Bernoulli series of length m with success probability q. Let $0 < \epsilon < 1$. Then*

$$Prob(X \leq (1 - \epsilon)mq) \leq e^{-\epsilon^2 mq/2}.$$

The second result is straightforward.

Exercise 6.2.5. Let X be the random variable denoting the number of successes in a Bernoulli series of length m with success probability q. Then, for all $0 \le k \le m$, we have $Prob(X \ge k) \le \binom{m}{k} q^k$.

The Chernoff bound can be used to establish the following result which is analogous to Proposition 6.2.3 but concerns the Byzantine fault scenario. Again it implies that logarithmic connectivity of a network is enough to guarantee almost safe broadcasting, provided that fault probabilities of components are sufficiently small to ensure that there are many paths between pairs of nodes, free of faults with probability larger than $1/2$.

Proposition 6.2.6. *Assume that components of the network are subject to Byzantine faults. Let d be any positive constant. There exists a constant $c > 0$ such that if fault-free nodes u and v are joined by $\lceil c \log n \rceil$ node-disjoint paths, each of which is free of faulty components with probability $\alpha > 1/2$, then it is possible to correctly transmit a message from u to v with probability at least $1 - n^{-d}$.*

Proof. Let $\epsilon = \frac{2\alpha - 1}{2\alpha}$. The assumption $\alpha > 1/2$ implies $0 < \epsilon < 1$. Then $(1 - \epsilon)\alpha = 1/2$. Let $c = \frac{2d}{\epsilon^2 \alpha \log e}$. This implies $d = c\epsilon^2 \alpha (\log e)/2$ and hence $e^{-\epsilon^2 (c \log n)\alpha/2} = n^{-d}$.

Let $m = \lceil c \log n \rceil$ and let $P_1, \ldots, P_m$ be node-disjoint paths joining u and v. Let E_i, for $i = 1, \ldots, m$, be the event that path P_i is free of faults. Since the paths are node-disjoint, events E_i are independent. Let X be the random variable denoting the number of paths P_i free of faults. Let E be the event that at most one-half of paths P_i are free of faults. By Lemma 6.2.4 we have:

$$P(E) = Prob(X \le m/2) = Prob(X \le (1 - \epsilon)m\alpha)$$
$$\le e^{-\epsilon^2 m\alpha/2} \le e^{-\epsilon^2 (c \log n)\alpha/2}$$
$$= n^{-d}.$$

Node v applies a voting mechanism considering the majority of the messages relayed along paths P_i as the correct message from u. If event E does not hold then this voting mechanism gives indeed the correct message sent by u. It follows that it is possible to correctly transmit a message from u to v with probability at least $1 - n^{-d}$.

□

We will also use the following combinatorial inequality

Exercise 6.2.7. For all $1 \le k \le n$,

$$\binom{n}{k} \le \left(\frac{en}{k} \right)^k.$$

6.3 The Bounded-Fault Model

In this section we assume that there are at most k crash faults in the network, with a worst-case distribution. In three subsections below we will consider three possibilities: at most k permanent link faults assuming fault-free nodes, at most k permanent node faults assuming fault-free links, and at most k transmission faults (i.e., transient link faults), assuming fault-free nodes, respectively. A broadcasting algorithm is k-tolerant if it guarantees broadcasting to all fault-free nodes whenever the number of faults of a given type (as specified above) does not exceed k.

6.3.1 Permanent Link Faults

We assume that there are at most k crash link faults in the network, and that all nodes are fault free. We seek fast k-tolerant broadcasting algorithms. Our first goal is to give lower and upper bounds on the time of such algorithms that differ only by a multiplicative constant.

Theorem 6.3.1. ([Li85]) *The running time of any k-tolerant broadcasting algorithm in an n-node graph G is at least $\log n + k$, for $n - 2 \geq k \geq 0$.*

Proof. Suppose that there is a k-tolerant broadcasting algorithm with running time smaller than $\log n + k$. Let v be the source and let $e_1, \ldots, e_k$ be the first k edges adjacent to v on which calls are made according to the algorithm. Suppose that all these edges are faulty. Then after time k only the source has the message. This implies that the source can broadcast to all other nodes in fewer than $\log n$ rounds, which is impossible.

□

We now establish an upper bound $\log n + O(k)$ on the time of k-tolerant broadcasting in complete graphs. To this end, we show a k-tolerant broadcasting scheme for the complete n-node graph, whose time is at most $\lceil \log n \rceil + 2k + 3$ [PS89]. We first describe a communication scheme called a round-robin matching procedure.

Given a set of m players, $1, \ldots, m$, a round-robin tournament RR on the set $1, \ldots, m$ is a set of matchings on the set $\{1, \ldots, m\}$ such that each pair of players is matched by at least one matching. Matchings of the tournament are called rounds, and, for all $i \leq m$, $RR(i, t)$ denotes the number of the player matched with player i in round t. If m is odd, we write $RR(i, t) = 0$ to mean that i is idle (not matched with anybody) in round t. Let $R(m)$ be the minimum number of rounds in a round-robin tournament on $1, \ldots, m$.

Exercise 6.3.2. $R(m) = m - 1$ for even m, and $R(m) = m$ for odd m.

We will use round-robin tournaments to establish a communication scheme called a round-robin matching procedure. Let $C_1, \ldots, C_m$ be disjoint sets of nodes of a complete graph. Enumerate all nodes in C_j by $v_{1,j}, v_{2,j}, \ldots$. Fix a round-robin tournament RR on the set $\{1, \ldots, m\}$. The round-robin matching procedure, corresponding to RR, is defined as follows. For any $i \le m$ and any time $t \le R(m)$, if $RR(i, t) = 0$ then all nodes in C_i are idle in time unit t. Otherwise, let $RR(i, t) = j$ and assume, without loss of generality, that $|C_i| \le |C_j|$. In time unit t, every node $v_{a,i} \in C_i$ calls $v_{a,j}$. Nodes $v_{b,j} \in C_j$, for $b > |C_i|$, remain idle in time unit t. The entire procedure takes time $R(m)$, i.e., at most m steps. In our application, the choice of the particular round-robin tournament RR on the set $\{1, \ldots, m\}$ is not important. We denote by $RRM(C_1, \ldots, C_m)$ the above-described round-robin matching procedure for any such (fixed) choice.

We will also use Algorithm BROADCAST-K_s described in Chapter 3. Recall that this algorithm broadcasts a message in a (fault-free) complete graph K_s in time $\lceil \log s \rceil$.

Let $s = \lfloor n/(k+1) \rfloor$ and $r = n \bmod (k+1)$. We first present a simple k-tolerant broadcasting scheme which works under the additional assumption that $s \ge r$. It presents the main ideas needed in the general case, in a simpler setting. Later we will present a more complex scheme in which this assumption will not be necessary.

Scheme SIMPLE k-TOLERANT BROADCASTING

Let S be the source. Partition all vertices into $k + 1$ *regular clusters* $C_1, \ldots, C_{k+1}$ of size s and one *special cluster* C_{k+2} of size $r \le s$. Let K denote the number of clusters ($K = k + 1$ if $r = 0$ and $K = k + 2$ otherwise). Enumerate all nodes in C_j by $v_{1,j}, v_{2,j}, \ldots$. Let $S = v_{1,1}$. $v_{1,j}$ be called the *root* of C_j. The scheme consists of the three following phases executed one after another.

1. In time unit t, where $1 \le t \le k$, S calls $v_{1,t+1}$.
2. Apply Algorithm BROADCAST-K_s in each regular cluster from its root in parallel for all regular clusters.
3. Apply Scheme $RRM(C_1, \ldots, C_K)$.

Lemma 6.3.3. ([PS89]) *Scheme SIMPLE k-TOLERANT BROADCASTING is k-tolerant.*

Proof. Let $v_{a,i}$ be any node different from S, situated in a regular cluster. We have to show $k + 1$ edge-disjoint calling paths from S to $v_{a,i}$. One such path uses the link between S and $v_{1,i}$ (if $i \ne 1$) and edges within C_i leading to $v_{a,i}$. Moreover, for each regular cluster $C_j \ne C_i$ there is a calling path from S to $v_{a,i}$ using the link from S to $v_{1,j}$, followed by edges within C_j, leading to $v_{a,j}$, followed by the link between $v_{a,j}$ and $v_{a,i}$. All these paths are edge-disjoint and there is a total of $k + 1$ of them.

Now consider a node $v_{a,k+2}$ in the special cluster C_{k+2} (if it exists). For each regular cluster C_j, there is a calling path from S to $v_{a,k+2}$ using the link between S and $v_{1,j}$, followed by edges within C_j, leading to $v_{a,j}$, followed by the link between $v_{a,j}$ and $v_{a,k+2}$. These $k + 1$ calling paths are edge-disjoint.

□

Lemma 6.3.4. ([PS89]) *Scheme SIMPLE k-TOLERANT BROADCASTING works in time* $\lceil \log n \rceil + 2k + 2$.

Proof. Phase 1 finishes by time k. Phase 2 takes time $\lceil \log s \rceil$, hence it finishes by time $\lceil \log n \rceil + k$. Phase 3 takes time at most $k+2$ (since $K \leq k+2$), hence it finishes by time $\lceil \log n \rceil + 2k + 2$.

□

We now present a second k-tolerant broadcasting scheme which, although slightly more complex than the preceding one, works without any extra assumptions and takes time $O(\log n + k)$ as well. Let $s = \lfloor (n-1)/(k+1) \rfloor$ and $r = (n-1) \bmod (k+1)$.

Scheme GENERAL k-TOLERANT BROADCASTING

Let S be the source. Partition all nodes other than S into $k + 1$ *regular clusters* $C_1, \ldots, C_{k+1}$ of size s and one *special cluster* C_{k+2} of size r. Now r may be larger than s but, by definition, $r \leq k$. Enumerate all nodes in C_j as follows: $v_{1,j}, v_{2,j}, \ldots$. Call $v_{1,j}$ the *root* of C_j. The scheme consists of the following four phases executed partially concurrently as described below.

1. In time unit t, where $1 \leq t \leq k+1$, S calls $v_{1,t}$.
2. For all $i \leq k+1$, in time units $i+1, \ldots, i + \lceil \log s \rceil$, Algorithm BROADCAST-$K_s$ is applied in regular cluster C_i from its root.
3. For all $i \leq k+1$, in time units $i + \lceil \log s \rceil + 1$, $i + \lceil \log s \rceil + 2$, ..., node $v_{1,i}$ calls nodes $v_{1,k+2}, v_{2,k+2}, \ldots$, in the special cluster, until all nodes of this cluster are called or until time $k + 2 + \lceil \log s \rceil$ has passed, whichever comes first.
4. Starting in time unit $k + 3 + \lceil \log s \rceil$ do in parallel:
   a) $RRM(C_1, \ldots, C_{k+1})$,
   b) $RRM(\{v_{1,k+2}\}, \ldots, \{v_{r,k+2}\})$.

Exercise 6.3.5. Scheme GENERAL k-TOLERANT BROADCASTING satisfies the requirements of the telephone model.

Lemma 6.3.6. ([PS89]) *Scheme GENERAL k-TOLERANT BROADCASTING is k-tolerant.*

Proof. We have to show $k+1$ edge-disjoint calling paths from S to any node. For nodes in regular clusters, the argument is the same as in the proof of Lemma 6.3.4. Consider any node $v_{j,k+2}$ in the special cluster. In phase 3 it receives calls from roots $v_{1,1}, v_{1,2}, \ldots, v_{1,k+2-j}$, thus establishing $k + 2 - j$ edge-disjoint calling paths from S. We call them calling paths of the first type. Moreover, for all $1 \leq i < j$, we also have calling paths of the second type, starting with the edge between S and $v_{1,k+2-i}$ (established in phase 1) then following the link between $v_{1,k+2-i}$ and $v_{i,k+2}$ (established in phase 3), and finally following the link between $v_{i,k+2}$ and $v_{j,k+2}$ (established in phase 4b). All calling paths of the second type are pair-wise edge-disjoint and edge-disjoint from paths of the first type. There are $j-1$ calling paths of the second type, which – together with the paths of the first type – gives a total of $k+1$ edge-disjoint calling paths from S to $v_{j,k+2}$.

$\square$

Lemma 6.3.7. ([PS89]) *Scheme GENERAL k-TOLERANT BROADCAST-ING works in time* $\lceil \log n \rceil + 2k + 3$.

Proof. Phases 1, 2, and 3 finish after at most $k+2+\lceil \log s \rceil$ time units. Since $r \le k$, phase 4 takes time at most $k+1$ and starts in time unit $k+3+\lceil \log s \rceil$, hence it must end by time $\lceil \log n \rceil + 2k + 3$.

$\square$

The following theorem summarizes the results of this subsection.

Theorem 6.3.8. ([PS89]) *For any $k \le n-1$, the optimal time of a k-tolerant broadcasting algorithm in the complete graph K_n, both adaptive and non-adaptive, is* $\log n + O(k)$.

6.3.2 Permanent Node Faults

In this subsection we assume that the source is fault free and at most k other nodes are subject to permanent crash faults. Faulty nodes are located in a worst-case manner. Links are assumed to be fault free. We assume that the network is a complete graph. As opposed to the previous subsection, our broadcasting algorithms will be adaptive. Hence messages sent by nodes may have another role apart from conveying the source message: they can contain some control information on the basis of which the receiving node makes a decision concerning its future calls. Thus we have to specify which nodes can make calls. We will consider two submodels: the *wake-up* model in which only (fault-free) nodes that already got the source message can make calls, and the *unrestricted* model in which any fault-free node can make a call at any time. The first model corresponds to the situation when the source has to wake up all other fault-free nodes which are dormant and cannot undertake any action before being waken up by the source message. On the other hand, the unrestricted model allows us to perform a preprocessing whose aim is to create a list of all fault-free nodes, and thus save time in subsequent broadcasting. It turns out that this preprocessing capability dramatically decreases broadcasting time: fault-tolerant broadcasting can be completed in logarithmic time in the unrestricted model (whenever the fraction of faulty nodes is bounded by a constant smaller than 1), while optimal broadcasting in the wake-up model takes time $k + \lceil \log(n-k) \rceil$ (and thus is linear in n under the same assumption). Most of the results discussed in this subsection are from [GP96] (results concerning the wake-up model are from a larger unpublished version of this paper).

We first describe precisely the format of the calls made in the broadcasting algorithms. The network is synchronous: all nodes use the same global clock measuring time units. A broadcasting algorithm is described using an elementary procedure call(u, p, u'), where u, u' are nodes and p is a packet. Packets are assumed to be sufficiently large to contain the source message m and a number from 1 to n, where n is the number of nodes. Apart from the source message, they will carry control messages about the number of nodes with a particular property, or the label of a particular node.

In every time unit, some of the fault-free nodes u (the *callers*) choose other nodes u' in such a way that pairs $\{u, u'\}$ form a matching. The action of the algorithm in this time unit consists of parallel executions of procedures call(u, p, u'), for all callers u, where the packet p contains a message computed by u in a way prescribed by the algorithm. It should be stressed that while preparing this message and choosing the node u', node u can only use information available to it locally at this time. (We do not assume any central monitor steering the broadcasting process.) We assume that (fault-free) nodes know labels of nodes which they call and by which they are called. The effect of procedure call(u, p, u') is the following. If node u' is fault-free, then u' receives packet p from u and u receives bit 1 from u' which means "I am fault-free". Thus each of u and u' learns that the other is fault-free (since it has sent a message). If u' is faulty then it does not receive packet p from u and does not send anything. Thus u learns that u' is faulty since it expected bit 1 from it and did not receive it.

Nodes which got message m are called *informed*. In the beginning only the source is informed and the goal of broadcasting is to inform all fault-free nodes. In the wake-up model we impose an additional restriction that all callers have to be informed.

The Wake-up Model

We present an optimal k-tolerant algorithm for the wake-up model and compute its running time. At each stage, every fault-free informed node u gets a list D of uninformed nodes, in increasing order of labels, called its *domain*. If the domain is not empty, u calls nodes of the domain in prescribed order until a fault-free node is found. In each call the packet sent by u contains the wake-up message m and a natural number equal to one-half of all nodes in the domain which have not yet been called. When the first fault-free node v is reached (u gets back 1 from v) and s is the number sent, the uninformed part of the domain of u is split. Each of nodes u and v computes its new domain: v gets the list $u + 1, \ldots, u + s$ and u keeps the remaining uninformed nodes from its previous domain in the same order. In the beginning, the domain of the source 1 is the list $(2, \ldots, n)$. Upon receiving a packet $[m, s]$, a fault-free node u wakes up and starts executing the following procedure.

procedure Call&Split(u, s);
$D := (u + 1, \ldots, u + s)$;
while $D \neq \emptyset$ **do**
    for consecutive nodes v in D **repeat**
        $D' :=$ the sequence of nodes in D after v;
        $l :=$ length of D';
        call$(u, [m, \lceil \frac{l}{2} \rceil], v)$;
    until v is fault-free;
    $D :=$ the final segment of D' of length $\lfloor \frac{l}{2} \rfloor$;
end.

Algorithm Wake-up
    Call&Split$(1, n - 1)$
end.

Theorem 6.3.9. *Algorithm Wake-up is an optimal k-tolerant algorithm for any $k < n$. Its running time is $k + \lceil \log(n - k) \rceil$.*

Proof. For any number $k < n$ of faulty nodes, every node is called by an informed fault-free node, hence Algorithm Wake-up is k-tolerant. Consider the procedure Call&Split $(1, n - 1)$ with at most $k \leq n - 1$ faulty nodes in the domain of u. Let $T(n, k)$ be the maximum time spent between the start of this procedure and the latest completion of its descendant in the tree of procedure calls, for any distribution of k faulty nodes. If $k = n - 1$ then $T(n, k) = n - 1$. Assume that $k < n - 1$ and the number of faulty nodes called by node 1 before reaching the first fault-free node is r. Then at most $k - r$ faulty nodes remain in the domain, and the total time is

$$r + 1 + T(\lceil \frac{n - r}{2} \rceil, \min(k - r, \lceil \frac{n - r}{2} \rceil - 1)).$$

It follows that

$$T(n, k) = \begin{cases} \max_{0 \leq r \leq k}(r + 1 + T(\lceil \frac{n-r}{2} \rceil, \min(k - r, \lceil \frac{n-r}{2} \rceil - 1))) & \text{if } k < n - 1 \\ n - 1 & \text{if } k = n - 1. \end{cases}$$

We prove the inequality

$$T(n, k) \leq k + \lceil \log(n - k) \rceil \tag{6.1}$$

by induction on n. Assume it holds for all $t < n$. If $k = n - 1$ then (6.1) is clearly satisfied. Hence assume $k < n - 1$. Fix $r \leq k$ and consider two cases:

Case 1: $k - r \leq \lceil \frac{n-r}{2} \rceil - 1$
We have

$$r + 1 + T(\lceil\frac{n-r}{2}\rceil, \min(k - r, \lceil\frac{n-r}{2}\rceil - 1))$$

$$\leq r + 1 + k - r + \lceil\log(\lceil\frac{n-r}{2}\rceil - k + r)\rceil$$

$$= k + 1 + \lceil\log(\lceil\frac{n-r}{2}\rceil - k + r)\rceil.$$

The right-hand side is maximized for $r = k$, hence we get:

$$T(n, k) \leq k + 1 + \lceil\log(\lceil\frac{n-k}{2}\rceil)\rceil \leq k + \lceil\log(n - k)\rceil.$$

Case 2: $k - r > \lceil\frac{n-r}{2}\rceil - 1$
We have

$$r + 1 + T(\lceil\frac{n-k}{2}\rceil, \lceil\frac{n-k}{2}\rceil - 1) = r + 1 + \lceil\frac{n-k}{2}\rceil - 1 < r + 1 + k - r = k + 1,$$

hence $T(n, k) \leq k \leq k + \lceil\log(n - k)\rceil$.

This proves inequality (6.1) by induction, and hence the running time of Algorithm Wake-up is at most $k + \lceil\log(n - k)\rceil$. In order to finish the proof, it is enough to show that any k-tolerant algorithm in this model has running time at least $k + \lceil\log(n - k)\rceil$.

Before the first fault-free node is called by the source, only the source can make calls. Suppose that the first k nodes called by the source are faulty. Then, after k time units, there is one informed fault-free node and $n - k - 1$ (fault-free) nodes to be informed. This requires $\lceil\log(n - k)\rceil$ additional time units, for a total of $k + \lceil\log(n - k)\rceil$ time units.

□

The Unrestricted Model

As previously mentioned, the unrestricted model allows preprocessing in which yet-uninformed nodes arrange themselves in a list which subsequently accelerates broadcasting. We will assume that the fraction $\frac{k}{n}$ of faulty nodes is bounded by a constant $\beta < 1$. Thus, for $\alpha = 1 - \beta > 0$, the number $n - k$ of fault-free nodes exceeds αn. Every segment S of the sequence $(1, \ldots, n)$, which contains at least $\alpha|S|$ fault-free nodes, is called *good*. Thus the entire sequence $(1, \ldots, n)$ is good. We construct a k-tolerant broadcasting algorithm which works in time $O(\log^2 n)$. It consists of two procedures. Procedure FormList is a preprocessing before actual broadcasting. It organizes all fault-free nodes in one list in time $O(\log^2 n)$. The time used for this preprocessing dominates the running time of the whole algorithm. When the list is already created, we use procedure Broadcast to broadcast the source message m from the source to all nodes in the list. The running time of this procedure is $O(\log n)$.

The computation structure for both procedures FormList and Broadcast is a complete binary tree of segments of the sequence $(1, \ldots, n)$, called the *partition tree PT*. The root of PT is the sequence $(1, \ldots, n)$. Every internal vertex S_p of PT has two children: the left child S_l is the initial segment of S_p of size $\lfloor \frac{|S_p|}{2} \rfloor$ and the right child S_r is the final segment of S_p of size $\lceil \frac{|S_p|}{2} \rceil$. Leaves of PT are on level 0, their parents on level 1, etc. The root is on level $h = \lfloor \log(\alpha n) \rfloor$. All vertices on the i-th level have size equal to $\lceil \frac{n}{2^{h-i}} \rceil$ or $\lfloor \frac{n}{2^{h-i}} \rfloor$, for $i = 0, \ldots, h$. In particular, each leaf has constant size (independent of n): either $\lceil \frac{n}{2^h} \rceil$ or $\lfloor \frac{n}{2^h} \rfloor$.

We will need the following easy fact.

Exercise 6.3.10. At least one child of an internal good vertex of PT is good.

A branch of PT (a path from the root to a leaf) is called good if all its vertices are good. The following fact is a consequence of Exercise 6.3.10.

Exercise 6.3.11. There exists a good branch in PT.

In our algorithm we use two procedures in doubly ranked lists: computing prefix and suffix sums, and broadcasting from the head of the list to all its elements. In a doubly ranked list, every element knows its predecessor and successor, and knows its head and tail ranks, i.e., the distance from the head and the distance from the tail of the list. (In the sequel, the head rank is simply called rank.) If every element of the list has a number called its value, computing prefix sums means computing at every element the sum of values of all elements preceding it in the list (see [CLR90]). Computing suffix sums is analogous. Both procedures can be performed in time $O(\log n)$ in an n-element list, using the well-known pointer jumping technique (see [CLR90]) in which distances between communicating elements of the list are doubled at every stage and new links, twice as long as in the preceding stage, are created.

Exercise 6.3.12. Computation of prefix (suffix) sums and broadcasting from the head to all elements of the list can be performed in a doubly ranked list in time logarithmic in the length of the list.

We now describe the first procedure FormList which will play the role of preprocessing in our broadcasting algorithm. It is divided into a logarithmic number of stages, corresponding to consecutive levels in the partition tree PT, preceded by the initial stage. The initial stage is performed at the leaf level in PT. During this stage, fault-free nodes in every leaf of PT are organized in doubly ranked lists. This is done as follows. Every fault-free node calls every other node in its leaf in a round-robin fashion, thus learning which nodes are fault-free. Then the list of all fault-free nodes ordered by increasing labels is created.

Further stages correspond to consecutive levels $i = 0, \ldots, h - 1$ in PT. During stage i, nodes in each vertex V on level i of PT communicate with nodes in the sibling of V, in the case when at least one of these sibling segments is good. The following invariant is preserved at all stages: all fault-free nodes in a good vertex of PT are organized in a doubly ranked list. If at least one of the siblings is good then, upon completion of stage i, all fault-free nodes in their union (which is their common parent) are organized in a doubly ranked list. To prevent communication conflicts, every stage is divided into two phases. The first phase is devoted to calls from fault-free nodes in left siblings in the partition tree PT. The second phase is reserved for fault-free nodes in the right siblings, and is performed by a node u only if it was not called in the first phase.

Define the domain $D_i(u)$ of a node u at the i-th stage of the procedure to be the vertex on the i-th level in the partition tree PT to which u belongs. $D_i(u)$ is *left* (*right*) if it is the left (right) child of its parent in PT. Let $D_i^*(u)$ be the sibling of $D_i(u)$. Every node u has a local Boolean variable *good* whose value is true iff $D_i(u)$ is good. Let $L(u)$ denote the current list which contains node u, and let r be the length of this list (computed by u from its head and tail ranks). Let j be the rank of u in $L(u)$. Define $Field(u)$ to be the j-th consecutive segment of $D_i^*(u)$ of size either $\lceil \frac{|D_i^*(u)|}{r} \rceil$, for $j \leq |D_i^*(u)| \bmod r$, or $\lfloor \frac{|D_i^*(u)|}{r} \rfloor$, otherwise. Thus every fault-free node u can compute $Field(u)$ at each stage. (Observe that $Field(u)$ is always of bounded size.) Also, u can compute the current value of variable *good*, as soon as it knows the length of the current list $L(u)$.

The computation is synchronized, which means that steps of the procedure are performed in the same time by all nodes of the network. If a step is not performed by a given node, this node waits for the prescribed period of time before going to the next step.

Here is a formal description of the procedure.

procedure FormList;

for all nodes u **in parallel do**
    1. Call all $v \in D_0(u)$ and form the list $L(u)$ of fault-free nodes of $D_0(u)$;
        $good := (|L(u)| \geq \alpha |D_0(u)|)$;
    for stage $i = 0$ to $h - 1$ **do**
       2. **if** $D_i(u)$ is left & *good* **then**
          A Compute $Field(u)$;
          B Call all $v \in Field(u)$;
             $n_u :=$ (number of fault-free nodes in $Field(u)$) $+ 1$;
          C Compute prefix and suffix sums
             of numbers n_u in the list $L(u)$;
          D Call all $v \in Field(u)$ inserting every fault-free node v
             into the list $L(u)$;
       3. **if** $D_i(u)$ is right & *good* & u was not called in step 3 **then**
          Perform all steps A, B, C and D;
       4. $good := (|L(u)| \geq \alpha |D_{i+1}(u)|)$;
end.

Lemma 6.3.13. *Procedure FormList organizes all fault-free nodes in the network into a doubly ranked list in time* $O(\log^2 n)$.

Proof. Every node u in the network knows the total number n of nodes. The partition tree PT is fixed, hence every node u can compute $D_i(u)$ and $D_i^*(u)$ for every level $i = 0, \ldots, h-1$ of PT. We will prove by induction that the following invariant holds:

INV

if the domain $D_i(u)$ is good then in the beginning of the i-th stage all fault-free nodes in it are organized in the doubly ranked list $L(u)$.

Thus every node u can compute $Field(u)$.

In step 1 (the initial stage) all fault-free nodes of each leaf $D_0(u)$ are organized in the doubly ranked list $L(u)$. Thus the invariant is satisfied for $i = 0$. Suppose inductively that it is satisfied for some $i \geq 0$.

A node u makes calls in stage i only if its domain $D_i(u)$ is good. To avoid communication conflicts, step 2 is performed only by nodes from left domains, nodes from right domains waiting until step 2 is finished. After computing $Field(u)$ in substep A, node u calls all nodes in $Field(u)$, and computes the number $n_u - 1$ of fault-free nodes in it, in substep B. Every element of list $L(u)$ has value n_u. The prefix and suffix sums of these values are computed in the list $L(u)$, according to Exercise 6.3.12, in substep C. The prefix (suffix) sum at node u is its head (tail) rank in the updated list $L(u)$. This list is updated in substep D by inserting all fault-free nodes from $D_i^*(u)$: u inserts all fault-free nodes from $Field(u)$ into $L(u)$, in increasing order of labels, immediately before itself. All list parameters (predecessor, successor and rank) of each inserted node are properly set by u. Thus, if $D_i(u)$ is good, all fault-free nodes from $D_{i+1}(u) = D_i(u) \cup D_i^*(u)$ are in the list $L(u)$ after completing step 2. Step 3 is performed only by good right domains which were not searched in step 2. All operations in this step are the same as in step 2, and the above property holds as well.

Now assume that $D_{i+1}(u)$ is good. By Exercise 6.3.10, either $D_i(u)$ or $D_i^*(u)$ is good, and hence, in the beginning of the $(i+1)$-th stage, all fault-free nodes in $D_{i+1}(u)$ are organized in the doubly ranked list $L(u)$. This concludes the proof of invariant **INV** by induction. Exercise 6.3.11 implies that after stage $h-1$ all fault-free nodes of the network are in the same doubly ranked list. (Note that if the parent W of a good vertex V on level i is not good, the list created by nodes in V contains all fault-free nodes of W before the $(i+1)$-th stage but is not used in this stage because nodes in this list do not make calls.)

It remains to compute the running time of procedure FormList. The initial stage (step 1) takes constant time, since leaves of PT are of constant size. Further stages $i = 0, \ldots, h-1$ are divided into steps 2, 3 and 4. Steps 2 and 3 consist of substeps A, B, C and D. Substeps A, B and D take constant time since the size of $Field(u)$ is constant. According to Exercise 6.3.12, substep C takes logarithmic time which dominates the running time of every stage. Step 4 takes constant time. The number of stages in procedure FormList (levels in PT) is logarithmic, and every stage takes $O(\log n)$ time, thus the running time of procedure FormList is $O(\log^2 n)$.

□

When all fault-free nodes are connected in one list, procedure Broadcast is called to perform actual broadcasting. The head of the list calls the source, asking for the message m. After it gets it, the message is broadcast from the head to all nodes in the list.

procedure Broadcast;
    1. The head of the list calls the source;
    2. The source transmits message m to the head;
    3. Broadcast message m from the head to all
        nodes in the list by pointer jumping;
end.

Exercise 6.3.12 implies:

Lemma 6.3.14. *Procedure Broadcast works in time $O(\log n)$.*

Algorithm FastBroadcast;
    FormList;
    Broadcast;
end.

Lemmas 6.3.13 and 6.3.14 imply:

Theorem 6.3.15. ([GP96]) *Assume that the fraction $\frac{k}{n}$ of faulty nodes is bounded by a constant smaller than 1. Algorithm FastBroadcast is a k-tolerant adaptive broadcasting algorithm for the complete n-node network, with running time $O(\log^2 n)$.*

As we have seen, the preprocessing part of Algorithm FastBroadcast, which organizes all fault-free nodes in a list, takes time $O(\log^2 n)$, thus dominating the running time of the entire algorithm. Hence it is natural to ask if a similar preprocessing could be done faster, ideally in logarithmic time. It turns out that this is, in fact, possible. In [DP00], a constant fraction of fault-free nodes are distributedly organized in a connected graph with constant degree and logarithmic diameter, and this preprocessing is done in time $O(\log n)$. The construction uses much more sophisticated methods than the above-described Procedure FormList: it relies on an expander construction. This permits us to obtain optimal k-tolerant broadcasting time in this model.

Theorem 6.3.16. ([DP00]) *Assume that the fraction $\frac{k}{n}$ of faulty nodes is bounded by a constant smaller than 1. There exists a k-tolerant adaptive broadcasting algorithm for the complete n-node network, with running time $O(\log n)$.*

6.3.3 Transmission Faults

Broadcasting in the presence of transmission failures will be presented in the following variation of the bounded-fault model, called the *linearly bounded-fault* model, introduced in [GP98]. Given a constant $0 < \alpha < 1$ we assume that at most αi faulty transmissions can occur during the first i time units of the communication process, for every natural number i. This assumption grasps the idea that more faults are possible when the algorithm runs longer (which is natural to suppose, since faults are transient) and, on the other hand, uses the worst-case approach, typical for the bounded-fault model. We assume that faults are of the crash type, i.e., faulty transmissions have no effect. Nodes are assumed fault-free. Notice that the assumption $\alpha < 1$ is necessary for broadcasting to be feasible. All algorithms considered in this subsection are non-adaptive, and lower bounds on time also concern non-adaptive broadcasting.

For a fixed parameter $0 < \alpha < 1$, a given network $\mathcal{N}$ and a source s, a broadcasting scheme is called α-*safe* if it broadcasts information from the source s to all nodes whenever the number of faulty transmissions during the scheme execution satisfies the above linearly bounded assumption with parameter α. By $B(\mathcal{N}, \alpha, s)$ we denote the least worst-case running time of α-safe broadcasting schemes from the source s. The maximum of $B(\mathcal{N}, \alpha, s)$ over all possible sources s is denoted by $B(\mathcal{N}, \alpha)$ and is called the α-*safe broadcasting time* of the network $\mathcal{N}$.

Networks whose α-safe broadcasting time is linear in their fault-free broadcasting time can be considered robust with respect to linearly bounded transmission faults, while those for which α-safe broadcasting time dramatically exceeds broadcasting time without faults are vulnerable to faulty transmissions. We will establish α-safe broadcasting time for important interconnection networks, and find out which of them are robust and which are vulnerable to faults.

We start with the study of α-safe broadcasting on the line of length n. It turns out that robustness of the line with respect to linearly bounded transmission faults heavily depends on the parameter α. We present a linear-time broadcasting scheme for the case $\alpha < \frac{1}{2}$, and show that for $\alpha \geq \frac{1}{2}$ any α-safe broadcasting scheme takes time

$$\Omega\left(\left(\frac{1}{1-\alpha}\right)^n\right).$$

For this range of the parameter α, a matching upper bound

$$\mathcal{O}\left(\left(\frac{1}{1-\alpha}\right)^n\right)$$

is also obtained by showing an α-safe broadcasting scheme working in this time. L_n denotes the line with nodes $v_0, v_1, \ldots, v_n$ and links $l_1, \ldots, l_n$.

First consider the situation when $\alpha < \frac{1}{2}$.

Assume that the source is in the leftmost node v_0. Consider the following simple broadcasting scheme.

Scheme Odd-Even (m)

> **for** m steps **do**
>> in odd (even) steps all pairs of nodes joined by odd (even) numbered links communicate.
>
> **end.**

Lemma 6.3.17. ([GP98]) *Assume that $\alpha < \frac{1}{2}$ and let $t = \lfloor \frac{n}{1-2\alpha} \rfloor$. Scheme Odd-Even (t) performs α-safe broadcasting in L_n.*

Proof. Assume that before a given step of the scheme, v_k is the rightmost informed node. Node v_{k+1} becomes informed in this step if link l_{k+1} is used in communication in this step and the transmission along this link is fault-free. Thus any single transmission fault can cause a delay of at most two steps in the broadcasting process. One step of delay is directly caused by the failure and the second one is caused by the fact that link l_{k+1} is not used in communication in the next step. Let T be the smallest m for which Scheme Odd-Even (m) broadcasts in L_n in an α-safe way. Since at most $\lfloor \alpha T \rfloor$ transmission failures can occur during the broadcasting process, every failure causes a delay of at most two steps, and the message has to traverse n links, the following inequality holds: $n + 2\lfloor \alpha T \rfloor \geq T$. This implies $T \leq \frac{n}{1-2\alpha}$.

$\square$

If the source is in an interior node, the above lemma holds as well. This implies the following result.

Theorem 6.3.18. ([GP98]) *For any fixed $0 < \alpha < \frac{1}{2}$*

$$B(L_n, \alpha) = O(n).$$

Next assume that $\alpha \geq \frac{1}{2}$. α-safe broadcasting time in the line dramatically changes in this case. We show an exponential lower bound $\Omega((\frac{1}{1-\alpha})^n)$ on the running time of all α-safe broadcasting schemes on L_n, and we present a scheme whose running time matches this bound.

In order to establish the lower bound, we use an adversary argument. Define the *adversary's account* $\mathcal{A}$ which changes during the scheme execution in the following way. At the beginning of the scheme, $\mathcal{A}$ is set to 0. After every step of the scheme, $\mathcal{A}$ is increased by 1. Whenever the adversary uses a transmission fault in a given step, the account $\mathcal{A}$ is decreased by $\frac{1}{\alpha}$. This corresponds to the fact that the adversary has to wait at least $\frac{1}{\alpha}$ time units (i.e., "earn" $\frac{1}{\alpha}$ units on the account) before "spending" one failure. The adversary can place failures in an arbitrary way, as long as the account remains non-negative at all times. Define the *head* to be the rightmost informed node.

Lemma 6.3.19. ([GP98]) *There exists an adversary for which every move of the head after the first step of the scheme increases $\mathcal{A}$ at least $\frac{1}{1-\alpha}$ times.*

Proof. After the first step, $\mathcal{A}$ has value 1. Assume that after a given step t_0 the account $\mathcal{A}$ is positive and v_k is the head. Consider t steps of the scheme following step t_0. Denote by $w(t)$ the number of steps among those t in which link l_{k+1} is used for communication and let $b(t) = t - w(t)$. Consider inequalities

$$w(t) \le \alpha(\mathcal{A} + t) \quad \text{and} \quad b(t) \le \alpha(\mathcal{A} + t). \tag{6.2}$$

As long as both of them hold, the adversary can put faults always on link l_{k+1}, preventing the head from moving right. Thus any α-safe scheme has to violate one of those inequalities. Let t_1 be the least positive integer for which one of the inequalities is violated. First assume that $w(t_1) > \alpha(\mathcal{A}+t_1)$. In this case we describe the behavior of the adversary as follows: in steps $t_0 + 1, \ldots, t_0 + t_1$ a failure is placed on link l_{k+2} whenever this link is used. In this way, after step $t_0 + t_1$ the head can move by at most one, to node v_{k+1}. Since the second inequality holds in all these steps, this is a legal behavior of the adversary. Denote $W = w(t_1)$ and $B = b(t_1)$. We have $W > \alpha(\mathcal{A} + t_1) = \alpha(\mathcal{A} + W + B)$, thus

$$W - \frac{\alpha}{1-\alpha}B > \frac{\alpha}{1-\alpha}\mathcal{A}. \tag{6.3}$$

Notice that

$$\frac{1-\alpha}{\alpha} \le \frac{\alpha}{1-\alpha}, \quad \text{for} \quad \alpha \ge \frac{1}{2}. \tag{6.4}$$

During t_1 steps, $\mathcal{A}$ is changed to at least $\mathcal{A} + W + B - \frac{B}{\alpha}$, because the account $\mathcal{A}$ is increased by $t_1 = W + B$ and decreased by using at most B transmission failures. The final value of the account after step $t_0 + t_1$ is at least

$$\mathcal{A} + W + B - \frac{B}{\alpha} = \mathcal{A} + W - \frac{1-\alpha}{\alpha}B.$$

Applying inequalities (6.4) and (6.3) we get:

$$\mathcal{A} + W - \frac{1-\alpha}{\alpha}B \ge \mathcal{A} + \left(W - \frac{\alpha}{1-\alpha}B\right) \ge \mathcal{A} + \frac{\alpha}{1-\alpha}\mathcal{A},$$

thus finally:

$$\mathcal{A} + W + B - \frac{B}{\alpha} \ge \frac{1}{1-\alpha}\mathcal{A}. \tag{6.5}$$

In steps $t_0 + 1, \ldots, t_0 + t_1$ the head moved by at most 1 and the value of the account increased at least by the factor $\frac{1}{1-\alpha}$.

The case when the second of the inequalities in (6.2) is violated first is handled analogously. In this case the adversary puts a failure in the link l_{k+1} whenever it is used and the head does not move at all, while the value of the account increases as before.

$\square$

After the first step of the algorithm, the value of $\mathcal{A}$ is 1. By Lemma 6.3.19, when the head gets to v_n, the value of $\mathcal{A}$ increases to $\Omega((\frac{1}{1-\alpha})^n)$. This means that the broadcasting time is also $\Omega((\frac{1}{1-\alpha})^n)$ because at each step the value of $\mathcal{A}$ can increase by at most 1. This proves the following result.

Theorem 6.3.20. ([GP98]) *For any constant* $\alpha \geq \frac{1}{2}$

$$B(L_n, \alpha) = \Omega((\frac{1}{1-\alpha})^n).$$

We now establish a matching upper bound on α-safe broadcasting time, for any constant $\alpha \geq \frac{1}{2}$. To this end we describe an α-safe broadcasting scheme with running time $O((\frac{1}{1-\alpha})^n)$. We present the scheme for the case when the source is in node v_0 (the general case is analogous). Let t_k, for $k = 1, \ldots, n$, be integers defined as follows:

$$t_0 = 0, \quad t_1 = 1, \quad \text{and} \quad t_k = \lfloor \alpha(t_1 + \cdots + t_k) \rfloor + 1, \quad \text{for } k > 1. \quad (6.6)$$

Scheme Exponential-Broadcasting-on-the-Line
    for $j := 1$ **to** n **do**
        if j is odd
            then in steps $t_{j-1} + 1, \ldots, t_j$, all pairs of nodes
                joined by odd numbered links communicate
            else in steps $t_{j-1} + 1, \ldots, t_j$, all pairs of nodes
                joined by even numbered links communicate
end.

First observe that the above scheme is α-safe. After t_1 steps node v_1 is informed. Assume that after $t_1 + \cdots + t_k$ steps node v_k is informed. In steps $t_1 + \cdots + t_k + 1, \ldots, t_1 + \cdots + t_k + t_{k+1}$ the link l_{k+1} is used for communication. Since $t_{k+1} > \alpha(t_1 + \cdots + t_k + t_{k+1})$, the adversary cannot put a failure in this link in all those steps and, consequently, after $t_1 + \cdots + t_{k+1}$ steps, node v_{k+1} becomes informed.

Since

$$t_{k-1} = \lfloor \alpha(t_1 + \cdots + t_{k-1}) \rfloor + 1 \quad \text{and} \quad t_k = \lfloor \alpha(t_1 + \cdots + t_k) \rfloor + 1,$$

we get

$$t_k = \lfloor \alpha(t_1 + \cdots + t_k) \rfloor + 1 \leq \lfloor \alpha(t_1 + \cdots + t_{k-1}) \rfloor + 1 + \alpha t_k + 1$$

and hence

$$t_k \leq t_{k-1} + \alpha t_k + 1. \tag{6.7}$$

This implies $t_n = O((\frac{1}{1-\alpha})^n)$ and hence $T_n = t_1 + \cdots + t_n = O((\frac{1}{1-\alpha})^n)$. Hence we get the following theorem:

Theorem 6.3.21. ([GP98]) *For any constant* $\alpha \geq \frac{1}{2}$

$$B(L_n, \alpha) = \Theta((\frac{1}{1-\alpha})^n).$$

We now turn attention to α-safe broadcasting in arbitrary trees. To this end, we first consider a very simple tree, the star S_n. This is the tree with central node v_0 and nodes $v_1, \ldots, v_n$ adjacent to it. Let l_i be the link joining v_0 and v_i. It turns out that the star is not robust with respect to linearly bounded faults for any value of α. We show that any α-safe broadcasting scheme for S_n has running time $\Omega((\frac{1}{1-\alpha})^n)$, for any constant $0 < \alpha < 1$. We also show an algorithm running in time $O((\frac{1}{1-\alpha})^n)$.

Lemma 6.3.22. ([GP98]) *For any constant* $0 < \alpha < 1$

$$B(S_n, \alpha) = \Omega((\frac{1}{1-\alpha})^n).$$

Proof. Without loss of generality we may assume that the central node v_0 is the source. (If a leaf is the source then after the first step the central node is informed and the situation is the same as if v_0 were the source.) Consider any α-safe broadcasting scheme $\mathcal{B}$. Let t_i^k be the number of steps in which link l_i is used for communication during the first k steps of scheme $\mathcal{B}$. Notice that for each $i = 1, \ldots, n$, there must exist a positive integer k such that $t_i^k > \alpha k$. Otherwise the adversary could always preclude informing node v_i. Let $k_i = \min\{k : t_i^k > \alpha k\}$. Notice that in step k_i of scheme $\mathcal{B}$, the link l_i is used, because of the minimality of k_i. This implies $k_i \neq k_j$ whenever $i \neq j$. Thus we may renumber all leaves of S_n in increasing order of k_i and assume, from now on, that $k_i < k_j$ for $1 \leq i < j \leq n$.

Since $k_i \geq k_j$, for all $j = 1, \ldots, i$, we get $k_i \geq t_1^{k_1} + \cdots + t_i^{k_i}$. Since $t_i^{k_i} > \alpha k_i$ we get

$$t_i^{k_i} > \alpha(t_1^{k_1} + \cdots + t_i^{k_i}).$$

Since $t_1^{k_1} = 1$ we obtain $t_n^{k_n} = \Omega((\frac{1}{1-\alpha})^n)$. There are at least $t_n^{k_n}$ steps of the scheme in which link l_n is used, hence the running time of scheme $\mathcal{B}$ is in $\Omega((\frac{1}{1-\alpha})^n)$.

$\square$

The following modification of Scheme Exponential-Broadcasting-on-the-Line shows that the above lower bound is tight. Let t_i have the same meaning as in the formulation of this scheme.

Scheme Star
    for $j := 1$ **to** n **do**
        in steps $t_{j-1} + 1, \ldots, t_j$, node v_0 communicates with v_j
end.

Exercise 6.3.23. Scheme Star is α-safe and its running time is for any constant $0 < \alpha < 1$

$$\mathcal{O}\left(\left(\frac{1}{1-\alpha}\right)^n\right).$$

Hence we get:

Theorem 6.3.24. ([GP98]) *For any constant* $0 < \alpha < 1$

$$B(S_n, \alpha) = O\left(\left(\frac{1}{1-\alpha}\right)^n\right).$$

Some of the techniques described above can be applied in the context of general trees. Since all edges of a tree of maximum degree d can be partitioned into d pairwise disjoint matchings, a generalization of Scheme Odd-Even can be used to obtain the following result.

Exercise 6.3.25. For any tree T with maximum degree bounded by a constant d, and for any constant $\alpha < \frac{1}{d}$, we have $B(T, \alpha) = O(D)$, where D is the diameter of T.

Thus, for $\alpha < \frac{1}{d}$, bounded-degree trees are robust with respect to linearly bounded transmission faults: their α-safe broadcasting time is linear in fault-free broadcasting time. However, for $\alpha \geq 1/2$, all trees become vulnerable to transmission failures. Theorem 6.3.20 implies that α-safe broadcasting time is exponential in the diameter of the tree, for such values of α. On the other hand, in view of Theorem 6.3.24, this time is exponential in the maximum degree of the tree, for any constant $0 < \alpha < 1$. Hence, if $\alpha \geq 1/2$, the α-safe broadcasting time of any tree largely exceeds its fault-free broadcasting time.

Since vulnerability of trees to linearly bounded transmission faults heavily depends on the value of parameter α, it is interesting to ask if there are networks which are robust with respect to transmission faults for any $0 < \alpha < 1$, more precisely, networks whose α-safe broadcasting time is linear in their fault-free broadcasting time for any α. It turns out that rings have this property.

Theorem 6.3.26. ([GP98]) *For any constant* $0 < \alpha < 1$

$$B(R_n, \alpha) = O(n).$$

Proof. Consider an n-node ring R_n with nodes $v_0, \ldots, v_{n-1}$ and links $l_0, \ldots, l_{n-1}$, link l_i joining v_i with $v_{(i+1) \bmod n}$. Let v_0 be the source. We show the argument for even values of n. The case of odd n is similar. Consider the analog of

Scheme Odd-Even on the line: in odd (even) steps all links of the ring with odd (even) indices are used for communication. After step 1, the informed nodes are v_0 and v_n. Fix an even positive integer j and let $v_a, v_{a+1}, \ldots, v_0, \ldots, v_b$ be the maximal segment of informed nodes before step j. Let $l' = l_{a-1}$ and $l'' = l_b$. Call two consecutive steps, j and $j+1$, a *phase*. (Thus the first phase consists of steps 2 and 3.) During phase $\{j, j+1\}$ each of links l', l'' is used once for communication. Thus, if x is the number of nodes which got the message in a given phase and f is the number of faults used in this phase then $x + f \geq 2$. The maximum number of faults that can occur during k phases is $\alpha(2k + 1)$ because one step was done before the first phase and no failure could occur in this step. Hence the number of nodes informed during k phases is at least $2k - \alpha(2k + 1)$. Whenever $2k - \alpha(2k + 1) \geq n - 2$, broadcasting is completed after time $T = 2k + 1$. (Two nodes were informed after the first step, before the first phase.) The latter inequality is satisfied for

$$k = \left\lceil \frac{\frac{n-1}{1-\alpha} - 1}{2} \right\rceil,$$

hence α-safe broadcasting is completed in time at most

$$2 \left\lceil \frac{\frac{n-1}{1-\alpha} - 1}{2} \right\rceil + 1.$$

$\square$

Although, as we have shown, rings are robust with respect to linearly bounded transmission faults for any value of parameter α, they are not good networks for broadcasting, even without faults, as their diameter is linear in the number of nodes. Hence the natural question is whether there exist networks whose α-safe broadcasting time is short for any α, ideally logarithmic in the number of nodes. It turns out that both hypercubes and complete graphs have this property.

Let us first consider the r-dimensional hypercube H_r and fix a labeling $1, \ldots, r$ of its dimensions. For any natural $t \geq r$ consider the following broadcasting scheme in H_r.

Scheme Cyclic(t)
    In time unit $i := 1, \ldots, t$ every node communicates with
    its neighbor in dimension $i \bmod r$.
end.

We will use the following lemma concerning k-tolerant broadcasting, for any number k of transmission faults.

Lemma 6.3.27. ([GP98]) *Scheme Cyclic (t) achieves k-tolerant broadcasting in H_r, from any source, for $t = r + k$.*

Proof. We prove the lemma by induction on r. For $r = 1$ and $r = 2$ it can be checked directly. Assume $r \geq 3$. For $k = 1$ the lemma is obvious, so assume

$k > 1$. Without loss of generality we may assume that the first transmission from the source (along dimension 1) is fault free: otherwise the number of faults and time both decrease by 1. Split H_r into two copies of H_{r-1} along dimension 1. Denote by L the copy containing the source and by R the other copy. Thus, after the first time unit, each of L and R contains an informed node. Define a *window* to be a time segment of length r in which transmissions in Scheme Cyclic are scheduled along consecutive dimensions $2, 3, \ldots, r, 1$. The number of windows during time $t = r + k$ is $x = \lfloor \frac{r+k-1}{r} \rfloor$. Since $r \geq 3$ and $k > 1$, we have $x \leq k/2$.

Let a be the number of transmission faults within L, b the number of transmission faults within R and z the number of transmission faults along dimension 1. First assume that $a \leq k - x$ and $b \leq k - x$. Ignore temporarily time units used by the scheme for communication in dimension 1. By the inductive assumption, all nodes of L and all nodes of R are informed after $(r-1)+(k-x)$ time units. At most $x+1$ time units of Scheme Cyclic are used for transmissions along dimension 1, hence both L and R become informed after a total time $(r - 1) + (k - x) + (x + 1) = r + k$. Next suppose that one of the numbers a or b exceeds $k - x$. Without loss of generality suppose that $b > k - x$. Hence $a < x - z$. Consequently, there exists a window in which all transmissions within L are fault free, followed by a time unit j used for communication in dimension 1, with all transmissions fault free as well. During time units of this window, L becomes informed, and in time unit j the information is passed to all nodes in R. By the inductive assumption, broadcasting is completed after time at most $(r-1)+a+(x+1) \leq r+2x \leq r+k$. $\qquad\square$

The following theorem establishes the exact value of α-safe broadcasting time in the hypercube, and shows that this time is proportional to the dimension of the hypercube, for any value of parameter α.

Theorem 6.3.28. ([GP98]) *For any* $0 < \alpha < 1$

$$B(H_r, \alpha) = \lfloor \frac{\alpha}{1 - \alpha}(r - 1) \rfloor + r.$$

Proof. We first prove the lower bound. Consider any α-safe broadcasting scheme in the hypercube H_r. After the first $r - 1$ time units, there remain some uninformed nodes. Pick one such node v. Assume that all transmissions involving node v, during x consecutive time units $r, r + 1, \ldots, r + x - 1$, are faulty. In order to satisfy the linear bound on the number of faults, it suffices to guarantee $\frac{x}{r-1+x} \leq \alpha$. This is satisfied for $x = \lfloor \frac{\alpha}{1-\alpha}(r - 1) \rfloor$. At least one more time unit is needed to inform v, thus the total time is at least $\lfloor \frac{\alpha}{1-\alpha}(r - 1) \rfloor + r$.

In order to prove the upper bound, we show that Scheme Cyclic (t) achieves broadcasting for $t \leq \lfloor \frac{\alpha}{1-\alpha}(r - 1) \rfloor + r$ against any adversary satisfying the linear bound with parameter α. Let $T = r+y$ be the minimum time t in which this can be done. Since time $T' = r+y-1$ is too short, more than $y - 1$ faults can occur in time T': otherwise, in view of the previous lemma, the hypercube could be informed in time T' even against a stronger adversary which can place faults arbitrarily. It follows that in time T' at least y faults can occur. This means that $y \leq \alpha T' = \alpha(r+y-1)$ and consequently $y \leq \lfloor \frac{\alpha}{1-\alpha}(r-1) \rfloor$. Finally we get $T \leq \lfloor \frac{\alpha}{1-\alpha}(r-1) \rfloor + r$, which concludes the proof.

□

We now turn attention to α-safe broadcasting time in the complete graph on n nodes. If n is a power of 2, this time is equal to $B(H_{\log n}, \alpha)$ and follows from Theorem 6.3.28. Otherwise, a modified approach is needed. Fix the parameter $0 < \alpha < 1$ and let $r = \lfloor \log n \rfloor$. Let b be the number of digits 1 in the binary representation of n and let $q = \lceil \log b \rceil$. Consider the hypercube H_q, called *basic*, one of whose nodes is the source. Represent n as a sum $2^{x_1} + \cdots + 2^{x_s}$, where $s = 2^q$ and $x_i \leq r$ are natural numbers, for all $i = 1, \ldots, s$. This partition is possible in view of $s \geq b$. At consecutive nodes of H_q attach a copy of H_{x_i}, for $i = 1, \ldots, s$. Let

$$T_0 = B(H_q, \alpha) = \left\lfloor \frac{\alpha}{1-\alpha}(q-1) \right\rfloor + q.$$

Let $T_1 = \lfloor \alpha(T_0 + T_1) \rfloor + r$. Consider the following scheme.

Scheme Cyclic-in-Many-Cubes
1. Run Scheme Cyclic (T_0) in the basic hypercube.
2. Run Scheme Cyclic (T_1) in all attached hypercubes in parallel.
end.

Exercise 6.3.29. Confirm that Scheme Cyclic-in-Many-Cubes is α-safe.

Exercise 6.3.30. Confirm that Scheme Cyclic-in-Many-Cubes runs in time at most

$$\frac{1}{1-\alpha}r + \frac{1}{(1-\alpha)^2}(q-1) + \frac{1}{1-\alpha}.$$

The same argument as in the proof of Theorem 6.3.28 shows that the number $B(H_r, \alpha)$ is a lower bound on the worst-case running time of any α-safe broadcasting scheme for the complete graph K_n on n nodes. Thus we get the estimates

$$\frac{1}{1-\alpha}(\lfloor \log n \rfloor - 1) \leq B(K_n, \alpha) \leq \frac{1}{1-\alpha}\log n + \frac{1}{(1-\alpha)^2}\log \log n + \frac{1}{1-\alpha},$$

which prove the following theorem:

Theorem 6.3.31. ([GP98]) *For any constant* $0 < \alpha < 1$

$$B(K_n, \alpha) = \frac{1}{1-\alpha}\log n + O(\log \log n).$$

We have shown that for hypercubes and for complete graphs, α-safe broadcasting time is logarithmic in the number of nodes, i.e., of optimal order of magnitude, for any $0 < \alpha < 1$. However, both these types of networks have degree growing with size, and hence are difficult and costly to construct for a large number of nodes. Hence a natural question arises. Is it possible to construct networks of bounded degree which have the same robustness property with respect to linearly bounded faults? The positive answer to this question was given in [CM98]. It turns out that butterflies have this property.

The following exercise [CM98] establishes a property of the butterfly important from the point of view of fault-tolerant broadcasting.

Exercise 6.3.32. Show that the butterfly BF_m, for $m > 1$, has the following properties:

1. There exist four edge-disjoint paths of length $O(m)$ joining nodes (i, w) and (j, w), for any levels $i \neq j$ and any w.
2. For any $w \neq 0^m$, there exist four edge-disjoint paths of length $O(m)$ joining node $(0, 0^m)$ with some nodes (i_1, w), (i_2, w), (i_3, w), (i_4, w).

We will also use the following easy fact.

Exercise 6.3.33. Show that all edges of the butterfly BF_m, for even $m > 0$, can be partitioned into four perfect matchings.

Consider a butterfly BF_m, for even $m > 0$. Let M_1, M_2, M_3, M_4 be the perfect matchings from Exercise 6.3.33, and let t be a positive integer. Consider the following broadcasting scheme.

Scheme Matching-Broadcast(t, M_1, M_2, M_3, M_4)
    repeat t times
        for $i = 1, \ldots, 4$ **do**
            for $\{x, y\} \in M_i$ **in parallel do**
                node x communicates with y
end.

The following theorem shows that Scheme Matching-Broadcast executes α-safe broadcasting in the butterfly BF_m, for even $m > 0$, in time $O(m)$, i.e., logarithmic in the number of nodes.

Theorem 6.3.34. ([CM98]) *Fix* $0 < \alpha < 1$. *Let* $D = O(m)$ *be the upper bound on the length of paths from Exercise 6.3.32, and let* $t = \lceil \frac{D}{(1-\alpha)^2} \rceil + \lceil \frac{D}{1-\alpha} \rceil$. *Scheme Matching-Broadcast(t, M_1, M_2, M_3, M_4) performs α-safe broadcasting in the butterfly BF_m, for even $m > 0$, in time $O(m)$.*

Proof. Scheme Matching-Broadcast(t, M_1, M_2, M_3, M_4) works in time $4t = O(m)$. It remains to show that it is α-safe. Let $T = 4\lceil \frac{D}{1-\alpha} \rceil$ and $T' = 4t$. By symmetry of the butterfly we may assume that node $(0, 0^m)$ is the source. Fix a node (i, w). It is enough to show that it gets the source message by step T' of the scheme.

First suppose that $w = 0^m$. Consider the first T steps of the scheme. We have $4D + T\alpha \leq T$. During the first T steps there are at most αT faults. Consider the four edge-disjoint paths joining node $(0, 0^m)$ with $(i, 0^m)$, which exist by part 1 of Exercise 6.3.32. On one of them there are at most $\alpha \lceil \frac{D}{1-\alpha} \rceil$ transmission faults during the first T steps. Each fault delays information progress on this path by at most 4 steps. Hence node (i, w) gets the source message after at most $4(D + \alpha \lceil \frac{D}{1-\alpha} \rceil) = 4D + T\alpha \leq T < T'$ steps.

Next suppose that $w \neq 0^m$. Again consider the first T steps of the scheme. During these steps there are at most αT faults. Consider the four edge-disjoint paths joining node $(0, 0^m)$ with some nodes $(i_1, w), (i_2, w), (i_3, w), (i_4, w)$, which exist by part 2 of Exercise 6.3.32. On one of them there are at most $\alpha \lceil \frac{D}{1-\alpha} \rceil$ transmission faults during the first T steps. By the same reasoning as above, one of these four nodes gets the source message after at most $4(D + \alpha \lceil \frac{D}{1-\alpha} \rceil) = 4D + T\alpha \leq T$ steps. Call this node (j, w). If $i = j$, we are done. Hence assume $i \neq j$. By part 1 of Exercise 6.3.32, there are four edge-disjoint paths joining nodes (i, w) and (j, w). During T' steps of the scheme there are at most $\alpha T'$ transmission faults, hence on one of these paths there are at most αt faults. Since each fault delays information progress on this path by at most 4 steps, we conclude that node (i, w) gets the source message by step $T + 4(D + \alpha t)$. We have $T'(1 - \alpha) \geq \frac{4D}{1-\alpha} + T \geq 4D + T$, hence $\alpha T' \leq T' - T - 4D$. It follows that $T + 4(D + \alpha t) \leq T'$, which concludes the proof.

$\square$

6.4 The Probabilistic Fault Model

In this section we assume that links and/or nodes of the network are subject to faults which occur independently with fixed and given probabilities. The source is fault free, all other nodes have the same probability of fault and all links have the same probability of fault. We seek *almost safe* broadcasting algorithms: all fault-free nodes must receive the source message with probability at least $1 - \frac{1}{n}$, if the number n of nodes is sufficiently large.

6.4.1 Crash Faults

Transmission Faults

We start with the study of almost safe broadcasting in the following model. All nodes are fault free and links are subject to transient crash faults occurring independently in all time steps, with fixed probability $0 < p < 1$. More precisely, for any two transmissions (both simultaneous and those occurring in different time units), the events that these transmissions are faulty are independent.

We start by considering broadcasting from one end of the line L_n with nodes $v_0, v_1, \ldots, v_n$, and links $l_1, \ldots, l_n$. Let c be a positive integer. Consider Scheme Odd-Even $(2cn)$, described in Section 6.3.3. Let $q = 1 - p$.

Lemma 6.4.1. ([DP92]) *Let E denote the event that Scheme Odd-Even $(2cn)$ does not accomplish broadcasting in the line L_n. Then $Prob(E) \leq e^{-dn}$, where $d = \frac{qc-1}{2qc}$.*

Proof. For a fixed step of Scheme Odd-Even $(2cn)$, node v_j is *critical* if at the beginning of this step, j is the largest index of an informed node. If v_j, for $j < n$, is critical, the transmission from v_j to v_{j+1} is called *critical*. For $i = 1, \ldots, cn$, define the i-th stage of Scheme Odd-Even $(2cn)$ to be the sequence of its two consecutive steps $2i - 1, 2i$. During each stage, the critical transmission is performed at least once. If n critical transmissions succeed, broadcasting in L_n is completed. For any step of the scheme, consider the event that the critical transmission succeeds. Since these events for distinct steps are independent, and the probability of each event is q, we have $E \subseteq F$, where F is the event that the number of successes in a Bernoulli series of length cn with success probability q is less than n. Using Lemma 6.2.4 with $m = cn$ and $(1 - \epsilon)qc = 1$, we get

$$Prob(E) \leq Prob(F) \leq e^{-\frac{(qc-1)^2}{(qc)^2} mq/2} = e^{-(qc-1)^2 n/(2qc)} = e^{-dn}.$$

$\square$

For $c > 4/q$, we have $d > 1$, and hence Lemma 6.4.1 implies: $Prob(E) \leq e^{-n} < 1/n$, which means that for such values of c, Scheme Odd-Even $(2cn)$ performs almost safe broadcasting in the line L_n, in time $O(n)$.

It is interesting to compare Lemma 6.4.1 to Theorem 6.3.20. Consider any broadcasting scheme for the line L_n, working in linear time and performing $\Theta(n^2)$ transmissions. Theorem 6.3.20 implies that if some $\Theta(n)$ among these transmissions are faulty, and faults are distributed in a worst-case manner by the adversary, then this scheme cannot correctly accomplish broadcasting in the line. However, Lemma 6.4.1 shows that even $\Theta(n^2)$ transmission faults distributed randomly do not prevent broadcasting in linear time with high probability.

Lemma 6.4.1 can be easily generalized to get the following result:

Theorem 6.4.2. ([DP92]) *Let G be a bounded-degree graph of diameter D and let the source be an arbitrary node of G. Assume that all transmissions fail independently with constant probability $0 < p < 1$. Then there is an almost safe broadcasting scheme in G working in time $O(D)$.*

Permanent Faults

We now turn attention to almost safe broadcasting in the presence of permanent crash faults. The network is a complete graph. Assume that all links fail with constant probability $0 < p < 1$, and all nodes, except the source, fail with constant probability $0 < q < 1$. All faults are independent. The source is assumed fault free. Our aim is to show an almost safe broadcasting scheme working in logarithmic time [CDP94].

Although the underlying graph is complete, our broadcasting scheme will use very few links. We will construct an n-node network of logarithmic degree in which almost safe broadcasting is possible. Let $r_1 = (1 - p)(1 - q)$ and $r_2 = 1 - (1 - p)^2(1 - q)$. Let $c = \max(\lceil \frac{-8}{\log r_2} \rceil, \lceil \frac{24}{r_1 \log e} \rceil)$. For $n \geq 2c$ we define an n-node network $G(n, c)$. Let $d = c\lfloor \log n \rfloor$ and $s = \lfloor n/d \rfloor$. For clarity of presentation, assume that d divides n and $s = 2^{h+1} - 1$, for some integer $h \geq 0$. It is easy to modify the construction and the proof in the general case. Partition the set of all nodes into subsets $S_1, \ldots, S_s$ of size d, called *supernodes*. In every supernode S_i, enumerate nodes from 0 to $d - 1$. For any $i = 1, \ldots, s$ and any $j = 0, \ldots, d - 1$, assign label (i, j) to the j-th node in the i-th supernode. We assume that node $(1, 0)$ is the source of broadcasting. Arithmetic operations on the second integers forming labels are performed modulo d.

Arrange all supernodes into a complete binary tree T with $h + 1$ levels $0, 1, \ldots, h$. Level 0 contains the root and level h contains leaves of the tree T. The supernode S_1 is the root of T. For every $1 \leq i \leq \lfloor s/2 \rfloor$, S_{2i} is the left child of S_i and S_{2i+1} is the right child of S_i, in the tree T. For every $1 < i \leq s$, supernode $S_{\lfloor i/2 \rfloor}$ is the parent of S_i. If a supernode is a parent or a child of another supernode, we say that these supernodes are adjacent in T.

The set of edges of the network $G(n, c)$ is defined as follows. If supernodes S_i and S_j are adjacent in T then there is an edge in $G(n, c)$ between any node in S_i and any node in S_j. These are the only edges in $G(n, c)$. The graph $G(n, c)$ has the following properties.

Exercise 6.4.3. Confirm that

1. For every $1 \leq i \leq s$, $|S_i| = O(\log n)$;
2. $G(n, c)$ has maximum degree $O(\log n)$;
3. $h < \log n$.

We now describe a broadcasting scheme working in networks $G(n, c)$. It will use the following procedure whose aim is communication between adjacent supernodes. The procedure uses one time unit.

procedure MultiCall (S_i, S_j, k)
    for all $0 \leq r < d$ **in parallel do**
        (i, r) calls $(j, r + k)$
end.

The scheme consists of three identical stages. The aim of the first stage is to transmit the source message in such a way that at least one node in every supernode gets it with high probability. Such a node will be called a *leader* of the supernode. (A supernode may have many leaders.) In stages 2 and 3, leaders transmit the source message to other fault-free nodes in their supernode. In order to do that, a leader of a supernode S_i transmits the message to all nodes of an adjacent supernode in stage 2, and then these nodes retransmit the message to all nodes of S_i in stage 3. In each stage, the calls are scheduled so as to respect the telephone model constraints.

Scheme Tree-Structured-Broadcasting

> **for** *stage* := 1 **to** 3 **do**
>   **for** *step* := 0 **to** $d - 1$ **do**
>   **for each** S_i on an even level in T, less than h, **do**
>   MultiCall $(S_i, S_{2i}, step)$;
>   MultiCall $(S_i, S_{2i+1}, step)$;
>   **for each** S_i on an odd level in T, less than h, **do**
>   MultiCall $(S_i, S_{2i}, step)$;
>   MultiCall $(S_i, S_{2i+1}, step)$;
> **end.**

Exercise 6.4.4. Every node of $G(n, c)$ is involved in at most one call per time unit in the execution of Scheme Tree-Structured-Broadcasting.

Exercise 6.4.5. Scheme Tree-Structured-Broadcasting works in time $O(\log n)$.

In order to prove that Scheme Tree-Structured-Broadcasting is almost safe, we will need the following lemmas.

Lemma 6.4.6. *Let E_1 denote the following event:*

> *upon completion of stage 1 of Scheme Tree-Structured-Broadcasting, every supernode has a leader.*

Then $Prob(\overline{E_1}) \le 1/n^2$.

Proof. The event $\overline{E_1}$ implies that there exists a branch of the tree T, such that some supernode of this branch does not have a leader after the first stage of the scheme. Fix such a branch $B = (S_{i_0}, S_{i_1}, \ldots, S_{i_h})$, where $S_{i_0} = S_1$, and let P be the event that some supernode of B does not have a leader after the first stage of the scheme. Every fault-free node in supernode S_{i_j} calls different nodes in supernode $S_{i_{j+1}}$ in d consecutive steps of the first stage. These attempts are independent and they have success probability r_1 (both the target node and the joining link must be fault free). Upon a successful call from a leader of S_{i_j}, some node of $S_{i_{j+1}}$ becomes a leader. Hence $Prob(P)$ does not exceed the probability of at most h successes in a Bernoulli series of length d with success probability r_1. Since $h < \log n$, $Prob(P)$ does not exceed the probability of at most $\log n$ successes in such a series.

Take $\epsilon = \frac{r_1 c - 1}{r_1 c}$ in Lemma 6.2.4. Since $c > 1/r_1$, we have $0 < \epsilon < 1$, and $(1 - \epsilon) r_1 d = \lfloor \log n \rfloor$. Hence Lemma 6.2.4 implies $Prob(P) \le e^{-\epsilon^2 r_1 c \lfloor \log n \rfloor / 2}$. Since there are fewer than n branches in the tree T, we get (for $n \ge 2$):

$$Prob(\overline{E_1}) \le n \cdot Prob(P) \le n \cdot e^{-\epsilon^2 r_1 c \log n / 4}$$

$$= n \cdot n^{-\epsilon^2 r_1 c \log e / 4} = n^{1 - (r_1 c - 2 + \frac{1}{r_1 c}) \log e / 4}$$

$$\le n^{1 - r_1 c \log e / 8},$$

because $c \ge \lceil \frac{24}{r_1 \log e} \rceil \ge \lceil 4/r_1 \rceil$ implies

$$r_1 c - 2 + \frac{1}{r_1 c} \ge \frac{r_1 c}{2}.$$

Since $r_1 c \log e / 8 \ge 3$, we finally get $Prob(\overline{E_1}) \le 1/n^2$.

$\square$

Lemma 6.4.7. *Let E_2 be the following event:*

between every pair of nodes in the same supernode, there exists a path of length 2 whose two links and the intermediate node are fault free.

Then $Prob(\overline{E_2}) \le 1/n^2$, for sufficiently large n.

Proof. Every supernode contains d nodes, and for sufficiently large n, $n/d \ge 2$, hence there are at least two supernodes. Between every pair of nodes in a supernode there exist at least d disjoint paths of length 2. The probability that in a single path $u - w - v$ the intermediate node or one of the links is faulty is r_2. Fix two fault-free nodes u, v in a supernode, and d disjoint paths of length 2 between them. The events that these paths contain a faulty component, are independent. Hence the probability that each of them does, is r_2^d. Since there are fewer than n^2 pairs of nodes in the network, we get $Prob(\overline{E_2}) \le n^2 r_2^d \le n^2 r_2^{c \log n / 2}$, for $n \ge 2$, and since $c \log r_2 \le -8$, we obtain

$$Prob(\overline{E_2}) \le n^2 \cdot n^{c \log r_2 / 2} \le 1/n^2.$$

$\square$

We are now ready to prove:

Theorem 6.4.8. ([CDP94]) *Assume that all links fail with constant probability $0 < p < 1$, all nodes, except the source, fail with constant probability $0 < q < 1$, and all faults are independent. Scheme Tree-Structured-Broadcasting performs almost safe broadcasting in the complete graph, in logarithmic time.*

Proof. In view of Exercise 6.4.5, it is enough to prove that Scheme Tree-Structured-Broadcasting is almost safe. Let E denote the event that the scheme broadcasts correctly. Then $E_1 \cap E_2 \subseteq E$. Indeed, in view of E_1, every supernode has a leader. In stage 2 of the scheme, a leader u of supernode S_i transmits the source message to all its fault-free neighbors, provided that the joining links are fault free. In stage 3, these neighbors transmit the message to every fault-free node v in S_i, provided that respective joining links are fault free. By event E_2, there is a path of length 2 between u and v, without faulty components, hence v obtains the source message upon completion of stage 3.

Hence $\overline{E} \subseteq \overline{E_1} \cup \overline{E_2}$ and we have

$$Prob(\overline{E}) \le Prob(\overline{E_1}) + Prob(\overline{E_2}) \le 2/n^2 \le 1/n,$$

for sufficiently large n. This implies that Scheme Tree-Structured-Broadcasting is almost safe.

$\square$

6.4.2 Byzantine Faults

In this section we assume that all nodes are fault free and all links are subject to independent faults occurring with probability $0 < q < 1/2$. Faults are permanent and Byzantine. Our aim is to present the result from [BDP97]: an almost-safe broadcasting scheme working in time $O(\log n)$ in the complete n-node graph. As always, we use the term "scheme" to denote a non-adaptive broadcasting algorithm. However, a fine point should be noticed here. Our algorithm is non-adaptive in the sense that all transmissions are *scheduled* in advanced, that is, it is known *a priori* which pairs of nodes will communicate in which time step. However, messages sent by each node are computed by it during the algorithm execution, and depend on what messages this node got in the previous steps. In schemes seen so far, the situation was much simpler: a node scheduled to transmit in a given time step transmitted the source message, if it had received it previously, and transmitted nothing otherwise. This particularity is due to Byzantine faults: because of them different (corrupted) versions of the source message circulate in the network.

Since our algorithm is non-adaptive, at each time step each node knows on which link it should expect a message and along what path this message should have travelled. This enables ignoring messages coming on faulty links at unexpected times, due to their Byzantine behavior. On the other hand, if at time t a node w expects a message coming from v via the path p, any message coming at this time on the final link of p is stored in w as the (possibly distorted) message coming from v via p. If no message comes on this link at the expected time, a default message is stored. For simplicity of presentation we assume that the source message consists of one bit. Our considerations can be easily generalized for arbitrary constant-size messages.

In the design of the algorithm we will use the notion of *expander*. A bipartite graph $G = (A, B, E)$ (with bipartition sets A, B and the set E of edges) is called a (α, β, n, d)-expander, if both sets A and B have size n, the degree of each node is d, and for every set $X \subset A$ of size at most αn, there are at least $\beta|X|$ neighbors of elements of X.

The following fact can be proven using a simple counting argument.

Exercise 6.4.9. For every integer $d > 1$ and for every $\beta < d - 1$, there exists $1 \geq \alpha > 0$ such that for all $n \geq d$ there exists a (α, β, n, d)-expander.

We start with a high-level presentation of the main idea of the scheme. Let $c > 2$ be a constant, and let A be the set of all nodes different from the source s. Let $m = \lceil c \log(n - 1) \rceil$. For simplicity of presentation assume that m is an even integer dividing $n - 1$ and that $k = (n - 1)/m = 2^{h+1} - 1$, for some integer $h \geq 0$. The design of the scheme and the arguments are easy to modify in the general case.

We define the following graph on all n nodes. Partition A into k subsets of size m, called supernodes, and arrange them into a complete binary tree, similarly as in Section 6.4.1. The supernode corresponding to the root of this tree is called ROOT, and supernodes which are children of supernode X are called $lc(X)$ and $rc(X)$, respectively. ROOT is at level 0 of the tree and leaves are at level h. As before, all nodes in a supernode other than ROOT are adjacent to all nodes of its parent. Additionally, all nodes of ROOT are adjacent to the source s, and all nodes of each supernode are adjacent to each other. The obtained graph has degree $O(\log n)$. Our scheme will work not only in the complete graph but also in the above sparse graph.

For $0 < \gamma \leq 1$, we say that the source message is γ-*dominating* in a supernode at some time of the scheme execution, if at least γm elements of the supernode have correctly computed the source message at this time.

The scheme works in three stages.

Stage 1. The source informs all nodes in ROOT. The aim of this stage is to make the source message $(1 - \lambda)$-dominating in ROOT, with probability at least $1 - 2^{-m}$, for some constant λ to be determined later.

Stage 2. The source message is transmitted along branches of T. The aim of this stage is to make the source message $(1 - \lambda)$-dominating in each supernode. In consecutive steps the source message is transmitted from a supernode to each of its children along edges of an appropriate bipartite expander. We guarantee the invariant that if the source message was $(1 - \lambda)$-dominating in a supernode then it becomes $(1 - \lambda)$-dominating in the child, with probability at least $1 - 2^{-m}$.

Stage 3. The source message is spread in parallel in all supernodes. If it was $(1 - \lambda)$-dominating in a supernode at the end of Stage 2 then it becomes 1-dominating in it, with probability at least $1 - 2^{-m}$, at the end of Stage 3. That is, all nodes in this supernode have correctly computed the source message, with this probability.

We now proceed with a formal description of the scheme. For every node u, X_u denotes the content of register X in this node. In the beginning, the source message is held in register M of the source, i.e., it is M_s. For every node u, the bit that it computes during the scheme execution, and considers at a given stage to be the source message, is stored in its register M. Each node changes the content of this register at most once during the scheme execution. Every node x outputs the final value of M_x as its version of the source message. We will use the elementary procedure $\text{SEND}(u, X, v, Y)$, for adjacent nodes u, v, which consists of the following action: u sends X_u to v, and v assigns the received value to its register Y. If the link joining u with v is fault free, the effect of $\text{SEND}(u, X, v, Y)$ is the same as the assignment $Y_v := X_u$. Otherwise, an arbitrary value is assigned to Y_v.

Our scheme uses the following three subroutines, corresponding to the three stages, informally described above. The first of them, executing Stage 1, is very simple: the source sends the message sequentially to all nodes in ROOT, in time $O(m)$.

procedure SOURCE-TO-ROOT
    for all $v \in$ ROOT **do**
            $\text{SEND}(s, M, v, M)$
    end.

The aim of the second subroutine, used in Stage 2, is transmitting the source message between two disjoint sets of equal size m, using an appropriate expander, to guarantee the invariant mentioned in the description of Stage 2. (In our scheme, the procedure will be used to transmit the source message from a supernode to its child.) We assume that constants α and d are such that there exists a $(\alpha, d/2 + 2, m, d)$-expander (see Exercise 6.4.9). Fix such a bipartite expander $G = (Y, X, E)$ and fix a partition $L_1, \ldots, L_d$ of all its edges into pairwise disjoint perfect matchings. The following subroutine transmits the source message from set X to set Y via the above matchings, in time d, upon which each node computes the majority of all received values. For any node $u \in X$, let $L_i(u)$ denote the node in Y adjacent to u in matching L_i.

procedure SUPERNODE-TO-SUPERNODE (X, Y)
    for $i := 1$ **to** d **do**
        for all $u \in X$ **in parallel do**
            $\text{SEND}(u, M, L_i(u), V[u])$
        for all $v \in Y$ **in parallel do**
$$M_v := \begin{cases} 0 \text{ if } |\{u \in \Gamma_G(v) : V_v[u] = 0\}| > d/2 \\ 1 \text{ otherwise} \end{cases}$$
end.

The third subroutine is used to spread the correct version of the source message in each supernode, by exchanging messages (stored in register M of each node) among all nodes of the supernode, in time $m - 1$, and subsequently computing the majority of all received values at each node. This value is stored in register M of each node, the node considers it as a correct version of the source message and outputs it at the end of the scheme. Let $F_1, \ldots, F_{m-1}$ be a fixed partition of all edges of the complete graph on an m-element set X into $m - 1$ pairwise disjoint perfect matchings (recall that we assume that m is even).

procedure SPREAD (X)
    for $i := 1$ **to** $m - 1$ **do**
        for all $\{x, y\} \in F_i$ **in parallel do**
            $\text{SEND}(x, M, y, W[x]); \text{SEND}(y, M, x, W[y])$
        for all $x \in X$ **in parallel do**
$$M_x := \begin{cases} 0 \text{ if } |\{y \in X \setminus \{x\} : W_x[y] = 0\}| > (m-1)/2 \\ 1 \text{ otherwise} \end{cases}$$
end.

Using the above subroutines, the entire scheme can be now succinctly formulated as follows.

Scheme Byzantine-Broadcasting
    SOURCE-TO-ROOT
    for $i := 0$ **to** $h - 1$ **do**
        for all supernodes X on level i **in parallel do**
            SUPERNODE-TO-SUPERNODE $(X, lc(X))$
            SUPERNODE-TO-SUPERNODE $(X, rc(X))$
        for all supernodes X **in parallel do**
            SPREAD (X)
end.

Exercise 6.4.10. The running time of Scheme Byzantine-Broadcasting is $O(\log n)$.

We now prove that if link failure probability q is smaller than some constant $q_0 < 1/2$ (to be defined later) then Scheme Byzantine-Broadcasting is almost safe. At the end of the section we will show how to modify the scheme and refine the analysis, in order to remove this additional assumption.

Take any even integer $d > 6$. Then $d/2 + 2 < d - 1$ and hence, by Exercise 6.4.9, there exists an $\alpha > 0$, such that a $(\alpha, d/2 + 2, m, d)$-expander exists for all $m \geq d$. Consider n sufficiently large to guarantee $m \geq d$. Let λ be a positive constant less than $\alpha/3$.

Lemma 6.4.11. *There exists a positive constant $q_1 < 1/2$ such that, if link failure probability q is less than q_1, then the source message becomes $(1 - \lambda)$-dominating in ROOT, with probability at least $1 - 2^{-m}$, for sufficiently large m, upon completion of procedure SOURCE-TO-ROOT.*

Proof. If a node in ROOT gets an incorrect source message, the link joining it to the source must be faulty. Exercise 6.2.7 and Exercise 6.2.5 imply that the probability that at least λm nodes in ROOT get the incorrect bit can be estimated as follows, for sufficiently large m:

$$\sum_{i = \lceil \lambda m \rceil}^{m} \binom{m}{i} q^i (1-q)^{m-i} \leq \binom{m}{\lceil \lambda m \rceil} \cdot q^{\lceil \lambda m \rceil} \leq ((e/\lambda)^{2\lambda} \cdot q^\lambda)^m,$$

which is less than 2^{-m} for $q < q_1 = (\lambda/e)^2/2^{1/\lambda}$.

$\square$

Lemma 6.4.12. *There exists a positive constant $q_2 < 1/2$ such that, if link failure probability q is less than q_2, and the source message was $(1 - \lambda)$-dominating in supernode X before the execution of procedure SUPERNODE-TO-SUPERNODE (X,Y), then, for sufficiently large m, it becomes $(1 - \lambda)$-dominating in supernode Y, with probability at least $1 - 2^{-m}$, upon completion of procedure SUPERNODE-TO-SUPERNODE (X,Y).*

Proof. During the execution of procedure SUPERNODE-TO-SUPERNODE (X,Y), dm operations SEND are performed. Let $\mu = \lambda/(2d)$. Similarly as in the proof of Lemma 6.4.11, the probability that at least μdm of these operations involve faulty links can be estimated as follows, for sufficiently large m:

$$\sum_{i=\lceil \mu dm \rceil}^{dm} \binom{dm}{i} q^i(1-q)^{dm-i} \leq \binom{m}{\lceil \mu dm \rceil} \cdot q^{\lceil \mu dm \rceil} \leq ((e/\mu)^{2\mu d} \cdot q^{d\mu})^m,$$

which is less than 2^{-m} for $q < q_2 = (\mu/e)^2/2^{1/(\mu d)} = (\lambda/(2de))^2/4^{1/\lambda}$.

Let E be the event that less than μdm of these operations involve faulty links. Denote by Z the set of those nodes in Y which assigned the incorrect bit to their register M during the execution of procedure SUPERNODE-TO-SUPERNODE (X,Y). Each of these nodes could get an incorrect bit either from a node in X which had such a bit before the execution of the procedure, or through a faulty link. Moreover, in order to assign a wrong bit to its register M, a node in Z must have gotten at least $d/2$ incorrect messages. Since the source message was $(1 - \lambda)$-dominating in supernode X, it follows that if the event E holds, the size of Z can be estimated as follows:

$$|Z| \leq \frac{\lambda dm + \mu dm}{d/2} = m(2\lambda + 2\mu) < 3\lambda m < \alpha m.$$

Suppose that $|Z| \geq \lambda m$. Let $|Z| = \delta m$, for $\delta \geq \lambda$. Every node $z \in Z$ gets the correct bit through at most $d/2$ links. Hence the total number of links through which the correct bit has been received in Z is at most $d\delta m/2$. On the other hand, $|\Gamma_G(Z)| \geq (d/2 + 2)\delta m$, by the definition of the expander. At most λm nodes from $\Gamma_G(Z)$ held the incorrect bit in their register M. The other nodes sent the correct bit to nodes in Z, each through at least one link. Since we assumed that event E holds, at most μdm of these links are faulty. Hence the total number of links through which the correct bit was received in Z is at least

$$(d/2 + 2)\delta m - \lambda m - \mu dm = d\delta m/2 + (\delta - \lambda)m + \delta m - \mu dm > d\delta m/2,$$

because $\delta \geq \lambda$ and $\mu = \lambda/(2d) < \lambda/d$.

This contradiction implies that if the event E holds then $|Z| < \lambda m$. Hence the source message becomes $(1-\lambda)$-dominating in supernode Y, upon completion of procedure SUPERNODE-TO-SUPERNODE (X, Y). Since $Prob(E) \geq 1 - 2^{-m}$, this concludes the proof of the lemma.

$\square$

Lemma 6.4.13. *There exists a positive constant $q_3 < 1/2$ such that, if link failure probability q is less than q_3, and the source message was $(1 - \lambda)$-dominating in supernode X, before the execution of procedure SPREAD(X), then, for every node in X, the probability that it assigns the correct bit to its register M upon completion of procedure SPREAD(X) is at least $1 - 2^{-m}$, for sufficiently large m.*

Proof. During the execution of procedure SPREAD(X), every node $x \in X$ gets messages from $m-1$ other nodes. At most λm of them send the incorrect bit. If less than $(m-1)/2 - \lambda m$ links incident to node x are faulty, then x receives the correct bit on the majority of links, and consequently assigns the correct bit to its register M. It remains to show that the probability that at least $(m-1)/2 - \lambda m$ links incident to node x are faulty is less than 2^{-m}.

For sufficiently large m, we have $m - \lambda m \geq \lceil (m-1)/2 - \lambda m \rceil \geq m/3 - \lambda m$, and hence, in view of Exercise 6.2.7 and Exercise 6.2.5, the above probability can be estimated as follows:

$$\sum_{i=\lceil (m-1)/2-\lambda m \rceil}^{m-1} \binom{m-1}{i} q^i (1-q)^{m-1-i}$$

$$\leq \binom{m}{\lceil (m-1)/2 - \lambda m \rceil} \cdot q^{\lceil (m-1)/2-\lambda m \rceil}$$

$$\leq \left(\frac{em}{\lceil (m-1)/2 - \lambda m \rceil} \right)^{\lceil (m-1)/2-\lambda m \rceil} \cdot q^{\lceil (m-1)/2-\lambda m \rceil}$$

$$\leq \left(\left(\frac{e}{1/3 - \lambda} \right)^{1-\lambda} \cdot q^{1/3-\lambda} \right)^m,$$

which is less than 2^{-m}, for

$$q < q_3 = \frac{1}{2^{3/(1-3\lambda)}} \cdot \left(\frac{1-3\lambda}{3e} \right)^{(3-3\lambda)/(1-3\lambda)}.$$

$\square$

Using the above lemmas we can now prove the following result.

Theorem 6.4.14. *There exists a positive constant $q_0 < 1/2$ such that, if link failure probability q is less than q_0, then Scheme Byzantine-Broadcasting is almost safe.*

Proof. Let $q_0 = \min(q_1, q_2, q_3)$. For sufficiently large n, the integer $m = \lceil c \log(n-1) \rceil$ is large enough to validate Lemmas 6.4.11, 6.4.12, and 6.4.13. By Lemma 6.4.11, the probability that upon completion of procedure SOURCE-TO-ROOT, the source message is not $(1 - \lambda)$-dominating in ROOT does not exceed 2^{-m}. By Lemma 6.4.12, the probability that upon completion of all SUPERNODE-TO-SUPERNODE procedures called by the algorithm the source message is not $(1 - \lambda)$-dominating in some supernode does not exceed $2^{-m} + (n-1)2^{-m}$. Finally, by Lemma 6.4.13, the probability that upon completion of all SPREAD procedures called by the algorithm some node does not assign the correct bit to its register M does not exceed

$$2^{-m} + (n-1)2^{-m} + n2^{-m} = 2n \cdot 2^{-\lceil c \log(n-1) \rceil},$$

which is less than $1/n$, for sufficiently large n, because $c > 2$.

$\square$

Recall that our goal is to construct an almost safe broadcasting scheme working in logarithmic time, for any link failure probability $q < 1/2$. In view of Theorem 6.4.14, so far we have achieved this goal only partially: Scheme Byzantine-Broadcasting satisfies this requirement under the extra assumption that link failure probability is small. In the sequel, we will show how to modify the scheme in order to remove this assumption. The modification uses a new procedure TRANSMIT with the following property. For any $r < 1$ and any link failure probability $q < 1/2$, the procedure enables us to transmit a message from node x to node y with probability of correctness r. The procedure works in time t and uses a set A of g intermediary nodes, where t and g depend only on q and r. We take $r = 1 - q_0$, where q_0 is the constant defined in the proof of Theorem 6.4.14. Since $q < 1/2$ is also constant, the new procedure works in constant time and uses a constant number of intermediary nodes.

Procedure TRANSMIT(x, H, y, G, A) has the following parameters: x is a node transmitting a bit which is held in its register H, and y is the node which has to receive this bit. It computes the version of H_x which it considers to be correct, and stores it in its register G. Nodes in the set A, such that $x, y \notin A$, serve as intermediaries in this transmission.

Using procedure TRANSMIT, which will be described below, Scheme Byzantine-Broadcasting can be modified as follows. Replace each instance of procedure SEND(x, H, y, G) in all subroutines of the scheme by procedure TRANSMIT(x, H, y, G, A), for some set A of nodes of constant size g to be determined later. By the property of this procedure, to be proved in Lemma 6.4.15, the probability that G_y becomes H_x upon completion of TRANSMIT(x, H, y, G, A) (i.e., that node y gets the bit sent by node x) is greater than r. Since $r = 1 - q_0$, the execution of procedure TRANSMIT(x, H, y, G, A) has the same effect as the execution of procedure SEND(x, H, y, G) with link failure probability smaller than q_0, and our previous analysis applies. Another modification is necessary to guarantee that the modified scheme satisfies the requirements of the telephone model. (There is a potential danger that procedures TRANSMIT, scheduled to work in parallel, use the same nodes as intermediaries.) To do that, consider any series of procedures TRANSMIT, scheduled to work in parallel. It replaces a corresponding series of procedures SEND, working in parallel, hence it corresponds to a matching in the underlying graph. Consequently, such a parallel series involves at most $n/2$ calls of TRANSMIT. Partition this series into g subseries, each of size at most $n/(2g)$, and assign pairwise disjoint sets of intermediary nodes, of size g, to each procedure in the subseries. We will use the name Scheme Modified-Byzantine-Broadcasting to denote the scheme resulting from these two modifications. Since TRANSMIT will be defined in a way satisfying the requirements of the telephone model, the new scheme satisfies these requirements as well. Since t and g are constant, its running time is still $O(\log n)$. Using Theorem 6.4.14, Scheme Modified-Byzantine-Broadcasting is almost safe.

It remains to specify the procedure TRANSMIT and compute its probability of correctness. Let $\epsilon = 1/2 - q$, and $x, y \notin A$.

procedure TRANSMIT(x, H, y, G, A)
    for $a \in A$ **do**
        SEND(x, H, a, W)
        SEND$(a, W, y, X[a])$
    for $a \in A$ **do**
        for $b \in A$ **do**
            SEND$(a, W, b, Y[a])$
            SEND$(b, Y[a], y, Z[a, b])$
    $g := |A|$

$$G_y := \begin{cases} 0 & \text{if there exist sets } Q, R \subset A \text{ of size } |Q| = |R| = \lceil g(1+\epsilon)/2 \rceil \\ & \text{such that, for all sets } Q^* \subset Q, R^* \subset R \text{ of size } |Q^*| = |R^*| = \\ & \lceil \epsilon g \rceil, \text{ either } Q^* \cap R^* \neq \emptyset \text{ and, for all } a \in Q^* \cap R^*, X[a] = 0, \\ & \text{or } Q^* \cap R^* = \emptyset \text{ and } |\{(a,b) \in Q^* \times R^* : Z[a,b] = 0\}| > \\ & \lceil \epsilon g \rceil^2/2 \\ 1 & \text{otherwise.} \end{cases}$$

end.

The following lemma shows that procedure TRANSMIT achieves correct transmission from node x to node y, with arbitrary probability $r < 1$, for a sufficiently large size g of the set A of intermediary nodes. Intuitively, Q is the set of nodes in A that receive the correct bit from x, and R is the set of nodes in A that send the correct bit to y. If Q and R are chosen wrongly, this should be revealed by correct subsets $Q^* \subset Q$ and $R^* \subset R$. As the size of A increases, so does the probability that these sets have the sizes required in the condition computed by y.

Lemma 6.4.15. *Let A be a set of size g and let x, y be distinct nodes outside of A. For every link failure probability $q < 1/2$ and every $r < 1$, the procedure TRANSMIT(x, H, y, G, A), working for $\epsilon = 1/2 - q$, satisfies $Prob(G_y = H_x) > r$, for sufficiently large g.*

Proof. Consider the following events.

E_z (for $z \in \{x, y\}$) – the event that at least $\lceil g(1 + \epsilon)/2 \rceil$ nodes in A are joined with node z by a fault-free link;

F_{Q^*, R^*} (for disjoint sets of nodes Q^*, R^* of size $\lceil \epsilon g \rceil$) – the event that the number of fault-free links joining nodes $v \in Q^*$ with nodes $w \in R^*$ exceeds $\lceil \epsilon g \rceil^2/2$.

By Lemma 6.2.4 with $\delta = \epsilon/(1 + 2\epsilon)$, we have

$$Prob(E_z) \geq 1 - e^{-\delta^2(1/2+\epsilon)g/2} = 1 - e^{-g\epsilon^2/(4+8\epsilon)},$$

and by the same lemma with $\delta = 2\epsilon/(1 + 2\epsilon)$ we have

$$Prob(F_{Q^*, R^*}) \geq 1 - e^{-\delta^2(1/2+\epsilon)\lceil \epsilon g \rceil^2/2} \geq 1 - e^{-g^2\epsilon^4/(1+2\epsilon)}.$$

Let $E = E_x \cap E_y$ and $F = \bigcap \{F_{Q^*, R^*} : Q^* \cap R^* = \emptyset, |Q^*| = |R^*| = \lceil \epsilon g \rceil\}$.
Hence

$$Prob(E) \geq 1 - 2e^{-g\epsilon^2/(4+8\epsilon)}$$

and

$$Prob(F) \geq 1 - \binom{g}{\lceil \epsilon g \rceil} \binom{g}{\lceil \epsilon g \rceil} e^{-g^2\epsilon^4/(1+2\epsilon)}.$$

Both $Prob(E)$ and $Prob(F)$ converge to 1 as g grows, hence, for sufficiently large g, we have $Prob(E \cap F) > r$. Now the lemma follows from the following:

Claim. If $E \cap F$ holds then $G_y = H_x$.

Proof. First suppose that $H_x = 0$. Let Q be a set of size $\lceil g(1+\epsilon)/2 \rceil$ of nodes in A joined by fault-free links with x, and let R be a set of size $\lceil g(1 + \epsilon)/2 \rceil$ of nodes in A joined by fault-free links with y. Such sets exist in view of E. Let $Q^* \subset Q$ and $R^* \subset R$ be any sets of size $\lceil \epsilon g \rceil$. If $Q^* \cap R^* \neq \emptyset$ then, for all $a \in Q^* \cap R^*$, links x—a and a—y are fault free, which implies $X[a] = 0$. If $Q^* \cap R^* = \emptyset$, event F implies that the majority of links a—b such that $a \in Q^*$ and $b \in R^*$, are fault free. Consequently, for the majority of pairs $(a, b) \in Q^* \times R^*$, all links x—a, a—b and b—y are fault free, and thus $Z[a, b] = 0$. According to the definition of G_y, this implies $G_y = 0$.

Next, suppose that $H_x = 1$. Take any pair of sets $Q, R \subset A$ of size $|Q| = |R| = \lceil g(1 + \epsilon)/2 \rceil$. Define Q' to be a set of size $\lceil g(1 + \epsilon)/2 \rceil$ of nodes in A joined by fault-free links with x, and define R' to be a set of size $\lceil g(1+\epsilon)/2 \rceil$ of nodes in A joined by fault-free links with y. The existence of Q' and R' is again guaranteed by event E. Thus $|Q \cap Q'| \geq \lceil \epsilon g \rceil$ and $|R \cap R'| \geq \lceil \epsilon g \rceil$. Choose $Q^* \subset Q \cap Q'$ and $R^* \subset R \cap R'$, such that $|Q^*| = |R^*| = \lceil \epsilon g \rceil$. If $Q^* \cap R^* \neq \emptyset$ and $a \in Q^* \cap R^*$ then links x—a and a—y are fault free; hence $X[a] = 1$. If $Q^* \cap R^* = \emptyset$, event F implies that the majority of links a—b, such that $a \in Q^*$ and $b \in R^*$, are fault free. Consequently, for the majority of pairs $(a, b) \in Q^* \times R^*$, all links x—a, a—b and b—y are fault free, and thus

$Z[a, b] = 1$. It follows that the criterion for $G_y = 0$ is not satisfied, and hence $G_y = 1$.

This proves that $G_y = H_x$, in both cases.

$\square$

Hence the claim is true, which completes the proof of Lemma 6.4.15.

$\square$

As discussed previously, Lemma 6.4.15 implies the following result:

Theorem 6.4.16. ([BDP97]) *For every link failure probability $q < 1/2$, Scheme Modified-Byzantine-Broadcasting is an almost safe broadcasting scheme working in time $O(\log n)$ in the complete n-node graph.*

Part II

Distributed Networks

7

Broadcast on Distributed Networks

It is better to be worn out
than to become like a rusty piece of iron.

Denis Diderot

7.1 Broadcasting on General Networks

This section presents distributed algorithms for two well-known basic graph traversal techniques, namely depth-first search and breadth-first search, and a basic lower bound on the message complexity of broadcast.

First we present a time-optimal distributed depth-first search algorithm, using $4m$ messages in the worst case on networks with m links. Then we present a distributed breadth-first search algorithm that operates within a factor $O(2^{\sqrt{\log N \log \log N}})$ on the lower bound for both time and communication, on networks with N nodes. Finally, we give a $\Omega(m)$ lower bound on the message complexity of broadcast on networks with m links.

7.1.1 Distributed Depth-First Search

Consider a distributed computing system consisting of a number of autonomous processors interconnected through communication links. The processors do not share a common memory, keep only local information and hence need to communicate frequently in order to coordinate any computation that has to be accomplished. The interconnection network can be modeled by an undirected communication graph $G = (V, E)$, where nodes correspond to the processors and the edges to bidirectional communication links. The processors have distinct identities, but each processor knows only the identities of its neighbors. Each processor performs a variety of local tasks, besides receiving messages from its neighbors, performing some local computation and sending messages to its neighbors. The exchange of messages between two neighboring processors is asynchronous in that the sender always hands over the message to the communication subsystem and proceeds with its own local task. The communication subsystem, we assume, will deliver the message at its destination, without loss or any alternation, after a finite but undetermined time lapse. The messages sent over any link also follow a FIFO rule. The messages received at any processor are stamped with the identity of the sender and transferred to a common queue before being processed one by one. Messages arriving at a node simultaneously from several neighbors may be placed in any arbitrary order in the queue. Since several computations may be in progress concurrently, we assume that the network has suitable mechanisms so that, at the receiving end, messages corresponding to any particular computation initiated by a particular node can be distinguished and separated out.

A distributed algorithm consists of the collection of identical node algorithms residing at the processors. These node algorithms specify the actions to be taken in response to the messages that may arrive at a node. It is assumed that the actions necessary for processing a message can all be performed in negligible computation time, uninterrupted by the arrival of other messages. Hence, the complexity measures used to evaluate the performance of distributed algorithms only relate to the communication aspect. The *message complexity* is the total number of messages transmitted during the execution of the algorithm. The *time complexity* is the time that elapses from the beginning until the termination of the algorithm, assuming that the delay in any link is exactly one unit of time. It must be recognized that this assumption of unit time delay in communication links is made only for the purpose of analysis, and the algorithm is expected to operate correctly under the previous assumption that the delay is finite, but cannot be bounded. Also, given a communication graph with $|V|$ nodes and $|E|$ edges, the actual performance of any distributed algorithm, in terms of its message and time complexities, will depend upon the structure of the graph, the degree and other characteristics of the node initiating the algorithm, the delays encountered in the links, etc., and hence we use only the worst-case analysis in comparing two algorithms, as well as in discussing the optimality of any algorithm.

Given an asynchronous communication network and a starting node s, the task is to construct a depth-first search (DFS shortly) tree, rooted at s, for a communication graph $G = (V, E)$. All messages are required to be of fixed length independent of the size of the graph. At the end of the computation, the DFS tree will be available in a distributed fashion, each node except the root knowing its parent in the tree. We assume that the algorithm is initiated by source node s and that no processor or link failure takes place during the entire execution.

Clearly, a DFS algorithm partitions the edges of an undirected graph into tree edges and back edges. This requires an exploration of the graph with the *center of activity* moving from one node to another in a systematic way. Initially, the start node s is the center of activity. When a node becomes the center of activity for the first time, it marks itself as *visited*. Also, whenever a node becomes the center of activity, it tries to identify a neighbor who is not yet visited and transfers the center of activity to that node. But if no such neighbor exists, i.e., if the node is completely scanned, then it shifts the center of activity to the parent node in the tree, or simply terminates if the node happens to be the start node itself.

If there are N nodes in the graph, then it is clear that there must be $2N - 2$ shifts of the center of activity. It is also easily seen that graphs for which the DFS tree constructed has a linear chain of N nodes are obvious cases requiring all shifts of the center of activity to proceed sequentially, no matter what algorithm is employed. Since each shift of the center of activity has to be accomplished by a transmission of a message, any distributed algorithm for DFS should have a worst-case time complexity of at least $2N - 2$.

We have also seen that a node can shift the center of activity to the parent node or terminate the algorithm only if it has ensured that each one of the neighbors has already been visited. If all the messages are required to be of fixed length, and cannot contain the number of nodes in the graph, the node identities, etc., then each node requires at least one message to arrive from each of its neighbors before it recognizes that it has been completely scanned. Thus, any distributed algorithm for DFS should have a message complexity of at least $2|E|$.

If we consider a straightforward distributed algorithm for finding a DFS tree, simulating a sequential DFS algorithm, its message and time complexities are $2|E|$. Though this algorithm is message optimal, it is time inefficient, as the messages are all transmitted one after another in sequence. By a slight increase of message complexity, one can achieve an algorithm for constructing DFS tree in $O(N)$ time. The improved algorithm requires four kinds of messages – DISCOVER, RETURN, VISITED, and ACK. DISCOVER messages are used to shift the center of activity from a visited node to an unvisited one. RETURN messages are used to shift the center of activity from a node to its parent in the tree. VISITED messages are used by a node to inform all its neighbors, except the parent, that it has been visited. ACK messages are sent in response to VISITED messages. In fact, when a node becomes the center of activity, VISITED messages are sent out to all neighbors, except the parent, and only after ACK messages have been received from all these neighbors is the center of activity shifted to another node. The basic idea is that by the time the center of activity is shifted to any node, every node knows exactly which of its neighbors have been visited. This ensures that DISCOVER messages are never sent to an already visited node. Thus the algorithm requires $N - 1$ DISCOVER, $N - 1$ RETURN, $2|E| - (N - 1)$ VISITED, and $2|E| - (N - 1)$ ACK messages, all adding up to $4|E|$. More importantly, the VISITED and ACK messages add two units of time at each node of degree greater than one to the time complexity. The DISCOVER and RETURN messages need $2N - 2$ time units. As a result, the algorithm has a time complexity of $4N - 2 - 2N_1$, where N_1 is the number of nodes of degree one.

The time complexity of the above algorithm can be further improved and made time optimal by simply eliminating the ACK messages and ensuring that VISITED messages are always transmitted in communication-time parallel to DISCOVER or RETURN messages. But this creates a new problem – more than one DISCOVER message could be sent to a node.

In the modified algorithm, a node marks itself as visited when it receives a DISCOVER message for the first time. It also tries to identify a neighbor which has not yet been visited and to send a DISCOVER message to it, at the same time informing, through VISITED messages, its status to all other neighbors except, of course, the parent. Since VISITED messages could suffer long delays in the communication links, a node trying to identify an unvisited neighbor may not have the correct and complete information regarding the status of all of its neighbors. It may, therefore, choose an already visited neighbor and send a DISCOVER message to it, simply because it has not received any message from that neighbor at that stage. But, luckily, it is possible to recover from such a mistake because a DISCOVER or VISITED message sent by that neighbor will eventually arrive at its intended destination. In order to enable this kind of recovery, the sender of a DISCOVER message always records the identity of the neighbor to whom such a message is sent, so that if a DISCOVER or VISITED message is received from that neighbor, an alternative neighbor, if any, can be found to shift the center of activity. This strategy also means that a DISCOVER message received at an already visited node can simply be ignored, except for the purpose of recognizing that the sender has also been visited already.

The Formal Presentation of the Algorithm

The algorithm uses messages and variables of the following form:
DISCOVER – sent to a neighbor who is not known to have been visited, for the purpose of visiting.
RETURN – returns the center of activity to the parent.
VISITED – informing neighbors about the status.
$neighbours(i)$ – set of neighbors of node i (input).
$parent(i)$ – parent of i in the DFS (output); initially, $parent(i) = i$ for all nodes; finally, $parent(i) = i$ only for the start node.
$nomessage(i)$ – subset of $neighbours(i)$ including those neighbors not known to have been visited; initially, $nomessage(i) = neighbours(i)$ for all nodes.
$visited(i)$ – Boolean flag set to true once visited, i.e., on receiving the DISCOVER message for the first time; initially, $visited(i)$ is false for all nodes.
$explore(i)$ – the identity of a neighbor in $nomessage(i)$ to whom a DISCOVER message has been sent; initially, $explore(i) = i$ for all nodes.

To trigger the algorithm, source node s delivers a DISCOVER message to itself. This message is not counted in the complexity.

Algorithm at node i
    for DISCOVER message from j **do**
        delete j from $nomessage(i)$;
        execute procedure recover;
        if node i has already been visited **then**

      do nothing

   else

     set $visited(i)$ to $true$;

     set $parent(i)$ to j;

     execute procedure shift-center-of-activity;

     for all $(p \in neighbours(i)) \wedge (p \neq parent(i)) \wedge (p \neq explore(i))$ **do**

       send VISITED to p

     end for

   end if

end for

for VISITED message from j **do**

   delete j from $nomessage(i)$;

   execute procedure recover

end for

for RETURN message from j **do**

   delete j from $nomessage(i)$;

   execute procedure shift-center-of-activity

end for

procedure recover;

if $explore(i) = j$ **then**

   execute procedure shift-center-of-activity

else

   do nothing

end if

procedure shift-center-of-activity;

if there exists $k \in nomessage(i)$ **then**

   set $explore(i)$ to k;

   send DISCOVER to k

else

   set $explore(i)$ to i;

   if $parent(i) = i$ **then**

     Terminate /* start node*/

   else

     send RETURN to $parent(i)$

   end if

end if

Time and Communication Complexity Analysis

Observe that the depth-first search built as a result of execution of the algorithm is really not dependent on the pattern of delays encountered in the communication links. It is only dependent on the order in which the neighbors are selected by an already visited node to send DISCOVER messages. On the other hand, the number of messages exchanged during the execution of the algorithm is dependent on the delays encountered in the communication links. The best-case situation arises when the communication delay in any link is one time unit, the same assumption to be made for the time complexity analysis. Recall that VISITED messages are always sent in communication-time parallel to DISCOVER or RETURN messages. The above assumption implies that by the time a node becomes a center of activity, all VISITED messages sent by its neighbors should have been received and processed. Thus, there will never be a mistake made of sending a DISCOVER message to an already visited node. In other words, exactly $N-1$ DISCOVER messages will be sent in the best case, consuming exactly $N-1$ units of time. Also, exactly $N-1$ RETURN messages will be sent, consuming an additional $N-1$ units of time. Hence, the time complexity of our algorithm is $2N-2$, which is optimal.

Now, in order to determine the number of messages exchanged in the best case we still have to account for VISITED messages. Every node i of degree d_i clearly sends out at least $d_i - 2$ VISITED messages. But, the start node s and these nodes which finally appear as leaf nodes in the DFS tree will send out one more VISITED message each. Thus, the total number of VISITED messages exchanged in the best case is $2|E| - 2N + l + 1$, where l is the number of leaf nodes in the DFS tree constructed. Including DISCOVER and RETURN messages, the total number of messages exchanged in the best case is $2|E| + l - 1$. Clearly, here $l \geq 1$. If the graph G is complete, then $l = 1$ so that in such a case the number of messages exchanged during the execution of the algorithm could be as low as $2|E|$, the optimum value.

In order to evaluate the number of messages exchanged by the algorithm in the worst case, we observe that the pattern of delays in communication links could be such that every DISCOVER or VISITED message that can prevent the mistake of sending a DISCOVER message to an already visited node is received at its destination only after such a DISCOVER message has been dispatched in the opposite direction. In other words, every node in the graph may have to send a DISCOVER message to every one of its neighbors other than the parent. Thus, the start node s will send as many as d_s DISCOVER and $d_s - 1$ VISITED messages. Any other node i of degree $d_i \geq 2$ will send $d_i - 1$ DISCOVER, $d_i - 2$ VISITED and one RETURN messages. A node of degree one can only send a RETURN message. Summing up, the total number of messages in the worst case is $4|E| - 2(N-1) + (N'-1)$, where N' is the number of nodes of degree one in the graph, excluding the start node s. Thus the message complexity of the algorithm could vary between $2|E|$ and $4|E| - (N-1) - 1$, depending on the structure of the graph and the pattern of delays in the communication links. Also, observe that if the communication graph G is a tree by itself, then $|E| = N - 1$ and $l = N'_1$, and thus in this case the algorithm will require exactly $2|E| + N'_1 - 1$ messages independent of the nature of delays in the communication links.

7.1.2 Distributed Breadth-First Search

By asynchronous network we mean a point-to-point communication network, described by a communication graph $G = (V, E)$, where the set of nodes V, $|V| = N$, represents processors and the set of edges E represents bidirectional non-interfering communication channels operating between neighboring processors. No common memory is shared by processors. Each processor receives messages from its neighbors, performs local computations, and sends messages to its neighbors, all in negligible time. The sequence of messages sent on any given edge in a given direction is correctly received in FIFO (First-In-First-Out) order, with finite but variable and unpredictable delay.

The following complexity measures are used to evaluate the performance of distributed algorithms. The communication complexity C is the worst-case total number of elementary messages sent during the algorithm, where an elementary message contains at most $O(\log N)$ bits. The time complexity T is the maximum possible number of time units from start to the completion of the algorithm, assuming that the inter-message delay and the propagation delay of an edge is at most one time unit. This is under the provision that the algorithm works correctly without this assumption.

Given an undirected graph $G = (V, E)$ and a source node $s \in V$, the *Breadth-First Search (shortly BFS) problem* is to find, for each node $i \neq s$, the length l of a shortest path in G (in terms of the number of edges) from s to i and the immediate predecessor (parent) of i on that path. The length l is called the "layer number" of node i. The edge leading from a node to its parent (child) is called the inedge (outedge).

In a distributed algorithm for the BSF problem, each node has a copy of an algorithm determining the response of the node to messages received at that node. Namely, the algorithm specifies which computations should be performed and which messages should be sent. Initially, each node is ignorant of the global network topology except for its own edges. The algorithm is initiated by the start node s. Upon completion of the algorithm, the start node enters a given final state. At that time, all other nodes in the same connected component of the network know their (correct) layer numbers, their inedges and their outedges.

Let us first outline the *basic coordinated algorithm*. It operates in successive iterations, each processing another BFS layer. The source node controls these iterations by means of a synchronization process, performed over the part of the BFS tree built in the previous iterations. At the beginning of a given iteration i, the BFS tree has been constructed for all nodes in layers $j < i$. The source node broadcasts a message over that tree, which is forwarded out to nodes at layer $i - 1$. These nodes send "exploration" messages to all neighbors, trying to discover nodes at layer i. When a node first receives such an exploration message, it chooses the sender as its parent and deduces that it belongs to layer i. It sends back an acknowledge (ack) with an indication of whether the sender was chosen as the parent or not. Each node at layer $i - 1$

waits until all exploration messages have been acknowledged and then sends an ack to its parent in the BFS tree, indicating whether any new descendents have been discovered. When an internal node gets such acks from all children, it sends an ack to its parent. Eventually, all the acks are collected by the start node and thus layer i has been processed completely. If any nodes have been discovered at that layer, the next iteration $i + 1$ is started. Otherwise, the algorithm terminates. This algorithm is quite inefficient in the number of messages and time used for synchronization purposes. If one considers a network with all nodes on a single path of length $N - 1$, one sees that the communication and time complexities are each $O(N^2)$.

The previous attempt can be improved by dividing the network into groups of successive layers, called *strips*, and processing these strips one-by-one, synchronizing only once per strip. We now present an approach that processes strips in another way. Before processing a new strip, assume that the BFS was constructed for all the previous strips. Each node in the last layer of this already constructed tree will be called a start node of the new strip. The problem of processing a strip is in a sense a generalization of the original BFS problem – instead of constructing a BFS tree from a single start node, we want to construct a forest of shortest paths from a set of start nodes.

We now want to distinguish between two types of BFS problems on a strip. The first problem, referred to as the *known-connectivity* BFS, is to construct a BFS forest on a strip with starting knowledge of a spanning forest of the strip, each tree spanning another connected component of the strip. The second problem, referred to as the *unknown-connectivity* BFS, is to construct a BFS forest on a strip without such knowledge.

We now present a simple algorithm solving the first problem, referred to as the *basic coordinated multiple-source algorithm*, which is a natural extension of the basic single-source algorithm above. The idea is that all the start nodes grow their BFS trees by applying the basic algorithm above. All the start nodes in the same connected component will synchronize through the spanning tree of the component so that they process at the same speed, i.e., perform the same iteration at a given time. In particular, the forest is constructed layer by layer. At the beginning of processing a new layer, say layer i, some leader among the start nodes in a given connected component propagates a message to the other start node via the spanning tree and such start nodes propagate the message through the already constructed layers of the forest. The nodes at layer $i - 1$ then send exploration messages as before, and acks are collected back to the start nodes and then through the spanning tree to the leader. This process is performed independently in such connected components. Note that the absence between components causes not harm.

With the above strategy, at most one exploration message is sent in each direction on each edge of the strip. Thus, the total number of exploration messages is linear in the number of edges. Also, each node in the strip (counting the strip's start nodes) receives at most one coordination message over the spanning tree from the leader and at most one message from a start node through the BFS tree forest during one iteration of the algorithm. Finally, there is one ack message for each broadcast and exploration message. Altogether there are 4 messages per node per iteration and 4 messages per edge of the strip.

However, what we really need to solve is the second problem. The coordinated algorithm above cannot be applied to solve the second problem, because the spanning forest inside the strip is not available. Conceptually, the second problem appears to be more complex than the first one because, since it is not clear a priori which nodes are in the new strip, there is no obvious method for coordination between start nodes. It turns out that there is no apparent way to reduce the second problem to the first one in an efficient way. One may suggest the following naive solution to the problem: construct a global spanning tree in the network and coordinate between start nodes of the strip through that tree. However, coordination over such a "big" tree is too expensive, and in fact we end up with an algorithm whose complexity may reach $O(N^2)$.

However, let us assume that spanning forests are available in each strip, and that we should only solve the known-connectivity BFS inside each strip. Then the BFS tree for the whole network can be generated by using the above coordinated multiple sources algorithm for each strip and by using the synchronization technique of the basic coordinated algorithm between strips. Each node then receives at most 2 synchronization messages per node for each strip being processed. Suppose that strips are chosen to contain $\sqrt{N}$ layers. Since there are at most $\sqrt{N}$ strips and $\sqrt{N}$ layers inside each strip, altogether we have $O(N^{1.5})$ synchronization messages, and in total $O(|E| + N^{1.5})$ messages have been sent; this is a considerable improvement compared to the basic algorithm above. This observation shows how nice it would be to reduce the unknown-connectivity problem to the known-connectivity one. Now we show how this reduction can be performed.

The next algorithm is referred to as the bootstrap algorithm. The idea is to perform BFS and the spanning tree algorithms interchangeably, each algorithm supporting the other. In general, the algorithm maintains a forest of trees, each of the start nodes belonging to some of these trees. Initially, each tree consists of a single start node. The set of start nodes in the same tree is called a cluster; one of these nodes is chosen as a leader of the cluster, which coordinates the operation of the whole cluster. The idea of the algorithm is to try to achieve a greater degree of coordination by merging together as many trees as possible.

The algorithm proceeds in iterations, each consisting of two stages. The purpose of the first stage is to grow independently BFS trees from each cluster while the second stage stitches together neighboring trees. Upon completion of the stage, all the start nodes of a certain cluster coordinate with all the other clusters back through the original start node. Thus, at a certain time, all the clusters perform the same stage of the same iteration.

The first stage is a slight generalization of the above coordinated multiple-source algorithm, which will be referred to as the generalized coordinated algorithm. It uses the above basic coordinated multiple-source algorithm to build an independent BFS forest from each cluster for a given number of layers. (This process is coordinated by the leader of the cluster via the tree.) Since there is no coordination between different clusters, and the network is asynchronous, some clusters might grow their forests much more quickly than others. As a result, some node might be improperly seized by a forest A which progresses quickly, while actually it might be closer to a start node in another cluster and thus should belong to a forest B grown by that cluster. This situation is discovered later when B eventually grows up and an exploration message sent by a node in B reaches one of the improperly seized nodes. In this situation, the edge carrying that message is remembered by the adjacent nodes as an inter-forest edge and both forests A and B terminate their part in the first stage, even though the required number of layers has not yet been explored.

Upon the completion of the first stage, the cluster and its BFS forest are called *active* if the BFS forest is adjacent to any inter-forest edges; otherwise they are called *inactive*. Note that an inactive cluster's BFS forest must contain all nodes whose true BFS paths must go through the start nodes of the cluster, since none of these nodes could be improperly seized by another forest. Thus, if upon completion of the first stage all the clusters are inactive, then the strip has been processed correctly and the bootstrap algorithm terminates.

Otherwise, the second stage is started, in which the spanning tree algorithm is applied to a super-graph whose nodes are active forests and whose edges are the inter-forest edges. The output of this algorithm is a spanning forest of the super-graph, each tree spanning another component; in addition a leader is chosen inside each tree. As a result, active forests which share common inter-forest edges are merged together into bigger forests.

In the successive iterations, the first stage is modified as follows. Whenever an active forest meets another inactive forest on its way, the former forest penetrates into the latter, absorbing all of its nodes which were improperly seized. (This could not occur in the first iteration since all the forests are initially active.) If an active forest is met, then, as before, the exploration process is stopped and an inter-forest edge is remembered. The idea behind this rather peculiar rule is to reduce the number of iterations without increasing too much the number of penetrations.

We now observe that after each iteration, each active cluster either becomes inactive or merges with at least one other active cluster. Thus, after i iterations, each cluster not yet inactive contains at least 2^i start nodes. Thus after at most $\log N$ iterations, each cluster is inactive and the resulting forest is a genuine BFS forest for the strip.

In summary, the bootstrap algorithm effectively solves the problem of processing a strip with unknown connectivity in at most $\log N$ iterations, each involving one application of the generalized coordinated algorithm, two global synchronization procedures and one application of the spanning tree algorithm. Each application of the spanning tree algorithm requires $O(\log N)$ messages per node plus $O(1)$ messages per edge of the strip. In terms of time, it requires $O(\log N)$ time per node of the strip. (Here, we use the fact that the subnetwork, which is the input to the spanning tree algorithm, belongs entirely to the strip.)

Let us now plug the bootstrap algorithm into the previous scheme, which divided the network into strips of size $\sqrt{N}$ and processed them serially. It is easy to show that the resulting algorithm has communication complexity $O((|E| + N^{1.5}) \cdot \log N)$. Since the exploration messages, which account for the term E above, are sent in parallel at each node (and similarly for the spanning tree algorithm) the time complexity is $O(N^{1.5} \cdot \log N)$. Note that this algorithm requires knowledge of N, however; both N and $|E|$ can be easily calculated with complexity $O(|E|)$ in communication and $O(1)$ in time.

$O(|E| \cdot 2^{\log N \log \log N})$ Algorithm

We now modify the generalized coordinated algorithm above, making it recursive. This modification will be referred to as the *main* algorithm. Recall that the input to this algorithm is a set of start nodes, or a cluster, and a spanning tree T of the cluster which belongs entirely to the current strip. In this algorithm, the strip is split into a set of substrips, each with a common number of layers. The algorithm proceeds in successive iterations, each processing another substrip by calling the bootstrap algorithm as a subroutine. The input to this subroutine is a set of start nodes for the substrip plus a structure for external coordination, which is needed in order to trigger the execution of the subroutine, to detect its termination and to synchronize between internal iterations of the subroutine. The structure consists of the BFS forest of the preceding substrips plus the above tree T. The start nodes of the present strips are the nodes on the final layer of the BFS forest. (For the first strip, this set of start nodes is simply the cluster being used by the main algorithm.)

The subroutine itself proceeds in at most $\log N$ internal iterations, each involving, among others, solving the known-connectivity problem for substrips. This problem will be solved by the same main algorithm (instead of the generalized coordinated algorithm, as before). Note that the main algorithm for strips calls the algorithm for smaller substrips, i.e., it is performed recursively, decreasing the depth of the strip as the recursion depth increases. At the bottom level of recursion, the basic coordinated multiple-source algorithm is used.

Let us now evaluate the complexity of the main algorithm. Consider the processing of a strip of size d with a known connectivity by the main algorithm above. For convenience, we will consider here the normalized complexities of the algorithm, denoted by $CE(d)$ and $TE(d)$, which are upper bounds on the number of messages sent per link of the strip and the time spent per node of the strip, respectively. Observe that the overall complexities of the BFS algorithm are $C = |E| \cdot CE(N)$ and $T = N \cdot TN(N)$. (We assume the worst case, i.e., that the diameter of the network is N.) We will provide here a recursive equation or evaluation for $CE(h)$; it turns out that $TN(h)$ satisfies the same equation and thus $CE(h) = TN(h)$.

Let us denote by d the depth of the strip being processed at the i-th level of recursion. $N = d_1 > d_2 > \cdots > d_r$, where r is the maximum depth of the recursion. The complexity at the bottom level is $CE(d_r) = TN(d_r) = O(d_r)$. The algorithm processing strips of depth d_i consists of $\frac{d_i}{d_{i+1}}$ iterations, each processing substrips of size d_i, and involving a logarithmic number of internal iterations. Each internal iteration involves one call to the algorithm processing strips of size d_{i+1}, one call to a synchronization procedure, and one call to the spanning tree algorithm. Taking these factors into account yields the following recursive formula

$$CE(d_1) = \log N \cdot \left(\frac{d_i}{d_{i+1}} + CE(d_{i+1}) \right).$$

It turns out that $TN(d)$ obeys the same recursive formula, and thus $CE(d) = TN(d)$.

Using dynamic programming, one can optimize the total complexities of the resulting algorithm, choosing properly the depth r of the recursion and the numbers d_i. Let us just mention that the optimum solution has the form

$$CE(N) = O(2^{\sqrt{\log N \log \log N}}).$$

Thus, the total communication and time complexities of the algorithm are bounded by

$$O(|E| \cdot 2^{\sqrt{\log N \log \log N}})$$

and

$$O(N \cdot 2^{\sqrt{\log N \log \log N}}),$$

respectively.

7.1.3 Basic Lower Bound on Message Complexity of Broadcast

In a common model of communication networks, in which each processor initially knows the identity of its neighbors, but does not know the entire network topology, we prove that the number of messages of bounded length required for broadcast is $\Omega(|E|)$, where $|E|$ is the number of links in the network. Using this result, one can demonstrate the optimality of the Distributed Breadth-First Search algorithm from the previous section for dense networks.

The Model

The communication model consists of a family $\mathcal{F}$ of undirected graphs, an ordered set (universe) S of possible identities (ID's), communication axioms, assumptions about the broadcast protocols, and a complexity measure. A graph $G = (V, E) \in \mathcal{F}$ represents the topological structure of a communication network. Initially, (unique) ID's are assigned to the processors (nodes) of the graph. An assignment $\phi : V \mapsto S$ is a one-to-one mapping from the set of nodes to the set of possible ID's.

For the sake of simplicity, we assume that communication is event-driven, i.e., except for the initial transition of the source a processor transmits a message only after receiving a message. For the sake of concreteness, we assume that the communication is synchronous, i.e., communication takes place in "rounds", where processors transmit only at the very beginning of a round and all messages are received by the end of the round. Clearly, the lower bound holds also if communication is asynchronous.

Consider broadcast protocols which operate on every network G in the family $\mathcal{F}$, and every assignment ϕ of ID's to the processors of G. The protocols are identical copies of a local program. An input to the local program consists of the ID of the processor P in which it runs and the ID's of the processors incident to each of the links of P. It should be stressed that the local program gets as input not only the set of ID's of neighboring processors, but also the correspondence between the communication links and these ID's.

The local computation of the program is restricted to comparing two ID's. Formally, the program has local variables of two types: variables *identity-type* (*ID-type*) and *ordinary* variables. Initially, the ID-type variables are empty, while the ordinary variables may contain some constants (e.g., 0 and 1). The local computations of the program are of two corresponding types:

1. Comparing two ID-type variables and storing the result of the comparison in an ordinary variable. The result of the comparison may be either of the three $<, =, >$. We should assume some standard encoding of the result of the comparison, e.g., $<$ is encoded as -1, $=$ as 0, and $>$ is encoded as $+1$.
2. Performing an arbitrary computation on ordinary variables and storing the result in another ordinary variable.

The communication instructions of the program are of two types: an unconditional "receive message", and a (possibly) conditional "send message". The condition in the "send" instruction may only be a comparison of two ordinary variables. The message sent consists of the values of some of the variables of the local program.

Assume that all messages sent by the protocol contain a bounded number of ID's; this bound is denoted by B. Further assume that all send instructions are of the form

if $vo_1 = vo_2$ **then** *send the message* $(id_1, id_2, \ldots, id_B; \overline{vo})$ *to processor* id_{B+1}

where id_i is the value of the i-th ID-type variable, vo_i is the value of the i-th ordinary variable, and $\overline{vo}$ is the sequence of values of all ordinary variables (all variables being of the sending processor).

The complexity measure will be the number of messages (containing at most B ID's) sent in the worst-case execution of the protocol on the network G. We stress that the protocol has to be correct, i.e., achieve broadcast not only when running on G, but rather when running on any network of the family $\mathcal{F}$.

The Lower Bound

Definition 7.1.1. *Two ID's* $x, y \in S$ *are called* close *if for all* $z \in S - \{x, y\}$ *the following holds:*

$$x < z \text{ if and only if } y < z.$$

Clearly, if $|S| \geq 2 \cdot |V|$, there exist two disjoint assignments ϕ_1, ϕ_2 (i.e., with disjoint range) of ID's to the nodes such that for every node v, the ID's $\phi_1(v)$ and $\phi_2(v)$ are close.

In the following definition we assume, w.l.o.g., that a processor sends at most one message to each of its neighbors in each round.

Definition 7.1.2. *An execution of a protocol on a graph* G *under ID-assignment* ϕ *is the sequence of messages sent during the corresponding execution, where messages sent appear in the sequence of the lexicographic order of triples* $(ROUND, SENDER, RECEIVER)$. *For simplicity, we append to each message in the sequence a* header *consisting of the corresponding* $(ROUND, SENDER, RECEIVER)$ *triple.*

It should be stressed that the sequence may contain "gaps", i.e., a message with header h must not be followed by a message with a header which is the successor of h in the lexicographic order.

Definition 7.1.3. *Let* σ_1 *and* σ_2 *be executions of protocol* Π *on the graph* G *with ID-assignments* ϕ_1 *and* ϕ_2, *respectively. We say that the executions* σ_1 *and* σ_2 *are* essentially the same *if, when substituting in* σ_1, *for every* $v \in V$, *the ID value* $\phi_1(v)$ *by* $\phi_2(v)$, *we get the sequence* σ_2.

Lemma 7.1.4. *Let* ϕ_1 *and* ϕ_2 *be two assignments so that for every* $v \in V$, *the ID's* $\phi_1(v)$ *and* $\phi_2(v)$ *are close. Then the execution of any protocol* Π *on the graph* G *with ID-assignments* ϕ_1 *and* ϕ_2, *respectively, is essentially the same.*

Proof. Immediate by observing that all the ordinary variables of all local programs will have the same values in both executions.

□

Corollary 7.1.5. *Let ϕ_1 and ϕ_2 be as in Lemma 7.1.4, and suppose that ϕ_1 and ϕ_2 have disjoint range. Let α be any N-long string of 1's and 2's, and let ϕ_α be an ID-assignment so that for every $i \in V$, $V = \{1, 2, \ldots, N\}$,*

$$\phi_\alpha(i) = \phi_{\alpha_i}(i), \text{ where } \alpha_i \text{ is the } i\text{-th element of } \alpha.$$

Then the execution of any protocol Π on the graph G with ID-assignments ϕ_α is essentially the same for every α.

We are now ready to introduce the family of networks, for which we will prove that any protocol achieving broadcast on each network in the family has a linear message complexity.

Definition 7.1.6. *For any given graph G and for every edge $e = (u, v)$ in E, we define the graph $G^{(e)} = (V^{(e)}, E^{(e)})$ so that*

$$V^{(e)} = V \cup \{u', v'\}, \text{ where } u', v' \in V,$$

$$e^{(e)} = (E - \{e\}) \cup \{(u, v'), (v', u'), (u', v)\}.$$

The family of graphs is

$$C_G = \{G\} \cup \{G^{(e)} \mid e \in E\}.$$

Note that the auxiliary graph $G^{(e)}$ is a copy of G, except that the edge $e = (u, v)$ is replaced by the path $u-v'-u'-v$. In our argument, we will concentrate on the graph G, passing when required to one of the auxiliary graphs $G^{(e)}$, and relying on the assumption that the protocol is also correct when run on $G^{(e)} \in C_G$. This is the underlying principle of the following important lemma which asserts that neighbors should "hear" of one another during any execution of a broadcast protocol.

Lemma 7.1.7. *Let G be a graph, S be a set of ID's ($|S| \geq 2|V|$), and Π be a protocol which achieves broadcast on each graph in G_G. Then there exists an ID-assignment $\phi : V \mapsto S$ such that on execution of Π on G with ID's ϕ, for every edge $(u, v) \in E$, at least one of the following three events takes place:*

(i) A message is sent on (u, v).
(ii) Processor u either sends or receives a message containing $\phi(v)$.
(iii) Processor v either sends or receives a message containing $\phi(u)$.

Proof. The intuition behind the proof is that in the case when the lemma does not hold for $e \in E$ no processor in the network can distinguish an execution on $G^{(e)}$. The only potential difference between these executions lies in whether u and v are neighbors or not, where $e = (u, v)$. But the neighborhood can not be tested if no messages bearing the ID of one processor are communicated from/to the other processor. This intuition needs the careful formalization sketched below.

Assume, on the contrary, that the lemma does not hold. That is, there exists a graph G such that during the execution of Π on G with any ID-assignment $\phi : V \mapsto S$, there is an edge $e = (u, v) \in E$ such that none of the above events $(i), (ii), (iii)$ takes place. Let ϕ_1 and ϕ_2 be as in the Corollary 7.1.5. Denote $x = \phi_1(u)$, $y = \phi_1(v)$, $\overline{x} = \phi_2(u)$, and $\overline{y} = \phi_2(v)$. For every $i \in \{u, v\}$, let $\psi^{(i)} : V \mapsto S$ be ID-assignments such that $\psi^{(i)}(i) = \phi_2(i)$ and $\psi^{(i)}(j) = \phi_1(j)$ for all $j \in V - \{i\}$. By the Corollary 7.1.5 the executions of Π on G are essentially the same with any of the three ID-assignments ϕ_1, $\psi^{(u)}$ and $\psi^{(v)}$. The later two executions will be useful in the sequel:

EX1: The execution of Π on graph G with Id's $\psi^{(u)}$.
EX2: The execution of Π on graph G with Id's $\psi^{(v)}$.

Now, we consider the execution of Π on the graph $G^{(e)}$. In particular, we will consider such an execution with ID-assignment $\psi : V^{(e)} \mapsto S$, where $\psi(v') = \overline{y}$, $\psi(u') = \overline{x}$, and $\psi(w) = \phi_1(w)$, for every $w \in V$. Denote

EX3: The execution of Π on graph $G^{(e)}$ with Id's ψ.

We recall the assumption that, except for the source in its initial transmission, a processor sends a message only after receiving one. Thus, unless processors u' and v' receive messages from v or u, they do not send any messages during $EX3$. Next, we claim that the executions $EX1$, $EX2$ and $EX3$ are essentially the same.

Claim. Let $EX1$, $EX2$ and $EX3$ be executions as defined above. Suppose that during $EX3$ processors u' and v' do not send messages before they receive a message. Then the executions $EX1$, $EX2$ and $EX3$ are essentially equal.

Proof. We show that in the above three executions all processors have the same values of ordinary variables, and essentially the same values of ID-type variables. Furthermore, during $EX3$ no processor has both the values $\psi(u)$ and $\psi(u')$ (resp. $\psi(v)$ and $\psi(v')$). We use induction on the number of rounds r. The induction base holds vacuously, u (resp. v) having $\psi(v')$ (resp. $\psi(u')$) during $EX3$. For the induction step, from round r to $r + 1$, we use the fact that during the $(r + 1)$-st round of both $EX1$ and $EX2$, processor u (resp. v) neither sends nor receives a message with the ID of v (resp. u) nor was a message sent over the link (u, v). By the induction hypothesis, the message sent in the $(r + 1)$-st round of the $EX3$ processor u (resp. v) does not send a message containing the ID of v' (resp. u') nor does it receive a message containing the ID of v (resp. u). Thus, u still does not have $\psi(v)$ and no other processor $w \in V - \{u\}$ has $\psi(v')$. It follows that the values of the ID-type variables of all processors are essentially the same in all three executions after round $r + 1$, and that the values of the ordinary variables are the same. The induction claim follows.

□

Proof of Lemma 7.1.7 continued. Once we have asserted that the three executions are essentially the same, we conclude that u (resp. v) does not send a message to v' (resp. to u') in the execution of Π on $G^{(e)}$ with Id's ψ. This contradicts the correctness of Π and the lemma follows.   □

Lemma 7.1.7 provides us with a way of charging messages sent during the execution of any broadcast protocol to the links of the network. We stress that if it is not that messages must be sent over every link, but rather that we can charge them essentially in this manner. We get:

Theorem 7.1.8. *Let G be an arbitrary graph, and Π a protocol achieving broadcast on every network of the family C_G. Then the message complexity of Π on G is $\Omega(|E|)$.*

Proof. Let ϕ be an ID-assignment such as in Lemma 7.1.7. We employ the following *charging role* to messages sent during the execution of Π on G with Id's ϕ:
For every message containing $\phi(w)$ sent from processor u to processor v, we:

(1) Charge the edge (u, v).
(2) Charge the pair (possibly edge) (u, w).
(3) Charge the pair (possible edge) (w, v).

We now show that the charge-count and the number of messages sent during the execution are closely related, and that this count is bounded below by $\Omega(|E|)$. Recall that B is the number of Id's contained in a single message.

Claim. The maximum number of links which get charged for a single message sent during the above execution is $2B + 1$.

Proof. A message sent from v to u is charged to the link (v, u) and to all the links (u, w) and (v, w), where $\phi(w)$ appears in the message. Since there are B Id's in each message, the claim follows.

$\square$

Claim. Each link is charged at least once.

Proof. For each edge (u, v), we use Lemma 7.1.7, and consider the following three cases:

(i) A message is sent on (u, v). Then the edge (u, v) is charged by (1).
(ii) Processor u either sends or receives a message containing $\phi(v)$. In the first subcase, the edge (u, v) is charged by (2). In the second subcase, the edge is charged by (3).
(iii) Processor v either sends or receives a message containing $\phi(v)$. Similar to case (ii), interchanging the roles of v and u.

The claim follows.

$\square$

Let C denote the total charge placed by the above rule on the above execution of Π, and let M denote the total number of messages sent during that execution. Combining the above claims, we get

$$|E| \leq C \leq (2B + 1) \cdot M$$

Recalling that B is a constant, Theorem 7.1.8 follows.

$\square$

7.2 Broadcast on Tori

7.2.1 Upper Bound

In this section we present a broadcasting algorithm used on asynchronous anonymous totally unlabeled $n \times m$ tori using $\frac{10}{7}nm + O(n + m)$ messages in the worst case. By anonymous tori we mean that the identification numbers of the nodes are not necessarily unique, and by totally unlabeled tori we mean that each node can distinguish its links by uninterpreted labels that have no topological meaning.

Informal Description of the Algorithm

Our algorithm $\mathcal{A}$ starts from the source s by sending an initial message in one direction until it returns back to s. This can be done using the *handrail* technique (introduced in the previous chapter) with $O(n)$ messages of size $O(1)$, where n is the size of the torus in the direction of the initial message. The created path circling around the tori is called the *equator*. From the handrail technique it follows that vertices on the equator can consistently distinguish between their *north* and *south* sides.

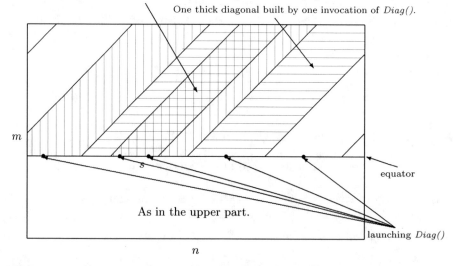

Fig. 7.1.

In the second phase, another message is sent along the equator eastward and at each 7-th vertex it launches northward the subroutine $Diag()$ until it returns back to the source. The first launch of $Diag()$, denoted as $Diag^1()$, is done using marked messages, so it will not interfere with the last one, which may overlap with it. The *Diag()* procedure broadcasts on the thick (7 vertex) diagonal in the north-east direction until it returns back to the equator.

The overall message complexity of the broadcasting algorithm is $\frac{10}{7}N + O(n+m)$ and it follows from these facts:

The cost of the start-up (building the equator and launching *Diag()*) procedures is $O(n)$.

Diag() uses $\frac{10}{7}$ messages of size $O(1)$ bits per each reached vertex.

The whole torus can be covered by disjoint thick diagonals built by *Diag()*. Since *Diag()* stops on the equator, different invocations of *Diag()* do not overlap. The only exception is the first and the last invocation of *Diag()*, which can overlap for n a non-multiple of 7, where n is the length of the equator. There are at most $6m$ vertices in the intersection, that means totally $\frac{60}{7}m$ additional messages. (See Figure 7.1.)

Detailed Description of the Algorithm

StartUp

At the source on start up:

0. For each incident link h:
   Send(S_x^1), where x is a label of h at the source vertex;

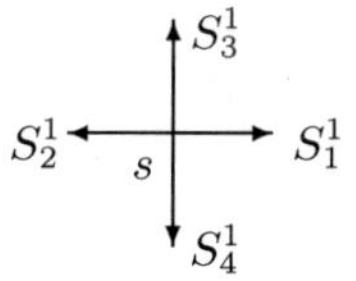

$$S_2^1 \longleftarrow \quad \begin{array}{c} S_3^1 \\ \uparrow \\ \underset{s}{\mid} \\ \downarrow \\ S_4^1 \end{array} \quad \longrightarrow S_1^1$$

Fig. 7.2.

At an arbitrary vertex:

	Upon receiving	Send	See Figure
1.	S_x^1	S_x^2 on all remaining links	7.3a
2.	S_1^2 on h and S_x^2	S_x^3 on h	7.3b
3.	S_x^3 on h_1 and S_y^3 on h_2, let $x < y$	U_1 on h_1, B_1 on h_2 and M_1 on link on which S_1^2 was sent, but nothing received.	7.3c

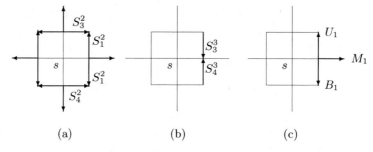

(a) (b) (c)

Fig. 7.3.

Building the equator:

	After receiving message(s)	Send	See Figure
4a.	U_1	U on unused links	7.4a
4b.	B_1	B on unused links	7.4a
4c.	M_1, not at source	M on unused links [1]	7.3c and 7.4d
4d.	M_1 at source	L on link with label 1 [2]	7.5
5a.	U on h and M	U on all links except h	7.4b
5b.	B on h and M	B on all links except h	7.4b
5c.	U and B	M_1 on links on which no M_1, U or B was received	7.4c

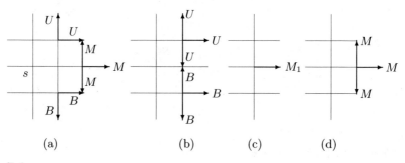

(a) (b) (c) (d)

Fig. 7.4.

---

[1] Links used only by S_x^y messages are considered unused.
[2] Starting launching phase.

Launching phase:

	After receiving message(s)	Send	See Figure
6.	L	D_0' where U_1 was sent L_1 where M_1 was sent	7.3c and 7.5
7.	L_i, not at source	$L_{(i+1) \bmod 7}$ where M_1 was sent if $i = 0$, send D_0 from where U message came	7.4c and 7.5 7.4b and 7.5

Diag() procedure:

	After receiving message(s)	Send to all unused links [3]
8a.	D_0	D_1 [4]
8b.	D_1 and not M received before	D_2
8c.	D_2 and (D_9 or S_x^2)	D_3
8d.	Two D_2	D_4
8e.	D_4	D_5
8f.	Two D_5	D_6
8g.	D_6	D_7
8h.	D_5 and D_7	D_3
8i.	Two D_7	D_8
8j.	D_8	D_9
8k.	Two D_9	D_1

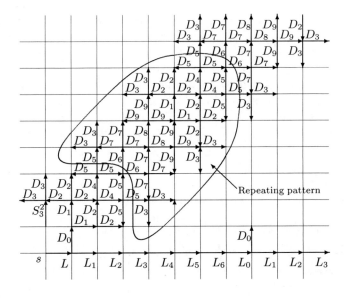

Fig. 7.5.

If D_0' message was received, D_i' messages will be used, to avoid possible interference between the first and the last invocation of $Diag()$.

It is easy to see that computation cyclically proceeds in cycle

$$8a \rightarrow 8b \rightarrow 8d \rightarrow 8e \rightarrow 8f \rightarrow 8g \rightarrow 8i \rightarrow 8j \rightarrow 8k \rightarrow 8b$$

with concurrent steps 8c and 8h. In one such cycle altogether 28 new vertices are added to the thick diagonal using 40 messages, resulting in overall $\frac{10}{7}$ messages per vertex.

$Diag()$ terminates when it returns back to the equator from the south – a vertex that has received M and B message will not send any $Diag()$ message.

Making Termination Explicit

The presented broadcasting algorithm terminates implicitly. One way to make the termination explicit within the same complexity bound is the following:

Vertices reached by messages of $Diag()$ terminate when they finish their work in Diag().

Vertices of the equator terminate when the launching token of the second phase has passed them.

The only problems are vertices of the first, $Diag^1()$, and the last, $Diag^q()$, thick diagonal. The problem is how to terminate in order to prevent blocking of broadcasting on the second thick diagonal. One possible solution is the following:

When $Diag^1()$ returns back to the equator, it returns k steps to the west and launches $Diag^q()$ to south-west. Vertices reached by $Diag^q()$ but not by $Diag^1()$ will terminate after finishing their work in $Diag^q()$. When $Diag^q()$ reaches the equator, it goes k steps eastward and launches $Diag^1()$ to the north-east. Now vertices can terminate after finishing their work in $Diag^1()$. k can be computed during the construction of the equator as the length of the equator modulo 7. This additional computation can be done using $O(m)$ messages.

7.2.2 Lower Bound

Now we prove the lower bound on the number of used links by a synchronous broadcasting algorithm on unlabeled $n \times m$ tori in the form $\frac{8}{7}nm - O(1)$. This lower bound is achieved using the adversary argument.

---

[3] Unused links – unused by D_i and M messages. No messages are sent from vertices that received a B and M message – termination of $Diag()$ when equator was reached from the south.

[4] Messages are sent along two links on which no U, D_0 or S_x^y message arrived.

Basic Adversary Strategy

We prove the lower bound by letting the broadcasting algorithm and the adversary behave according to the following strategy.

On one side, we consider the broadcasting algorithm as a synchronous algorithm which maintains its *domain D* (informally, the domain is the set of already informed vertices together with used links and it is represented as a graph). Initially, the domain of the broadcasting algorithm consists of a single vertex – the source (of the information). The goal of the algorithm is to extend its domain to span the whole torus (i.e., to spread the information from the source to all vertices). The algorithm will be aware of the graph D representing its domain, but it does not know how it is embedded into the torus. Here, by embedding we mean the mapping of D to an $n \times m$ torus such that each node of D is mapped to a vertex of the torus and each edge of D is mapped to a link of the torus.

On the other side, the goal of the adversary strategy is to maximize the number of explored links. The basic adversary can achieve this aim by choosing the embedding that leads to the currently smallest possible new domain (measured by the number of vertices).

The game between the algorithm and the adversary proceeds in rounds. Each round begins with a step of the algorithm, which is followed by a step of the adversary. The algorithm (during its step) specifies for each vertex of its domain D the number of yet unused links it wants to explore. The adversary (during its step) chooses an embedding of D into a torus and (considering numbers specified by the algorithm) it decides which of the yet unused links at a given vertex will be explored next. At the end of each round all recently explored links are added to the domain D of the algorithm. The game terminates when the domain D spans the whole torus.

The *basic adversary* tries all possible ways of choosing explored links and for each alternative it tests all possible embeddings. The embedding (with the choice of explored links) which leads to the smallest new domain is chosen. (See Figure 7.6.) If there are more than one possibility, it decides on an arbitrary one.

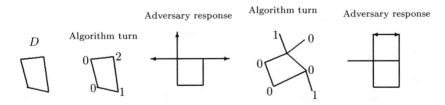

Fig. 7.6.

Structure of the Domain

Consider the game between the broadcasting algorithm and the adversary introduced in the previous subsection. Let $D(i)$ be the domain of the algorithm after round i. $D(i)$ can be divided into the *core graph* $C(i)$ and *hanging trees* $T(i)$. The *core graph* can be defined as the maximal subgraph $C(i)$ of $D(i)$ such that $\forall v \in C(i)$, v has at least two neighbors in $C(i)$. $T(i)$ is the rest of the domain. $T(i)$ can be viewed as a forest of trees with roots in $C(i)$. (These roots are not in $T(i)$.) We denote these graphs after the termination of the game by D, C and T.

Let $i \geq 0$ be a round of the game. The following facts follow directly from the definitions of $D(i)$ and $C(i)$.

$D(i)$ is a subgraph of $D(i + 1)$.

$C(i)$ is a subgraph of $C(i + 1)$.

The following lemma is crucial in the structural characterization of the domain under the basic adversary strategy.

Lemma 7.2.1. *For each round $i \geq 0$ of the game between the algorithm and the adversary, if $C(i - 1) \neq \emptyset$ then the depth of the hanging trees $T(i)$ in the domain is limited by 2.*

Proof. By contradiction. Let $t > 0$ be the first round of the game in which a hanging tree (say T''') of depth 3 appears. It means that in the previous round $t - 1$ there was a hanging tree T'' of depth 2 and from one of its leaves at depth 2 (say from a vertex v) a new link was explored. Let the root of T'' be a vertex $r \in C(t-1)$. The vertex r has two neighbors in $C(t-1)$. Take the embedding $\mathcal{E}$ used by the basic adversary at round $t - 1$ and modify it locally to embed T'' such as is prescribed in Figure 7.7. Such a modification is indeed possible, because the place for the vertex v is either free or it is occupied by another branch of T'' (which can be exchanged with the branch leading to v). If this place belongs to $C(t - 1)$ or to another tree, then in the previous round the adversary should have directed the growth of T'' to this place, thus constructing a smaller domain.

The resulting embedding $\mathcal{E}'$ (together with the choice of the explored link (v, w)) results in a smaller domain compared with $\mathcal{E}$, which is a contradiction with respect to our choice of adversary.

$\square$

Using similar arguments as in the proof above it can be shown that $T(i)$ cannot contain a path of length 4. A simple case analysis yields the following lemma:

Lemma 7.2.2. *If $C(i) = \emptyset$ then $|T(i)| \leq 8$.*

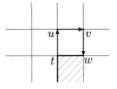

Fig. 7.7.

So far we have shown that the basic adversary can limit the depth of hanging trees by looping their branches back to the core graph. Another way to reduce the depth of hanging trees is to connect two trees, thus eliminating trees that grow deep. We will refine the basic adversary by making it prefer the following option:

If there are several possibilities (for embedding and choice of explored links) resulting in domains of the same size, prefer the option in which a tree branches loop back to the core graph rather than the option in which trees come into contact. (See Figure 7.8.)

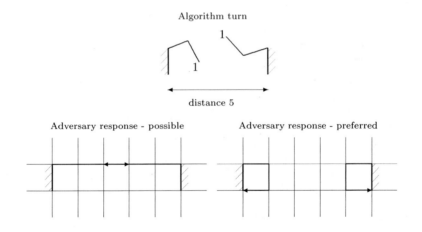

Fig. 7.8.

7.2.3 The Growth of the Core Graph

The following lemma states that $C(i+1)$ can be obtained from $C(i)$ by successive adding of short (at most 4 edges) ears.

Lemma 7.2.3. *If $C(i) \neq \emptyset$ there exists a sequence of graphs*

$$C(i) = C_{k_i}, C_{k_i+1}, \ldots, C_{k_{i+1}} = C(i+1)$$

such that for each j, $k_i \leq j < k_{i+1}$, C_{j+1} is obtained from C_j by adding a path π_j of length at most 4 which starts and ends in C_j.

Proof. We show: If $C_j \neq C(i+1)$ for $j < k_{i+1}$, we can always find such a path (ear) π_j in $C(i+1) - C_j$.

First note that we can always find an ear of length at most 6. See that new vertices are added to the core graph only when branches of some tree loop back to the core graph or when two trees meet. (The third possibility occurs when two branches of the same tree meet. But this case is not possible under a refined adversary, because the adversary prefers to loop branches back to the core graph.) In the first case an ear of length at most 3 is formed (since trees are of depth at most 2, plus a newly explored link), in the second case each tree can contribute by a path of length at most 3, bounding the overall length of the newly created ear by 6.

To form a *long ear* of length 5 or 6, at least one tree must contribute with a path of length 3. The refined adversary prefers looping back branches of trees to the core graph. It may happen that the adversary cannot loop back a branch of length 3, because that will increase the size of the domain. One such situation is shown in Figure 7.9.

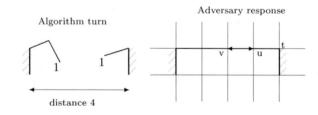

Fig. 7.9.

However, that is possible because the tree of the vertex u contributed only with 2 links $((t, u), (u, v))$ and there were overlapping links from different trees $((u, v)$ and $(v, u))$. If there are not overlapping links (there is no request for exploration from u), looping back to the core graph is possible, because it will not enlarge the domain. Similarly, looping back is possible if both trees contribute by paths of length 3, as shown in Figure 7.8.

That means that if there is still a long ear, then it must also have small loop(s) at its end(s). We can first add an ear formed by this (these) small loop(s), decreasing the overall length of the long ear to at most 4. (It is easy to see from the definition of the refined adversary that if an ear of length 6 was formed, it must have small loops at both of its ends.)

□

Estimating the Number of Explored Links

Let $E_{C(i)}$ be the set of links in the core graph $C(i)$. We prove the lower bound on the number of used links by proving a lower bound on the expression

$$\frac{|E_{C(i)}| + |T(i)|}{|C(i)| + |T(i)|} \tag{7.1}$$

which relates the number of the explored links and the number of reached vertices after the i-th round of the game.

Let $C_0 = C(s)$, where s is the first round such that $C(s) \neq \emptyset$. Let C_j, $j > 0$, be the core graph after adding j ears according to Lemma 7.2.3 and let $|E_{C_j}|$ be the number of links in C_j. Let T'_j be the set of vertices at a distance 1 or 2 from C_j. Define $T_j = T(i)$ for $k_{i-1} < j \le k_i$. Due to Lemma 7.2.1 $|T'_{k_i}| \ge |T_{k_i}|$.

Let us rewrite (7.1) for round $i \ge s$:

$$\frac{|E_{C(i)}| + |T(i)|}{|C(i)| + |T(i)|} \ge \frac{|E_{C_{k_i}}| + |T'_{k_i}|}{|C_{k_i}| + |T'_{k_i}|}$$

$$= \frac{|E_{C_0}| + |T'_0| + \sum_{j=1}^{k_i}(|E_{C_j}| - |E_{C_{j-1}}| + |T'_j| - |T'_{j-1}|)}{|C_0| + |T'_0| + \sum_{j=1}^{k_i}(|C_j| - |C_{j-1}| + |T'_j| - |T'_{j-1}|)}. \tag{7.2}$$

Denote $|E_{C_j}| - |E_{C_{j-1}}|$ by e_j, the number of links in the j-th ear. Similarly $|C_j| - |C_{j-1}| = v_j$, the number of inner vertices in the j-th ear. Clearly $e_j = v_j + 1$. Let t_j be the number of vertices which are at distance 1 or 2 from C_j, but they are at a greater distance from C_{j-1}. $|T'_j| - |T'_{j-1}|$ can be estimated as $t_j - v_j$, since v_j inner vertices of the j-th ear are transferred from T'_{j-1} to C_j. (Note that all inner vertices of the j-th ear are inside T'_{j-1}, because the ear has length at most 4.)

We can rewrite (7.2):

$$\frac{|E_{C_0}| + |T'_0| + \sum_{j=1}^{k_i}(e_j + (t_j - v_j))}{|C_0| + |T'_0| + \sum_{j=1}^{k_i}(v_j + (t_j - v_j))} = \frac{|E_{C_0}| + |T'_0| + \sum_{j=1}^{k_i}(t_j + 1)}{|C_0| + |T'_0| + \sum_{j=1}^{k_i} t_j}. \tag{7.3}$$

Let $t = \max_{1 \le j \le k} t_j$. Replacing each t_j in (7.3) by t we get

$$\frac{|E_{C(i)}| + |T(i)|}{|C(i)| + |T(i)|} \ge \frac{|E_{C_0}| + |T'_0| + k_i(t+1)}{|C_0| + |T'_0| + k_i t}. \tag{7.4}$$

Let $k'_i = k_i + (|C_0| + |T_0|)/t$. Then $|C_0| + |T_0| + k_i t$ can be expressed as $k'_i t$ and $|E_{C_0}| + |T_0| + k_i(t+1) = k'_i(t+1) + |T_0|/t + |C_0|(t+1)/t - |E_{C_0}| \in k'_i(t+1) - O(1)$, because $|C_0|$, $|E_{C_0}|$, $|T_0|$ and t are in $O(1)$ (see Lemma 7.2.2). Applying to (7.4) and taking into account $|C_k| + |T_k| \in O(k't)$ we get

$$|E_{C(i)}| + |T(i)| \geq \frac{k_i'(t+1) - O(1)}{k_i't}(|C(i)| + |T(i)|)$$

$$= \frac{t+1}{t} \cdot (|C(i)| + |T(i)|) - O(1). \tag{7.5}$$

Note that this expression is in a more general form than the simple lower bound for ceased broadcasting. It says that any algorithm that has reached r vertices must have used $r(t+1)/t - O(1)$ links.

All we need now is to bound t:

Lemma 7.2.4. *t can be bounded by 7 for ears of length at most 4.*

Proof. First note that there is only a finite (and not really large) number of possible cases which can be tested by computer. We perform the case analysis.

Consider an ear of 4 links and 3 vertices u, v and w being added to the current core graph C_j. These three vertices either lie on a line or not. In the first case, all possible situations (up to the symmetry) are shown in Figure 7.10.

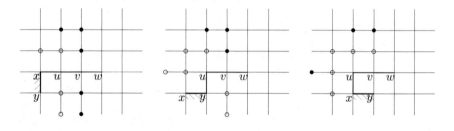

Fig. 7.10.

The vertices x and y are in C_j. Full circles represent vertices potentially added to the 2-neighborhood of C_{j+1} by adding the ear uvw to C_j. Only these vertices may contribute to t. Empty circles represent vertices in the 2-neighborhood of uvw which also belong to the 2-neighborhood of C_j and thus do not contribute to t.

Due to the symmetry only the left part from the vertical axis passing through v is shown. t is in these cases bounded by 6, 4 and 5, respectively. (If the right part is a mirror image of the left part. Otherwise t is even smaller.)

If u, v and w don't lie on a line, the four possible cases are shown in Figure 7.11. Again it is sufficient to consider only one half (left bottom) of the situation.

The value of t is in these cases bounded by 4, 2, 7 and 7, respectively.

Ears of smaller length are handled similarly. All possible cases can be drawn on previous figures, just with a smaller number of t-vertices.

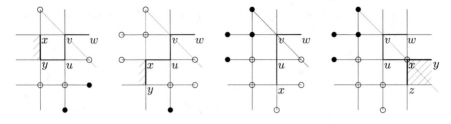

Fig. 7.11.

☐

Now we can apply (7.5):

Proposition 7.2.5. *Any broadcasting algorithm on unoriented tori that reached r vertices must have used at least $\frac{8}{7}r - O(1)$ links in the worst case.*

Corollary 7.2.6. *Broadcasting on unoriented N-node tori requires the use of at least $\frac{8}{7}N - O(1)$ links, even in the synchronous case.*

7.3 Broadcast on Hypercubes

In this section we present a time- and bit-optimal broadcast on hypercubes, without using any sense of direction (i.e., without using any node and edge labeling comprising the network topology).

7.3.1 Preliminaries

An n-dimensional hypercube is an undirected graph Q_n consisting of $N = 2^n$ vertices and $1/2Nn$ edges.

Vertices are represented by binary strings of length n. For two vertices, their *Hamming distance* is the number of positions on which they differ. There is an edge between two vertices iff their Hamming distance is 1. Q_n is regular of degree n and has diameter n.

The following notation will be used:

e_i for $i = 1, \ldots, n$ is a bit vector representing $0^{i-1}10^{n-i}$. Vectors e_i are called *unit vectors*.

v_i denotes the i-bit of v, where v is an n-bit binary string.

and, or and $\oplus$ denote bitwise *and, or* and *exclusive or* on n-bit binary strings, respectively.

$Q(v; e_{i_1}, \ldots, e_{i_l}), 1 \leq l \leq n$, is a sub-hypercube containing vertices $v \oplus \alpha_1 e_{i_1} \oplus \cdots \oplus \alpha_l e_{i_l}$, where $\alpha_i \in \{0, 1\}$, and e_{i_j} are distinct unit vectors. $e_{i_1}, \ldots, e_{i_l}$ are called *generators* of $Q(v; e_{i_1}, \ldots, e_{i_l})$.

Let E be any subset of the set of unit vectors and E^C be its complement, then $Q(v; E)$ and $Q(v; E^C)$ are called orthogonal.

Proposition 7.3.1. *The following properties on Q_n hold:*

1. *Every pair of distinct vertices of Q_n have either 0 or 2 common neighbors.*
2. *Vertices u and v in Q_n have 2 common neighbors iff their Hamming distance is 2. If u and v have the same number of 1's in their representation, these neighbors are u **and** v and u **or** v.*
3. *Each vertex in Q_n is unambiguously determined by any triple of its neighbors.*
4. *For all v in Q_n and all l-tuples $(e_{i_1}, \ldots, e_{i_l})$, $1 \leq l \leq n$, of distinct unit vectors there exists a unique isomorphism σ of $Q(v; e_{i_1}, \ldots, e_{i_l})$ to $Q(0^n; e_1, \ldots, e_l)$ such that v is mapped to 0^n and each e_{i_j} is mapped to e_j.*

A *dominating set* of a graph G is a set D of vertices such that every vertex of G belongs to D or is adjacent to a vertex from D. We say that the dominating set D is *perfect* iff each vertex has exactly one neighbor (including itself) in D.

Proposition 7.3.2. *If $n = 2^k - 1$ then there exists a perfect dominating set of size $2^n/(n+1)$ in Q_n.*

Proof. Let X be the identity of a vertex x written as the binary row matrix $n \times 1$. Take

$$D_n = \{x | AX = (0, \ldots, 0)^T\},$$

where A is the $k \times n$ binary matrix, in which the i-th column is the binary representation of the number i, for $i = 1, \ldots, n$. All computations are in Z_2 (modulo 2).

Let $AX = Y$ for some vertex x. Interpret Y as the binary number y. If $y = 0$, then $x \in D_n$, otherwise consider $x' = x + e_y$. Then x' is the neighbor of x that lies in D_n:

$$AX' = A(X + E_y) = AX + AE_y = Y + Y = (0, \ldots, 0)^T.$$

$\square$

Proposition 7.3.3. *For each n there exists a dominating set D_n of size at most $2^{n+1}/(n+1)$ in Q_n.*

Proof. Take

$$D_n = D_n^0 \cap D_n^1,$$

where

$$D_n^0 = \{x | AX = (0, \ldots, 0)^T\} \text{ and } D_n^1 = \{x | AX = (1, 0, \ldots, 0)^T\}.$$

A is the matrix from the previous proposition.

Let $AX = Y$ for some vertex x. Interpret Y as the binary number y. If $y < n$, then follow as in the previous proposition. If $y > n$, then flip the most significant bit of y to get y'. Now $y' < n$, since $y < 2n$. Consider $x' = x + e_{y'}$. Then

$$AX' = AX + AE_{y'} = Y + Y' = (1, 0, \ldots, 0)^T,$$

thus x has the neighbor x' in D_n^1.

□

7.3.2 Partial Broadcasting and Orientation

Definition 7.3.4. *Let M be a subset of vertices of Q_n and s be some vertex in M. Define $M_i^s = \{v \mid v \in M, d(v, s) = i\}$, where $d(x, y)$ is the Hamming distance of x and y.[1] Let $\mathcal{L}_M = \max\{i \mid M_i \neq \emptyset\}$.*
We say that M is a computable mask from s if and only if for all $v \in M$ at least one of the following conditions is true:

1. *$v \in M_0$ ($v = s$),*
2. *$v \in M_1$ (v is a neighbor of s),*
3. *$v \in M_k$ and v has at least three neighbors in M_{k-1},*
4. *$v \in M_k$ and v has two neighbors x and y in M_{k-1}. Let v' be the second common neighbor of x and y. Then $v' \in M_{k-2}$.*

If $s = 0^n$, we say that M is a computable mask.

The following statements follow easily from the previous definition:

The union of computable masks from s is a computable mask from s.
$Q(s; E)$ is a computable mask from s for any set of generators E.

The Basic Partial Broadcasting and Orientation

The following algorithm $\mathcal{A}1$ is used to perform partial broadcasting and orientation on a given computable mask M.

Algorithm $\mathcal{A}1$:

Mask M and source vertex s are fixed and known to all vertices.

Messages used:
(*You are:*, x) and (*I'm*, x)
where x is an n-bit binary string.

Variables used at vertex v:

$Name_v$: n-bit binary string containing the identity of v. Initial value is *empty*, only s knows its identity.
$Label_v(h)$: labels of links incident to v, initial value is *empty*. Exceptions are links from s leading to vertices in M, which have their correct labels.

---

[1] We will use M_i instead of M_i^s when s is obvious from the context.

Algorithm in the starting vertex s:

For all links h:
    if $s \oplus Label_s(h) \in M$ **then**
        Send(*You are:*, $s \oplus Label_s(h)$) on link h;

Algorithm in a vertex v:

Upon receiving a message (*You are:*, x) on link h:

$Name_v := x$;
$Label_v(h) := x \oplus s$;
if $Name_v \in M$ **then**
    Send(*I'm*, $Name_v$) on all links except h;

Upon receiving a message (*I'm*, x) on link h_1:

if $Name_v = empty$ **then**
    Wait for the second message (*I'm*, y); {arrived on link h_2}
    Compute identities w_1 and w_2 of the two common neighbors
        of x and y.
    if $x \in M_{k-1}$ and $y \in M_{k-1}$ and $w_1 \in M_{k-2}$ and $w_2 \in M_k$ **then**
        $Name_v := w_2$;
    else
        Wait for the third message (*I'm*, z); {arrived on link h_3}
        $Name_v :=$ the unique vertex that has neighbors x, y and z;
    fi

    if $Name_v \in M$ **then**
        Wait for *I'm* messages from all your predecessors in M;
        Send(*I'm*, $Name_v$) to all neighbors;
    fi
fi

Each vertex that has computed its name computes labels of links within M:

    For each link h on which (*I'm*, x) has been received:
        $Label_v(h) := x \oplus Name_v$;

Proposition 7.3.5. *If M is computable from s and s knows labels of all links leading to its neighbors in M, then the algorithm $\mathcal{A}1$ computes*

  1. identities of all vertices in M,
  2. link labels for each link between vertices in M.

Proof. 1. We will prove Proposition 7.3.5 by induction on the distance from s.

A vertex v computes its identity in the following cases:

After receiving the message *You are*: In this the case the identity is correctly computed because the source s knows the labels of the links to vertices in M_1. This is the first step of induction.

After receiving two messages (*I'm*, x) and (*I'm*, y) from x and y in the same layer: This case applies only if w_1, the common neighbor of x and y in the previous layer, is also in M. By induction, the computed names of x and y are correct, and because each vertex waits for messages from all its predecessors before announcing its name, w_1 has already computed its name (or $w_1 = s$). It follows that w_2 is the correct name for v.

After receiving three *I'm* messages: Its identity is unambiguously given in this case. Identities that came in these messages are correct by the induction hypothesis.

We have proved that vertices correctly compute their identities. The fact that indeed each vertex $\in M$ computes its identity can be easily proved by induction on the distance from s from the computability of M.

2. This follows from the fact that each vertex in M computes its identity and announces it, together with the handling of these messages by the algorithm.

$\square$

Proposition 7.3.6. *The algorithm $\mathcal{A}1$ uses at most $n|M|$ messages of size $O(n)$ bits.*

Proof. Each vertex in M sends at most n messages of size $O(n)$ bits each. Vertices outside M do not send messages.

$\square$

Proposition 7.3.7. *The time complexity of $\mathcal{A}1$ is $\mathcal{L}_M + 1$.*

Proof. By induction on the distance from the s: Vertex $v \in M_k$ computes its name not later than at the time k. The $+1$ term stands for computing the labels of the last links.

$\square$

7.3.3 Bit-Efficient Partial Broadcasting Algorithm

The algorithm $\mathcal{A}1$ does not communicate efficiently, because each vertex v sends its whole identity to all its neighbors. During the computation only some vertices really need the identity of v. The basic scheme is to send just *Hello* messages, and only vertices which are really interested in your full identity will ask you for it. A vertex v needs to learn the identities of only two or three of its neighbors to be able to compute its identity.

Algorithm $\mathcal{A}2$:

Messages used – *Hello, Who are you?* and messages of $\mathcal{A}1$

Variables used at vertex v – as in $\mathcal{A}1$

Algorithm in the starting vertex s is the same as in $\mathcal{A}1$

Algorithm in a vertex v:

Upon receiving a message (*You are:*, x) on link h:

$Name_v := x$;
if $Name_v \in M$ **then**
    Send(*Hello*) on all links except h;

Upon receiving the first *Hello* message on link h_1:

Wait for the second *Hello* message; {arrived on link h_2}
Send *Who are you?* on h_1 and h_2;
Wait for (*I'm*, x) and (*I'm*, y) on h_1 and h_2, respectively.
Compute identities w_1 and w_2 of the two common neighbors of x and y.
if $x \in M_{k-1}$ and $y \in M_{k-1}$ and $w_1 \in M_{k-2}$ and $w_2 \in M_k$ **then**
    $Name_v := w_2$;
else
    Wait for the third *Hello* message; {arrived on link h_3}
    Send *Who are you?* on h_3;
    Wait for (*I'm*, z) on h_3;
    Compute $Name_v$ as the unique vertex that has neighbors x, y and z;
fi

if $Name_v \in M$ **then**
    Wait until you receive *Hello* messages from all your predecessors in M;
    (Wait until you have received k *Hello* messages, where k is the number
    of your predecessors in M.)
    Send(*Hello*) to all neighbors;

Upon receiving *Who are you?* message on link h:

    Send(*I'm*, $Name_v$) on h

Proposition 7.3.8. *If M is a computable mask from s, and s knows the labels of all links leading to its neighbors in M, then the algorithm $\mathcal{A}2$ correctly computes the identities of the vertices in M.*

Proof. Follows the same line as the proof of $\mathcal{A}1$. The only substantial difference is that $\mathcal{A}2$ at vertex v tests the condition that it received *Hello* messages from all its predecessors in M by counting these messages, while $\mathcal{A}1$ can do this by testing identities. Note that $\mathcal{A}2$ can't be fooled by *Hello* messages from successors of v, because the successor must receive a *Hello* from v to proceed and send its *Hello*.   □

Proposition 7.3.9. *The algorithm $\mathcal{A}2$ uses $O(n|M|)$ messages containing totally $O(n(|M|+|M'|))$ bits, where M' is the set of all vertices of $Q_n - M$ that have at least two neighbors in M.*

Proof. We charge constant-bit messages (*Hello* and *Who are you?*) to the sender and long ((*I'm*, x) and (*You are:*, x)) messages to the receiver. Each vertex in $M \cup M'$ is charged $O(n)$ bits, because each vertex in M sends $O(n)$ messages and each vertex in $M \cup M'$ receives at most 3 long messages. $\qquad\square$

Note that if $M = Q(v; E)$ then $M' = \emptyset$.

Proposition 7.3.10. *Time complexity of $\mathcal{A}2$ is $1 + 3(\mathcal{L}_M - 1)$.*

Proof. By induction on the distance from s, similarly as for $\mathcal{A}1$. $\mathcal{A}2$ will proceed with at most 3 time units per one layer, with the exception of the first layer which takes at most one time unit. $\qquad\square$

7.3.4 Applications – Optimal Computable Masks for Special Target Sets

The general scheme of the broadcasting from the vertex s to some target set of vertices T is the following:

Choose a set M, $M \supset T$, such that M is computable from s. It is desirable to make M as small as possible.
Make sure that s knows the labels of the links leading to its neighbors in M.
Use $\mathcal{A}1$ ($\mathcal{A}2$) to broadcast on M.

One particular case is to inform a single vertex v at the distance d about its identity. (See, for example, the algorithm FarSend() from the previous chapter.) Because of Proposition 7.3.1 it is sufficient to consider the case $s = 0^n$, $v = 1^d 0^{n-d}$.

Lemma 7.3.11 (FarSend). *There exists a computable mask M of size $1 + d(d+1)/2$ which contains the vertex $v = 1^d 0^{n-d}$.*

Proof. It is easy to verify that

$$M = \{0^i 1^k 0^{n-i-k} \mid 0 \leq k \leq d, i + k \leq d\}$$

is such a mask. $\qquad\square$

Let $M_{|d}$ denote

$$M \cap \{0,1\}^d 0^{n-d}.$$

The mask used in the previous lemma is indeed optimal.

Lemma 7.3.12. *Let M be any computable mask that contains vertex $1^d 0^{n-d}$, then*

$$|M_{|d}| \geq 1 + d(d+1)/2.$$

Proof. By induction on d. For $d = 1$ the condition trivially holds, because M must contain both the source and the target vertex.

General case: The vertex $1^d 0^{n-d}$ must have at least two predecessors x and y in M_{d-1}. We may assume

$$x = 1^{d-1} 0^{n-d+1} \text{ and } y = 01^{d-1} 0^{n-d}.$$

From the induction hypothesis applied on x we get

$$|M_{|d-1}| \geq 1 + d(d-1)/2.$$

It holds that $y \notin M_{|d-1}$, because y has 1 at position d. Since the identity of a vertex is determined bitwise **or** according to its predecessors (this holds only for the source 0^n), y has a predecessor y' with 1 at position d. We may apply this argument on y' and so on until we get down to $0^{d-1} 1 0^{n-d}$. This means that there is a chain of $d - 1$ vertices

$$y, y', y'', \ldots, 0^{d-1} 1 0^{n-d}$$

that are not in $M_{|d-1}$. However, all these vertices must be in $M_{|d}$, otherwise $y \notin M_{|d}$. Hence we get

$$|M_{|d}| \geq 1 + |M_{|d-1}| + d - 1 \geq 1 + 1 + d(d-1)/2 + d - 1 = 1 + d(d+1)/2.$$
$\square$

Another useful task is to orient a vertex v consistently with the source s. (See, for example, the algorithm FarOrient() from the previous chapter.) While this can be trivially done using a mask of size $O(nd^2)$, the following lemma shows how to do it using a mask of size $O(nd)$.

Lemma 7.3.13 (FarOrient). *There exists a computable mask M_{Orient} of size $nd + n - 2d + 2$ containing the target set $S(1^d 0^{n-d}; 1)$. Here $S(x; r)$ denotes the sphere with center x and radius r.*

Proof. Take

$$M_{Orient} = M' \cup M'' \cup M'''$$

where

$$
\begin{aligned}
M' &= \{0^i 1^k 0^{n-i-k} \mid 0 \leq i + k \leq d\}, \\
M'' &= \{1^k 0^{d-k} 0^j 1 0^{n-d-j-1} \mid 0 \leq k \leq d, 0 \leq j < n - d\}, \text{ and} \\
M''' &= \{1^i 0^j 1^{d-i-j} 0^{n-d} \mid i > 0, j > 0, i + j < d\}.
\end{aligned}
$$
$\square$

This mask is close to optimal.

Lemma 7.3.14. *If M is a computable mask containing $S(1^d 0^{n-d}; 1)$, then*

$$|M| \geq nd + n - d(d+1)/2 + 1.$$

Proof. The proof uses the same ideas as the proof of Lemma 7.3.12.

□

This lower bound differs from the upper bound only by the term $|M'''|$.

Combining the previous results with $d \leq n$, we get that any vertex can be reached (or oriented) using $O(n^3)$ messages and $O(n^4)$ bits, which is optimal when using $\mathcal{A}1$.

7.3.5 The Broadcasting Algorithm

We will follow the outline of the algorithm from [DDKP+98]. The main difference lies in the application of $\mathcal{A}1$ and $\mathcal{A}2$ for partial broadcasting on sub-hypercubes instead of the "jo-jo" technique from [DDKP+98].

The broadcasting algorithm $\mathcal{BR}$ works in two stages:

STAGE I: $\mathcal{A}1$ is launched with the mask $M1 = \cup_{v \in D_k} M_{Orient}(v)$ from the source, where D_k is a perfect dominating set of $Q(0^n; e_1, \ldots, e_k)$. k is chosen to be of the form $2^r - 1$ for some integer r, while being as close as possible to $n/2$. Clearly $n/3 \leq k \leq 2n/3$.

STAGE II: $\mathcal{A}2$ is launched with the mask

$$M(v) = \{x | x \in Q_n, x_1 x_2 \ldots x_k = v_1 v_2 \ldots v_k\}$$

from each $v \in D_k$.

$\mathcal{A}1$ and $\mathcal{A}2$ were presented using implicit knowledge of M and s. Now we use more invocations, so we must specify what is implicit now and how we use it. Each vertex knows (from knowing n) $M1$, $M = \cup M(v)$ and D_k. Messages of different stages are marked by a stage mark (i.e., by $O(1)$ bits). We cannot afford to mark messages of different invocations of $\mathcal{A}2$ in Stage II, but this is not a problem since these invocations do not interfere.

There are no vertices receiving messages from two different invocations of $\mathcal{A}2$, because there is no vertex in $Q(0^n; e_1, \ldots, e_k)$ with two neighbors in D_k. (D_k is a perfect dominating set.)

Each vertex of $M(v)$ can decide to which invocation of $\mathcal{A}2$ it belongs simply by looking at the first k bits of the first ($I'm$, x) message it received, learning which v and $M(v)$ to use.

Proposition 7.3.15. *(Correctness) The algorithm $\mathcal{BR}$ is a broadcasting algorithm on unoriented hypercubes.*

Proof. First, note that the mask $M1$ is computable from 0^n and each $M(v)$ is computable from v.

It is sufficient to show that each vertex v has a neighbor in some $M(v)$ – that means that it receives a *Hello* message.

Let $v \in Q_n$ be an arbitrary vertex. Take $v' = v_1 v_2 \ldots v_k 0^{n-k}$. Since $v' \in Q(0^n; e_1, \ldots, e_k)$, it has a neighbor $x \in D_k$. Therefore v has a neighbor $x' = x_1 \ldots x_k v_{k+1} \ldots v_n \in M(x)$.

$\square$

Proposition 7.3.16. *The algorithm $\mathcal{BR}$ broadcasts on unoriented N-vertex hypercubes using $O(N)$ bits.*

Proof. The total bit cost of Stage I can be bounded by $|D_k|$ times the bit cost of orienting one vertex, which is

$$O\left(\frac{2^k}{k+1} \cdot n^4\right) \in o(N).$$

The total bit cost of Stage II can be bounded by the number of invocations of $\mathcal{A}2$ times the cost of one such invocation:

$$O\left(\frac{2^k}{k+1} \cdot n \cdot 2^{n-k}\right) = O(nN/(k+1)) = O(N).$$

$\square$

The time complexity of Stage I can be bounded by $k + 1$. Similarly, Stage II works in time $3(n - k) - 1$, summing to the total time $3n - 2k < 7/3n$.

7.3.6 Bit-Optimal Broadcasting in Time n

Note that we can apply $\mathcal{A}1$ instead of $\mathcal{A}2$ to achieve the algorithm that reaches optimal time n, but its bit complexity will be $N \log N$.

However, we could choose $k = n - 4$ and use the previous approach. This would result in an algorithm running in time

$$k + 1 + 3(n - k) - 1 = n + 8,$$

being still bit optimal. If the root v at Stage II is at the distance $d < n - 12$ from the initiator, all vertices in $M(v)$ and their neighbors will be informed in time

$$d + 1 + 3(n - k) - 1 = n - 12 + 12 = n.$$

There are

$$O(n^{12} \cdot (n + 1) \cdot 2^4) = O(n^{13})$$

vertices that would be informed later than in time n. These vertices can be informed directly in time n, using $\mathcal{A}1$ with a mask containing them all, in cost $O(n^4)$ per vertex, with the total added cost

$$O(n^{17}) \in o(N).$$

The problem is that there may not be a perfect dominating set D_{n-4} of Q_{n-4}. But we do not need a perfect dominating set as we get a dominating set which can be partitioned into the constant number of sets $D^0, \ldots, D^c$ such that invocations of $\mathcal{A}2$ from vertices of D_i do not interfere. All we need is to mark an invocation of $\mathcal{A}2$ from v by the index of D^i in which v lies.

One such dominating set is the dominating set from Proposition 7.3.3.

8

Leader Election in Asynchronous Distributed Networks

*I have no high opinion of a man
who is not cleverer today
than he was yesterday.*

Abraham Lincoln

8.1 Distributed Point-to-Point Network Model

In asynchronous distributed networks, there are inherently three sources of nondeterminism. Processes (performing within network nodes) are inherently asynchronous since there is no universally available clock for synchronizing their actions. The uncertainties in communication delays imply nondeterministic behavior in the sending and receiving of messages. And if FIFO requirements on links are not necessary, messages sent via the same link can overtake each other. All these three forms of nondeterminism are allowed in the model.

Processes are allowed to communicate only by sending messages out of a finite number of output ports. Output ports are connected (by a fixed point-to-point communication network defined separately from the processes) to the input ports of other processes. Thus, a process need not know the identity of the processes with which it is communicating.

Indeterminate delays between the sending and receiving of messages are explicitly allowed. Messages sent between a given pair of ports may even arrive out of order. The only requirement is that if a process continues to try to receive messages from a certain input port, then any message sent to this port will be received within a finite number of steps. Communication has no implied synchronization. A process always continues to execute after sending a message. If a process attempts to receive a message when none is ready, the process may continue with other tasks or wait for a message by explicit looping. The given mechanism can be used to implement a virtual system in which processes exchange messages in pairs, but this must be done explicitly in the programs of the individual processes.

Now we present a formal model for describing message passing systems on a general network topology.

A process is envisioned as an algorithm (protocol), residing at a node of a distributed network. A finite set of ports connects each process to the network. Every step, or transition, of a process can be identified as a *send*, *receive* or *local* transition. One of the ports is associated with each send or receive.

Formally, a *process* is an 8-tuple

$$p = (States_p, Input_p, Output_p, M_p, X_p, T_p, Msg_p, Port_p)$$

where $States_p$ is a set of states of p, $Input_p$ is a finite set of input ports, $Output_p$ is a finite set of output ports, M_p is a set of messages, X_p is an initial state of p, $X_p \in States_p$, T_p is the state transition function of p, Msg_p is the message output function of p, and $Port_p$ is the port designation function of p. $States_p$ is partitioned into three sets: $Send_p$, $Receive_p$ and $Local_p$. The transition function, T_p, is total on two distinct domains,

$$T_p \ : \ Send_p \cup Local_p \mapsto States_p$$

and

$$T_p \ : \ (Receive_p \cup NONE) \times M_p \mapsto States_p$$

(where $NONE$ is a unique value, distinct from any message). The message output function,

$$Msg_p \ : \ Send_p \mapsto M_p$$

is also total. Finally, the total function

$$Port_p \ : \ Send_p \cup Receive_p \mapsto Input_p \cup Output_p$$

is such that if $x \in Send_p$, then $Port_p(x) \in Output_p$, and if $x \in Receive_p$, then $Port_p(x) \in Input_p$.

The basic unit of communication in the model is a "message". To explicitly allow indeterminate delays between the sending and receiving of a message, a "delay factor" is associated with each message sent. The following definition allows manipulation of messages which are being transmitted.

For any process p, a *message pair* of p is an element of $M_p \times N$, where N is the set of non-negative integers. The first element of a message pair is the *message*, and the second is the *delay factor*. An *input queue* of p is a finite list of message pairs of p. Let

$$mq = (m_1, d_1), (m_2, d_2), \ldots, (m_k, d_k)$$

be an input queue of p. Message m_i of mq is said to be *ready* if $d_i = 0$. For any message pair (m, d), define

$$add(mq, (m, d)) = (m_1, d_1), (m_2, d_2), \ldots, (m_k, d_k), (m, d).$$

Define $first(mq)$ to be m_i if there exists a least i, $1 \leq i \leq k$, such that $d_i = 0$, and to be $NONE$ otherwise. Define $remove(mq)$ to be mq if $first(mq) = NONE$, and otherwise to be

$$(m_1, d_1), (m_2, d_2), \ldots, (m_{i-1}, d_{i-1}), (m_{i+1}, d_{i+1}), \ldots, (m_k, d_k),$$

where i is the least positive integer such that $d_i = 0$. Finally, define

$$decr(mq) = (m_1, d'_1), (m_2, d'_2), \ldots, (m_k, d'_k),$$

where $d'_i = 0$ if $d_i = 0$, and $d_i - 1$ otherwise, for $i = 1, 2, \ldots, k$.

Let P be a set of processes. Define

$$Input_P = \cup_{p \in P} Input_p,$$

$$Output_P = \cup_{p \in P} Output_p$$

and

$$M_P = \cup_{p \in P} M_p.$$

P is said to be *compatible* if for every $p, r \in P$ such that $p \neq r$ it holds that $M_p = M_r$, and

$$Input_p \cap Input_r \neq \emptyset$$

and

$$Output_p \cap Output_r \neq \emptyset.$$

An *instantaneous description* (shortly *id*) q of a compatible set of processes P specifies the state $p(q)$ of p for each $p \in P$, and the input queue $i(q)$ at input port i for each $i \in Input_P$. If q and q' are both *id*'s of P and $p(q) = p(q')$ for each $p \in P$ and $i(q) = i(q')$ for each $i \in Input_P$, then $q = q'$. Let Q_P be the set of all *id*'s of P. Define the *id* $initid(P) \in Q_P$ to be the unique *id* such that $p(initid(P)) = X_p$ for all $p \in P$ and such that $i(initid(P))$ is empty for each $i \in Input_P$; $initid(P)$ is called the *initial id* of P.

A *message system* is a pair $S = (P, C)$, where P is a compatible set of processes, and C is a subset of $Output_P \times Input_P$. C is called the *communication relation* of S. The *transition function of* S

$$T_S : Q_P \times P \times N \mapsto Q_P$$

is defined as follows. Assume that q, q' are in Q_P, $p \in P$, $j \in N$ and $T_S(q, p, j) = q'$. Then, for every $r \in P$, $r \neq p$ implies $r(q') = r(q)$. (That is, no process can affect the state of another process.) In addition, the appropriate one of the following three cases must hold.

Case 1. $p(q) \in Send_p$. Then $q \mapsto q'$ is called a send transition of p.

Let $I = \{i \in Input_p \mid (Port_p(p(q)), i) \in C\}$.

1. For every $i \in Input_p - I$, $i(q') = i(q)$.
2. $p(q') = T_p(p(q))$.
3. For every $q'' \in Q_p$ such that $p(q'') = p(q)$ and every $k \in N$,

$$p(T_S(q'', p, k)) = p(q').$$

4. For every $i \in I$, $i(q') = add(i(q), (Msg_p(p(q)), j))$.

Case 2. $p(q) \in Receive_p$. Then $q \mapsto q'$ is called a receive transition of p.

1. For every $i \in Input_p - Port_p(p(q))$, $i(q') = i(q)$.

2. $p(q') = T_p(p(q), first(i(q)))$, where $i = Port_p(p(q))$.
3. For every q'' in Q_p such that $p(q'') = p(q)$ and $first(i(q'')) = first(i(q))$
   for $i = Port_p(p(q))$, and for every $k \in N$, $p(T_S(q'', p, k)) = p(q')$.
4. $i(q') = decr(remove(i(q)))$, where $i = Port_p(p(q))$.

Case 3. $p(q) \in Local_p$. Then $q \mapsto q'$ is called a local transition of p.

1. For every $i \in Input_p$, $i(q') = i(q)$.
2. $p(q') = T_p(p(q))$.
3. For every $q'' \in Q_p$ such that $p(q'') = p(q)$ and every $k \in N$,

$$p(T_S(q'', p, k)) = p(q').$$

By 1., a send does not affect the input queues that are not connected to its output port, a receive does not affect any input queue but the one it is receiving from, and a local transition does not affect any input queue. 2. requires that T_S conform to the transition functions of the individual processes. 3. states that the action taken depends only on the current state of the process and (for receive only) on the first ready message, if any. Finally, 4. says that a send causes its message to be appended (with delay factor j) to the input ports connected to its output ports, and a receive removes a message (if one is ready) from its input queue and decrements the waiting time of all of the remaining, unready messages.

A message system thus allows messages to be sent with arbitrary, but finite, delay factors. (If a message is sent with delay factor j, then at least j receive transitions must be made on the input port before the message can be read. Note that the FIFO handling of ready messages guarantees that no message will be delayed forever, assuming that the receiving process continues to execute receive transitions on the appropriate input port.) Messages from the same process may arrive in a different order by an appropriate restriction on the choice of delay factors. For example, all the delay factors could be forced to be zero.) Also note that a communication relation can be chosen to allow broadcasts or one-to-one process communication.

Let $S = (P, C)$ be a message system. A *schedule* h of S is a sequence whose elements are chosen from the set $P \times N$. Schedule h is *admissible* if each $p \in P$ occurs infinitely often as a first component in h. Admissible schedules are those in which no process in the system ever stops.

If q is an *id* of S and $h = (p_1, d_1), (p_2, d_2), \ldots$ is a schedule of S, then the *computation* from q by h is the sequence of *id*'s $q_1, q_2, \ldots$ such that $q = q_1$ and for $j \geq 1$, $T_S(q_j, p_j, d_j) = q_{j+1}$. Id q' is *reachable* from q by schedule h if q' occurs in the computation from q by schedule h. If schedule h is finite, then $final(S, q, h)$ is the last *id* in the computation from q by schedule h. Let $h = (p_1, d_1), (p_2, d_2), \ldots$ be a schedule of S and $q_1, q_2, \ldots$ be the computation from q_1 by h. Then the *number of message transmissions* from q_1 by schedule h is

$$msgs(S, q_1, h) = |\{j \geq 1 \mid p_j(q_j) \in send(p_j)\}|.$$

An *id* is called *quiescent* if for all schedules h, $msgs(S, q, h) = 0$. The *worst case number of messages* for a system S is given by

$$MSGS(S) = \max\{\ msgs(S, initid(S), h)\ |\ \text{h is a schedule of S}\ \}.$$

8.2 Leader Election on Distributed Networks

In this section we discuss the problem of *election* on asynchronous networks both on general as well as on specific topologies (such as rings, tori, hypercubes and complete networks).

Assuming the configuration of the network where initially each processor is in the same state, the election procedure computes the configuration where exactly one processor is in a distinguished state called *leader* and the others are in the state called *non-leader*. The processor in the state *leader* is then elected by the algorithm. Note that some algorithms require that exactly one processor is in the state *leader*; it is then easily extended by the additional procedure in which all remaining processors are informed about the chosen leader.

8.3 Leader Election on Unidirectional Rings with Optimal Average Performance

Suppose we are given a protocol that works on a unidirectional ring. The protocol is message driven, that is, it is composed of atomic steps. In such a step, one processor receives a message, does local computation, and then it might send a message. Also, we assume that each message is eventually delivered. So, all processors will eventually execute their first step, which will be identical in all possible executions. Using induction, it can be shown that all processors will also execute their second step, which will be identical in all executions, and so on (complete the proof!). In other words, in all executions of a protocol in a unidirectional ring, all messages sent are the same; in particular, all executions have the same number of messages.

Now we present a Chang and Robert's algorithm solving leader election on a unidirectional ring with N processors. This algorithm, though being $O(N^2)$ in the worst-case, will prove to be the best possible within the multiplicative constant in terms of average performance.

In this algorithm, at each processor *id* is a constant holding the identity of the processor, v_{id} is a variable containing identities received by the processor, and *state* is a variable that will initially be set to *candidate* by all processors, and eventually be *leader* in one processor and *non-leader* in all others.

    state := *candidate*;
    send(*id*);
    repeat

```
receive(v_id);
if v_id > id then
    send(v_id);
    state := non-leader;
end if
until v_id = id;
state := leader;
```

One can easily show by induction that each processor eventually sends its first message, and eventually receives its first message, and, in a similar manner, that each processor that receives an identity larger than its own eventually sends its i-th message, and eventually receives its i-th message, for every $2 \leq i \leq N$. Due to the FIFO ordering, it follows that each processor will eventually receive the maximal identity, and thus will become a *non-leader*, the processor that holds this identity will receive it and terminate the algorithm. Note that each other processor will not be aware of termination, though it will reach a stage where no more messages will be sent or received (or observed from the outside). We have thus proved that

Theorem 8.3.1. *Every execution of Chang and Robert's algorithm terminates, with only the processor with the maximal identity being aware of termination, and such that exactly one processor stays in a leader state.*

The worst-case message complexity of Chang and Robert's algorithm is $O(N^2)$, and its time complexity is clearly N.

Lemma 8.3.2. *The average message complexity of Chang and Robert's algorithm is $O(N \log N)$.*

Proof. Assume a ring with N designated positions, and a set of identities $\{1, 2, \ldots, N\}$ such that the $N!$ permutations are equally probable. Let $P(i, k)$ denote the probability that a message carrying the identity i will be sent exactly k times. This equals the probability that the first $k - 1$ neighbors of the processor holding identity i will have identities smaller than i, and the k-th neighbor is holding an identity greater than i.

Hence

$$P(i, k) = \frac{\binom{i-1}{k-i}}{\binom{N-1}{k-i}} \cdot \frac{N-i}{N-k}.$$

Therefore, the expected number α of the messages is

$$\alpha = N + \sum_{i=1}^{N-1} \sum_{k=1}^{i} k \cdot P(i, k)$$

and, by changing the order of the summation, we can derive

$$\alpha = N + \sum_{k=1}^{N-1} \frac{N}{k+1} = N(1 + \frac{1}{2} + \frac{1}{3} + \cdots + \frac{1}{N})$$
$$= N \cdot H_N \approx 0.69 N \log N + O(1).$$

Note: The following is an alternative proof: We count the total number of messages sent over all $N!$ possible rings, and then divide it by $N!$ to get the average behavior. Observe a given processor P_i of the ring; its identity certainly makes one step. This counts for a total of $N!$ steps. In order for P_i's identity to make its second step, it must be the largest among the P_i's and P_{i+1}'s identities, which happens in exactly $\frac{N!}{2}$ cases. In order for P_i's identity to make its k-th step, it must be the largest among the k processors $P_i, P_{i+1}, \ldots, P_{i+k-1}$, which happens in exactly $\frac{N!}{k}$ cases. We thus get a total of

$$N!(1 + \frac{1}{2} + \frac{1}{3} + \cdots + \frac{1}{N})$$

messages, as in the previous alternative.

$\square$

Now we show that on a unidirectional asynchronous ring, whose size N is unknown to the processors, any algorithm for the asynchronous ring requires at least $\Omega(N \log N)$ messages on average.

Assuming the processors $P_0, P_1, \ldots, P_{N-1}$ have identities $id_0, id_1, \ldots, id_{N-1}$. Being in a complete asynchronous environment, we assume that the algorithm is message-driven; that is, we assume that any non-empty subset of the processors starts the algorithm, which includes atomic steps of receiving a message, doing local computation and sending (maybe more than one message) to neighbors (in the unidirectional ring there is exactly one neighbor to each processor); no other messages are sent; specifically, no "spontaneous" messages are generated in the system. Being on a unidirectional ring, we can assume that the messages sent during each execution of a given algorithm are the same specific execution. Also, we make the assumption that each message contains only the identity of the sender, concatenated to the message it received. Thus, each message is of the form $(id_i, id_{i+1}, \ldots, id_j)$, by which we mean that message (id_i) was initiated by P_i to P_{i+1}, which, as a response, sent the message (id_i, id_{i+1}) to P_{i+2}, etc., and eventually P_j sent the message $(id_i, id_{i+1}, \ldots, id_j)$ to P_{j+1}. With each execution of the algorithm we can thus associate a set of such messages.

Each ring we associate with the sequence of identities of its processors. For a sequence $s = (s_1, s_2, \ldots, s_t)$, whose length $length(s)$ is t, any sequence $s = (s_1, s_2, \ldots, s_r)$, $1 \le r < t$, is termed a prefix. For two sequences s and r, their concatenation st is the sequence obtained by writing the elements of s followed by those of r. If $s = rtu$, then we say that t is a subsequence of s. $C(s)$ is the set of all cyclic permutations of s. Clearly $|C(s)| = length(s)$.

Another important assumption is that in any execution of a maximum-finding algorithm A, at least one processor must see its own value; in other words, in a ring labeled s, at least one message $C(s)$ is sent by A.

Let D denote the set of all finite sequences of distinct integers; that is, if Z denotes the set of integers, then

$$D = \{(s_1, s_2, \ldots, s_k) \mid k \geq 1, \forall i \; s_i \in Z, \forall i \neq j \; s_i \neq s_j\}.$$

For $s \in D$, $E \subset D$, $|E| < \infty$, let

$$M(s, E) = |\{t \in E \mid t \text{ is a prefix of some } r \in C(s)\}| \tag{8.1}$$

and

$$M(s, E) = |\{t \in E \mid t \text{ is a prefix of some } r \in C(s) \wedge length(t) = k\}| \tag{8.2}$$

Clearly

$$M(s, E) = \sum_{k=1}^{length(s)} M_k(s, E). \tag{8.3}$$

Definition 8.3.3. *A set $E \subseteq D$ is exhaustive if it has the following two properties:*

Prefix property: if $s \in E$ then any prefix of s is also in E $\qquad$ (8.4)

Cyclic permutation property: $\forall s \in D, \; C(s) \cap E \neq \emptyset.$ $\qquad$ (8.5)

Lemma 8.3.4. *Let $s, t, u \in D$, such that u is a subsequence of both s and t, and let A be a maximum-finding algorithm. If in the execution of A on ring s a message u is sent, then in the execution of A on ring t a message u is sent.*

Proof. By induction on the length of u, and using the assumption of the message-driven model.

$\square$

The essence of the proof is that a processor can tell about its environment only through the messages it receives; hence, if in two scenarios it gets the same messages, it must react identically. This idea is widely used in lower bounds arguments in distributed computing.

Theorem 8.3.5. *Given a maximum-finding algorithm A for unidirectional rings, there exists an exhaustive set $E(A)$ such that for every ring s, A sends at least $M(s, E(A))$ messages on s.*

Proof. Define

$$E(A) = \{s \in D \mid \text{a message } s \text{ is sent when } A \text{ executes on ring } s\}. \tag{8.6}$$

1. *Prefix property:* if $tu \in E(A)$, $length(t) \geq 1$, then message tu was sent on the ring labeled tu by (8.6), hence by Lemma 8.3.4, t was sent on the ring t, hence $t \in E(A)$.

2. *Cyclic permutation property:* for a ring s, by the above assumption, at least one processor must send a message in $t \in C(s)$, thus $t \in E(A)$.
3. Let $t \in E(A)$ and let t be a subsequence of s. A message t was sent on ring t by (8.6), hence by Lemma 8.3.4 the message t was also sent on s. Thus, by (8.1) at least $M(s, E(A))$ messages were sent on s.

This completes the proof of the theorem.

□

For the lower bound, we assume that, given a set of N distinct identities I, all the $N!$ permutations of its elements are equally probable. (Note that we assume here N distinct locations on the ring; assuming only $(N-1)!$ permutations yields similar results; this modification is not discussed here). Let $ave_A(I)$ and $worst_A(I)$ denote the average-case and worst-case message complexities of a given algorithm A for a set of initial identities I. Following Theorem 8.3.5, we get:

Theorem 8.3.6. *For a given algorithm A for maximum-finding in unidirectional rings, and set of N identities I, we have*

$$ave_A(I) \geq \frac{1}{N!} \sum_{s \in perm(I)} M(s, E(A)) \tag{8.7}$$

and

$$worst_A(I) \geq \max_{s \in perm(I)} M(s, E(A)). \tag{8.8}$$

Theorem 8.3.7. *For a given algorithm A for maximum-finding in unidirectional rings, and set of N identities I, we have*

$$ave_A(I) \geq \Omega(N \log N).$$

Proof. By (8.7)

$$ave_A(I) \geq \frac{1}{N!} \sum_{s \in perm(I)} M(s, E(A))$$

$$= \frac{1}{N!} \sum_{s \in perm(I)} \sum_{k=1}^{N} M_k(s, E(A))$$

$$= \frac{1}{N!} \sum_{k=1}^{N} \sum_{s \in perm(I)} M_k(sE(A)).$$

There are $N \cdot N!$ prefixes of size k among all the permutations in $perm(I)$. We can group them into $\frac{N \cdot N!}{k}$ groups, with exactly k cyclic permutations in each. But, by the *Cyclic permutation property*, at least one of these must be counted in our summation, thus

$$ave_A(I) \geq \frac{1}{N!} \sum_{k=1}^{N} \frac{N \cdot N!}{k} = N \cdot H_N \simeq 0.69 N \log N.$$

$\square$

We now recall that the Chang and Robert's algorithm uses exactly $N \cdot H_N$ messages in the average case, and this is clearly best possible by Theorem 8.3.7; namely:

Theorem 8.3.8. *In a unidirectional ring whose size N is unknown to the processors, the Chang and Robert's algorithm has an optimal average-case performance of $N \cdot H_N$ messages.*

8.4 Leader Election on Unidirectional Rings – Worst Case Analysis

We present a simple but efficient deterministic algorithm for leader election on asynchronous unidirectional rings of size N with worst-case message complexity of $1.271 N \log N + O(N)$. We start with a basic algorithm for leader election on unidirectional rings called *BASIC* and then add two enhancements.

We start with presenting the basic leader election algorithm. Initially each processor creates a message $\langle label, round \rangle$, called an *envelope*, containing a *label* initially set to its own identifier, and a *round* initially set to 1, and forwards the envelope to its neighbor. Note that the label of an envelope remains unchanged as long as the envelope survives. Each processor keeps only the label and the round of the last envelope it sent. When an envelope with an odd (respectively, even) round number arrives at a processor, it is destroyed only if it has the same round number as the processor and a larger (respectively, smaller) label. If the round numbers and the labels both match, then the algorithm terminates and the receiving processor is the leader. If the round numbers match and the envelope is not destroyed, then its round number is incremented by one and the updated envelope is forwarded. Finally, if the round numbers do not match then the envelope is forwarded unchanged.

Algorithm BASIC(proc-id)

$id \leftarrow proc - id;$
$rnd \leftarrow 1;$
$fwd\text{-}id \leftarrow -\infty;$
$fwd\text{-}rnd \leftarrow -1;$
repeat
  if not *Causality-test* **then**
    if $fwd\text{-}rnd = rnd$ **then**
      $rnd \leftarrow rnd + 1;$
    end if

$fwd\text{-}rnd \leftarrow rnd;$
$fwd\text{-}id \leftarrow id;$
$send(id, rnd)$
end if
until Leader-test

The following two tests are employed when an envelope $\langle id, rnd \rangle$, containing a label id and a round rnd, reaches a processor that has recorded a label $fwd\text{-}id$ and a round number $fwd\text{-}rnd$. The functions $odd(\cdot)$ and $even(\cdot)$ return the obvious Boolean values.

Leader-test: $(id = fwd\text{-}id) \wedge (rnd = fwd\text{-}rnd)$
Causality test: $(fwd\text{-}rnd = rnd) \wedge [(odd(rnd) \wedge (id > fwd\text{-}id)) \vee (even(rnd) \wedge (id < fwd\text{-}id))]$.

The protocol for each processor is parameterized by its identifier *proc-id*. The correctness of $BASIC$ follows immediately after establishing:

safety: the algorithm never deletes all message envelopes,
progress: eventually only one envelope remains, and
correct termination: the algorithm elects a leader when exactly one envelope remains.

Because the ring is unidirectional and the algorithm is message-driven with messages processed in FIFO order, the messages received by each processor and the order in which each processor processes its messages is entirely determined by the initial input. The scheduler (i.e., the adversary that assigns delays to messages) is powerless to influence the outcome of the computation. For message driven-algorithms on unidirectional rings, correctness under any fixed scheduler implies correctness under all schedulers. We are free, therefore, to adopt any convenient scheduler to establish correctness. We choose a scheduler that processes envelopes in such a way that all undestroyed envelopes have the same round number.

Safety and progress are consequences of the following property of $BASIC$. In odd- (respectively, even-) numbered rounds, the envelopes that are eliminated are exactly those envelopes except the one with the smallest (respectively, largest) label in any maximal chain of successive envelopes with descending (respectively, ascending) labels. The *Leader-test* is passed only when a processor receives the envelope that is last sent. This can happen only when only one message envelope remains, thus ensuring correct termination.

The execution of this algorithm can be described as a sequence $A_1, A_2, \ldots, A_t$, where A_i is the cyclic list of envelopes that are alive in round i. Algorithm $BASIC$ guarantees that:

(P1) given any list A_i, and any envelope $b \in A_i$, either b is destroyed or b is promoted to A_{i+1}. Thus $A_{i+1} \subset A_i$, and
(P2) A_i is not the empty list.

Note that even when processors are permitted to arbitrarily promote any envelope to its next round, properties $(P1)$ and $(P2)$ remain true. Our improvement to $BASIC$ employs a strategy that we call *early promotion* that helps keep the message complexity low.

Next, we present the improved algorithm. The first improvement is to promote an envelope when it can be verified that the envelope would be promoted by $BASIC$ anyway. Denote an envelope with label a and round i by $\langle a, i \rangle$. Suppose an envelope $\langle b, i \rangle$, where i is even, encounters a processor, say x, that sometime in the past sent an envelope $\langle a, i - 1 \rangle$, where $a < b$. We claim that the first processor with round number i, say w, that $\langle b, i \rangle$ next encounters necessarily will have a label no bigger than a. As a consequence, $\langle b, i \rangle$ would be promoted to $\langle b, i+1 \rangle$ at w, so instead it receives *early promotion by witness* at x, and x is a *witness for round* i.

To see the claim, consider the fate of the last envelope, $\langle a, i-1 \rangle$, forwarded by x. When $\langle a, i - 1 \rangle$ arrives at a processor z with round number $i - 1$, either it is promoted to round i because a is less than the label of z, or it is destroyed because a is greater than the label of z. In the first case, z will have recorded a as its last label and i as its round, so z is the claimed processor w. In the second case, some chain of envelopes with decreasing labels is destroyed and the lowest in the chain is promoted to round i. The processor that promoted the lowest envelope is the claimed w and necessarily has recorded a label even smaller than a. A symmetric argument can be made for an envelope with an odd round number i that encounters a processor with a round number $i - 1$ and a larger label.

Suppose in the algorithm $BASIC$ an envelope is promoted from round i to round $i+1$ by a processor y with label a and round i. Then, if early promotion by witness is incorporated into round i, and $i \geq i$, there will necessarily be a witness for round i that promotes the envelope before it reaches y. In particular, the processor z that promoted a to round $i - 1$ is a witness for round i, and the envelope meets z before y.

Let F_t denote the t-th Fibonacci number defined by

$$F_0 = 0, F_1 = 1,$$

and for $t \geq 2$,

$$F_{t+1} = F_t + F_{t-1}.$$

The technique of *early promotion by distance* promotes an envelope from round i to round $i + 1$ if it has travelled a distance of F_{i+2} without encountering any processor with a matching round number. To implement this technique a counter is added to each envelope. When an envelope is promoted to round i, its counter is set to F_{i+2}. The counter is decremented each time the envelope is forwarded without promotion, and if the counter reaches zero, then the envelope is promoted, before being forwarded.

Our final leader election algorithm, called $ELECT$, consists of $BASIC$ augmented with early promotion by distance in odd-numbered rounds, and early promotion by witness in even-numbered rounds.

Algorithm ELECT(proc-id);

$id \leftarrow proc\text{-}id$;
$rnd \leftarrow 0$;
$cnt \leftarrow 0$;
$fwd\text{-}id \leftarrow -\infty$;
$fwd\text{-}rnd \leftarrow -1$;
repeat
   if not Causality-test **then**
     if Promotion-test **then**
        $rnd \leftarrow rnd + 1$;
        $cnt \leftarrow F_{rnd+2}$
     end if
     $fwd\text{-}rnd \leftarrow rnd$;
     $fwd\text{-}id \leftarrow id$;
     $send(id, rnd, cnt - 1)$;
   end if
   $receive(id, rnd, cnt - 1)$;
until Leader-test

Envelopes $\langle id, rnd, cnt \rangle$ now contain three fields, consisting of a label id, a round number rnd, and a counter cnt. The following test is employed in addition to those used by $BASIC$.

Promotion-test.

$$(even(rnd) \wedge (fwd\text{-}rnd = rnd - 1) \wedge (id > fwd\text{-}id)) \vee$$

$$(odd(rnd) \wedge ((cnt = 0) \vee ((fwd\text{-}rnd = rnd) \wedge (id < fwd\text{-}id))))$$

Now we concentrate on proving the correctness of the algorithm $ELECT$. Recall that safety is guaranteed by properties $(P1)$ and $(P2)$.

Suppose, contrary to progress, that after some point k envelopes, $k \geq 2$, continue to circulate around the ring. Then eventually all envelopes will receive a count (the value of cnt) at least as large as the ring size. At this point each envelope has a large enough count to allow it to travel to the processor that forwarded the successor envelope. So if the round number is odd (respectively, even) the envelope with the maximum label (respectively, minimum label) must be destroyed.

The algorithm cannot prematurely elect a leader or fail to elect a leader because a processor will receive an envelope with id equal to its $fwd\text{-}id$ if and only if there are no other envelopes, thus confirming correct termination.

Now proceed with the analysis of the algorithm.

For an envelope with label a and round number i, let $host_i(a)$ denote the processor that promoted the envelope from round $i - 1$ to round i. For an envelope with label a that is eliminated in round i, let $destroyer_i(a)$ denote the processor that eliminated this envelope. Let $\delta(x, y)$ denote the distance, travelling in the direction of the ring, from processor x to processor y. Since algorithm $ELECT$ never changes the label of an envelope for the duration of its existence, let *envelope a* be an abbreviation for the envelope with label a, and say that an envelope is *in round i* if its round number is i. Envelope b is the *immediate successor in round i* of envelope a if the first round i envelope encountered after envelope a in round i, travelling in the direction of the ring, is envelope b.

Lemma 8.4.1. *(a) If envelope a reaches round $i + 1$, and i is odd, then*

$$\delta(host_i(a), host_{i+1}(a)) \geq F_{i+1}.$$

(b) If envelope a is destroyed in round i, and i is odd, then

$$\delta(host_i(a), destroyer_i(a)) \geq F_i.$$

Proof. The proof is by induction on the odd round number. The basis, round 1, holds trivially because each envelope travels $1 = F_1 = F_2$ link. So suppose that the lemma holds for round $i - 2$ where $i \geq 3$ is odd.

First observe the following two facts for odd round i.

Claim 1. A witness for round $i - 1$ that promotes an envelope to round i was a host for round $i - 2$.

Proof. An envelope with label c and round $i - 1$ is promoted to round i at its first encounter with a processor that has round number $i - 2$ and label, say d, less than c. The processor that promoted envelope d to round $i - 2$, $host_{i-2}(d)$, has label d and round number $i - 2$, and all other processors with label d and round number $i - 2$ must follow $host_{i-2}(d)$. Envelope c could not have reached $host_{i-2}(d)$ in round $i - 2$ since otherwise it would have been destroyed by $host_{i-2}(d)$, so it encounters $host_{i-2}(d)$ in round $i - 1$. □

Claim 2. For any envelope b that is promoted to odd round i,

$$\delta(host_{i-2}(b), host_i(b)) \geq F_i.$$

Proof. Consider b's travel in odd round $i - 2$. If b was promoted to round $i - 1$ after travelling F_i links then the claim is immediate. Otherwise, b was promoted by some processor, say

$$host_{i-2}(c) = host_{i-1}(b) \text{ and } b < c,$$

and, by the induction hypothesis,

$$\delta(host_{i-2}(b), host_{i-1}(b)) \geq F_{i-1}.$$

Since b reaches round i, in round $i - 1$, b travels from $host_{i-1}(b)$ to some witness w for round $i - 1$ with label $d < b$ that promotes b to round i. By the inequalities, $c \neq d$, and hence w must be some processor not in the interval travelled by c in round $i - 2$. Hence

$$\delta(host_{i-1}(b), host_i(b)) \geq \delta(host_{i-2}(c), destroyer_{i-2}(c)) \geq F_{i-2}$$

by the induction hypothesis. The combined distance is therefore at least $F_{i-1} + F_{i-2} = F_i$.

$\square$

Now we are ready to prove the lemma. Let a and b be labels of two envelopes in round i where envelope b is the immediate successor in round i of envelope a. We distinguish two cases.

Case 1: *Suppose a survives round i.*

According to the algorithm, envelope a travels F_{i+2} links in round i unless it reaches $host_i(b)$ before travelling this distance. If a survives because it travels a distance of F_{i+2} without encountering a processor with round number i then the lemma holds trivially for round i. Otherwise, a travels to $host_i(b)$ and is promoted because $a < b$. Now, since $i-1$ is even, a was promoted from round $i-1$ to round i by some witness for round $i-1$. By Claim 1, this witness must be $host_{i-2}(g)$ for some envelope with label g. Also, since a is promoted, $g < a$. Consider the first processor, say $host_{i-1}(h)$, after $host_{i-2}(g)$ that is a host for round $i - 1$. Then by the algorithm $h \leq g$. So, by the inequalities $h \neq b$.

Thus

$$
\begin{aligned}
&\delta(host_i(a), host_{i+1}(a)) \\
={}& \delta(host_{i-2}(g), host_i(b)) \\
={}& \delta(host_{i-2}(g), host_{i-1}(h)) + \delta(host_{i-1}(h), host_i(b)) \\
\geq{}& \delta(host_{i-2}(h), host_{i-2}(h)) + \delta(host_{i-2}(b), host_i(b)).
\end{aligned}
$$

By the induction hypothesis,

$$\delta(host_{i-2}(h), host_{i-1}(h)) \geq F_{i-1}.$$

By Claim 2,

$$\delta(host_{i-2}(b), host_i(b)) \geq F_i.$$

Therefore,

$$\delta(host_i(a), host_{i+1}(a))$$
$$\geq \delta(host_{i-2}(h), host_{i-1}(h)) + \delta(host_{i-2}(b), host_i(b))$$
$$\geq F_{i-1} + F_i = F_{i+1}.$$

Case 2: *Suppose a is eliminated in round i.*

Since i is odd, $a > b$. Since a reached round i, a must have been promoted by a witness for round $i - 1$. By Claim 1, the witness for round $i - 1$ that could promote a and most closely precedes $host_i(b)$ is $host_{i-2}(b)$.

Thus,

$$\delta(host_i(a), destroyer_i(a)) = \delta(host_i(a), host_i(b))$$
$$\geq \delta(host_{i-2}(b), host_{i-2}(b)) \geq F_i$$

by Claim 2.

□

We examine the consequences of Lemma 8.4.1. For any round i of the algorithm $BASIC$, all envelopes in round i travel a total of N links because each envelope travels from its host in round i to the host of its immediate successor in round i. The algorithm uses fewer than N messages per round (except for the last round) because an envelope does not always travel all the way to the next host before being promoted. We now bound the savings due to early promotion. Let a and b be the label of two envelopes in round i where envelope b is the immediate successor in round i of envelope a. We say that envelope a *saves k links in round i* if, in round i, the distance envelope a travels is $\delta(host_i(a), host_i(b)) - k$.

Corollary 8.4.2. *Every envelope that reaches round $i + 1$ where i is even saves at least F_i links in round i.*

Proof. Let a be the label of an envelope that remains alive after an even round i, and let b be the label of the envelope in round i of envelope a. Then $a >_, b$ because a survives an even round. According to algorithm $ELECT$, if envelope a has not already been promoted to round $i + 1$ before reaching $host_{i-1}(b)$, it will achieve early promotion by witness at $host_{i-1}(b)$ thus saving at least $\delta(host_{i-1}(b), host_i(b))$. But, by Lemma 8.4.1,

$$\delta(host_{i-1}(b), host_i(b)) \geq F_i.$$

□

Theorem 8.4.3. *Algorithm $ELECT$ sends fewer than $1.271N \log N + O(N)$ messages on rings of size N.*

Proof. To bound the number of messages, we bound

(1) the total number of rounds, and
(2) the number of messages, in any *block* of two consecutive rounds consisting of an even round followed by an odd round.

By the previous lemma, if round number i is odd, then the distance travelled by any round i envelope is at least F_i. However, no round i envelope travels beyond the host of its successor in round i. Hence, if i is odd, then the distance between any two hosts in round i is at least F_i. Thus, in round i, where i is odd, there can be at most n/F_i remaining envelopes. Denote by $F^{-1}(x)$ the least integer j such that $F_j \geq x$. It follows that the algorithm $ELECT$ uses at most $F^{-1}(N) + O(1)$ rounds for rings of size N.

To estimate the number of messages sent in a block, consider even round i followed by an odd round $i + 1$. Assume that there are x envelopes in round

$i+1$. Then, by Corollary 8.4.2, the total number of links travelled by envelopes in round i is at most $N - xF_i$. Clearly, the total number of links travelled by envelopes in round $i + 1$ is at most N. Since, in odd round $i + 1$, each envelope travels at most F_{i+1+2} before promotion, this number is also at most xF_{i+3}. Thus the total number of messages in round $i + 1$ is at most $\min(xF_{i+3}, N)$ and the total number of messages in the block is bounded by $\min((N + x(F_{i+3} - F_i)), 2N - xF_i)$. Since $N + x(F_{i+3} - F_i) = 2N - xF_i$ for $x = N/F_{i+3}$, this bound is at most $2N - NF_i/F_{i+3}$.

Recall that ψ denotes $(1 + \sqrt{5})/2$, and let ψ_1 denote $(1 - \sqrt{5})/2$. Observe that for even i:

$$\frac{F_i}{F_{i+3}} = \frac{\frac{1}{\sqrt{5}}(\psi^i - \psi_1^{-i})}{\frac{1}{\sqrt{5}}(\psi^{i+3} - \psi_1^{i+3})} = \frac{\psi^i - \psi^{-i}}{\psi^{i+3} + \psi^{-(i+3)}}$$

$$> \frac{\psi^{2i+3} - \psi^3 - \psi^{-3}}{\psi^{2i+6}} > \psi^{-3} - \psi^{-2i}.$$

Therefore, the number of messages in a block starting with even round i is at most:

$$2N - N\frac{F_i}{F_{i+3}} < N(2 - \psi^{-3} + \psi^{-2i})$$

$$= N(2 - \frac{8}{(1 + \sqrt{5})^3} + \psi^{-2i} = N(4 - \sqrt{5} + \psi^{-2i}).$$

Hence there are at most

$$\sum_{even(i),\ 2 \leq i \leq F^{-1}(N)} N(4 - \sqrt{5} + \psi^{-2i}) + O(N)$$

$$= \frac{4 - \sqrt{5}}{2} N \log_b N + O(N)$$

$$< 1.271 N \log N + O(N)$$

messages sent by any computation of $ELECT$ on a ring of size N.

$\square$

Note that the appropriate lower bound for this problem of $0.69N \log N$ proved in the previous section holds even for the average message complexity when the ring size is unknown.

8.5 Leader Election on Bidirectional Rings – Worst Case Analysis

Consider an asynchronous bidirectional ring of N processors. We assume that each processor has separate queues for the incoming messages from each direction on the ring, and that the links and queues preserve the FIFO order in which messages are sent. A processor can send a message in one or two directions simultaneously on the ring. While every processor can distinguish between the two directions on the ring, there is no global sense of orientation at the outset. The distributed election algorithm is based on a series of message exchanges between all processors on the ring, and repeatedly generates new information when messages "meet". As conceptually messages should meet in a processor and not bypass each other in a link, we make the simplifying assumption that the links are "half-duplex", i.e., the links carry at most one message at a time regardless of their direction. At the end of this section we will rid ourselves of this assumption by a suitable modification of the algorithm.

The distributed election algorithm operates in the following manner. At all times the processor maintains a register ID that contains the name (identifier) of a "large" processor that is still in the game, and a (Boolean) register DIR that contains a "direction" on the ring in which there are processors that still have a "smaller" processor up for election. Typically the "large" candidate will exist on one side of the processor, and DIR will point towards the other side. Messages are generated that contain the name of a "large" candidate, and are send out (or passed on) in the direction where a "smaller" candidate is known to be still alive. The idea of the chase is to eliminate the smaller candidate, and force agreement on the larger candidate. Processors that initiate a chase are termed *active*, and the remaining processors are termed *observant* ("passive"). After the initial phase of the algorithm (phase 0), the processors that are local minima are active and all other processors are observant.

After the current active processors have initiated a chase, the observant processors basically relay messages onwards unless they notice an "unusual" situation on the ring. Active processors immediately go to the observant state after initiating a chase, and some of the observant processors will become active again as a result of their conclusion from the "unusual" situation that was observed. The new active processor will initiate another chase, and the same procedure repeats. In order to distinguish new chases from old, the algorithm will be organized in phases and each processor will maintain a register PNUM containing the phase number of the most recent phase in which it participated. Initially all PNUM registers are set to 0. Through the phase numbers we keep track of the logical ordering of the chases (as if they were scheduled ring-wide and strictly separated in time). All messages are stamped with the phase number of the generating (active) processor, to separate the phases and to synchronize the processors that did not receive messages for a while and are unaware of the current phase.

There are two "unusual" situations that can arise at the site of an observant processor as the algorithm proceeds (recall that the links are assumed to be half-duplex).

1. The processor receives a message of the current phase, say via its left link, that contains a value that is less than the current value in its ID regis-

ter. (This happens when the message has completed its chase of smaller processors and now bumps into a larger one that should take over.) The processor turns active, increments its phase number by 1, and initiates a chase with the value in its current ID in the direction of the message that was received, i.e., out over its left link.

2. The processor receives two messages of the same phase from opposite directions (a "collision"). The processor turns active, increments its phase number by 1, and initiates a chase with the largest value contained in the two messages in the direction of the smallest. It will be shown later that the number of active processors that can arise in a phase rapidly decreases as the algorithm proceeds, and that in the end precisely one processor will be left. This processor will know that it is elected because either it receives a message of the current phase with a value identical to the one it sent out (and stored in its ID register) or it receives two messages of the same phase from opposite directions that hold identical values. Thus the processor elected by the algorithm is not necessarily the processor with the largest identification number. The correctness of the algorithm will be demonstrated as well.

In the remainder of this section we will describe the algorithm performed by every participating processor in more detail. We assume that each processor starts with the initialization phase (phase 0) and then alternates between the active and the observant state as dictated by the algorithm. Messages are of the form $\langle v, p \rangle$ where v is a "value" (a processor identification number) and p a phase number.

8.5.1 Election Algorithm Due to van Leeuwen and Tan

The following algorithm describes the actions of an arbitrary processor on a bidirectional ring with half-duplex links as required for electing a leader.

Let u be the executing processor's identification number. It can either begin the algorithm spontaneously or be induced into it by one of its neighbors.

Algorithm VAN LEEUWEN – TAN
(1) INITIALIZATION
send message $\langle u, 0 \rangle$ **to** both neighbors;
$PNUM := 0$;
wait for the corresponding messages $\langle u_1, 0 \rangle$ and $\langle u_2, 0 \rangle$ to come in from the two neighbors;
if $(u_1 > u) \wedge (u_2 > u)$ **then**
    $ID := \max(u_1, u_2)$;
    $DIR :=$ the direction of $\min(u_1, u_2)$;
    go to active
else

    go to observant;
end if

This ends the initialization phase. A processor continues in either the active or the observant state, to perform the further steps of the election algorithm.

(2) ELECTION

(a) Active

A processor enters the active state with some value v stored in its ID-register and a phase number p. The phase number p is either stored in $PNUM$ or it is an "update" stored in a temporary register.

$PNUM := p + 1$;

send message $\langle v, PNUM \rangle$ **to** the direction DIR;

go to observant;

(b) Observant

In this state a processor receives messages and passes them on, unless an "unusual" situation is observed that enables it to initiate a new phase.

receive message(s) **from** one or both directions;

Discard any message $\langle v, p \rangle$ received with $p < PNUM$.

Discard any message that does not have the highest p-value among the messages;

case "The number of messages left"

0: **go to** observant;

1: Received message $\langle v, p \rangle$, $p \geq PNUM$

if $p = PNUM$ **then**

    if $v = ID$ **then**

        go to inaugurate

    else if $v < ID$ **then**

        $DIR :=$ direction from which the message was received

    else if $v > ID$ **then**

        go to observant

    else

        $PNUM := p$;

        $ID := v$;

        $DIR :=$ the direction in which the message was going;

        send message $\langle ID, p \rangle$ **to** the direction DIR;

        go to observant

    end if

end if

2: Received messages $\langle v_1, p \rangle$ and $\langle v_2, p \rangle$, $p \geq PNUM$;

if $v_1 = v_2$ **then**

    $PNUM := p$;

    go to inaugurate;

else if $v_1 \neq v_2$ **then**

    $ID := \max(v_1, v_2)$;

$DIR :=$ the direction of $\max(v_1, v_2)$;
    go to active
end if

8.5.2 Correctness and Message Complexity Analysis

In this subsection we show that the algorithm "terminates" and that it is correct, i.e., upon termination precisely one processor has been elected as a leader. Then we prove that the algorithm uses at most $1.44N \log N + O(N)$ messages, by showing that it terminates within $1.44 \log N$ phases and each phase requires at most N messages total (except perhaps the first).

Clearly every processor that wishes to begin the election must begin with the "initialization phase" (phase 0). The message $\langle u, 0 \rangle$ of processor u will awaken its two neighbors to the election process as well, if they aren't aware of it yet. Eventually every processor on the ring is woken up and is passing through its "phase" 0, and thus every processor will receive two messages with p-value 0 in this phase (one from each neighbor). It follows that all processors follow suit as soon as one processor starts the election, and that precisely $2N$ messages are exchanged in the initialization phase. The initialization ends by declaring the processors that are "local minima" to be active for the next phase (phase 1) and all other processors to be observant. Clearly at least 1 and at most $\frac{N}{2}$ processors will be local minima and thus are active in phase 1. Active processors will be separated by observant ones.

As the algorithm proceeds messages are generated and relayed around the ring, and the phase number (PNUM) of observant processors continues to be updated (increased) whenever necessary. If an observant processor turns active, its phase number is incremented again. Messages with lower phase numbers than the current one at the receiving processor are ignored and (hence) are not passed on. It follows that at every processor the messages that come in from the same direction in fact come in with increasing p-values! (Note that all links and queues preserve the order of the messages.) It implies that for the analysis we may as well assume that the algorithm proceeds synchronously, in "increasing" phases. This gives the worst possible bounds on the message complexity, because it automatically means that no messages will be discarded in this regime. (In the asynchronous case one part of the ring could be "faster" and proceed to higher phase numbers, and thus "overtake" the ongoing process in the other part of the ring in its lower phase. Thus the asynchronous algorithm may, in fact, use fewer message exchanges to complete the election.) Clearly, for the termination and correctness proofs no such restriction can be made, although some aspects of the synchronous regime will be implicit in any version of the algorithm.

Let $PNUM(u)$ be the contents of the PNUM register of processor u at some moment, and $ID(u, p)$ be the contents of the ID register of processor u at some moment while $PNUM(u) = p$. Let $p \geq 1$.

Lemma 8.5.1. *If a processor u receives a message $m \equiv \langle u, p \rangle$ with $p = PNUM(u)$, then m must have arrived from a direction (i.e., a link) opposite to the one over which u has last sent out a message.*

The processors u with $PNUM(u) = p$ are said to be in phase p. Phase p is reached either in the active state (by incrementing $PNUM$ to p) or in the observant state (by updating $PNUM$ to p while relaying a message of the form $\langle v, p \rangle$). In the former case we say that u is "active in phase p". For every processor u we call the first value of $ID(u, p)$ in phase p the "color" of u in phase p. (Observe that $ID(u, p)$ will be constant during phase p. Only when a message $\langle v, p \rangle$ is received with $v > ID$ could an update of u's color to v be made, but this is omitted from the given version of the algorithm.)

Definition 8.5.2.

1. A_p *is the set of processors u that are active in phase p.*
2. $\tilde{A}_p$ *is the set of colors of processors u that are active in phase p (shortly the "active colors").*

The idea of the colors is that we keep track of the identification numbers of outstanding candidates for election.

Lemma 8.5.3. *For all $p \geq 1$, $\tilde{A}_p \supseteq \tilde{A}_{p+1}$.*

For every phase p we will distinguish between *c-active* processors, which became active because of the collision of two messages of the same phase, and *d-active* processors, which became active because of the receipt of one message of the same phase with smaller value (the "non-collision" case). It is easily seen from the algorithm that there can be no other types of active processors.

Lemma 8.5.4.

1. *Between every two c-active processors in the phase p there are at least two non-active processors in the same phase.*
2. *A c-active processor in phase p cannot be c-active in the next phase.*
3. *A d-active processor in phase p cannot be c-active in the next phase.*
4. *If two adjacent processors are active in the same phase p, then they do not send messages to each other directly in this phase (i.e., over the joining link and simultaneously).*

Lemma 8.5.4 shows that there is always "room" during a given phase for messages to collide. The activities of the distributed election algorithm in all phases are triggered by the active processors. Our first concern is to show that the algorithm does not "die", i.e., fall silent before a leader is found (if ever).

Proposition 8.5.5. *For all $p \geq 1$, if $A_p \neq \emptyset$ and a leader is not found in phase p, then $A_{p+1} \neq \emptyset$.*

Our concern is to show that the algorithm is "partially correct", i.e., if a processor enters the inauguration state ("declares itself a leader") then it is the only processor to do this and the algorithm terminates. Let $p \geq 1$.

Lemma 8.5.6. *Suppose two active processors $u, u' \in A_p$ have the same color α. Then*

1. *u and u' must be adjacent-active, i.e., u and u' are separated only by non-active processors;*
2. *all processors in between u and u' (in one direction) have the same value α in their ID register, and*
3. *u and u' send their phase-p messages out in opposite directions, i.e., "out away from the edges of the α-colored interval".*

Proposition 8.5.7. *If a processor moves to the inauguration state in phase p, then*

1. *it is the only one doing so,*
2. *$|\tilde{A}_p| = 1$, and*
3. *all processors on the ring have the same value in their ID register.*

By Proposition 8.5.7, any processor that moves to the inauguration state knows that the election algorithm has terminated. In particular, Proposition 8.5.7 implies that the leader is unique whenever the algorithm terminates. We now prove that the algorithm indeed terminates, by showing that $|\tilde{A}_p|$ gradually decreases as p increases. Let $p \geq 2$.

Lemma 8.5.8. *There is a one-to-one mapping from the active colors in phase p to the active colors in phase $p - 2$ that are no longer active in phase $p - 1$.*

Proposition 8.5.9. *The algorithm always terminates in finitely many phases and hence it is a correct distributed election algorithm.*

Finally, we analyze the message complexity of the algorithm for arbitrary bidirectional rings of N processors.

Theorem 8.5.10. *The algorithm terminates within $1.44 \log N$ phases and uses at most $1.44N \log N + O(N)$ messages.*

Proof. Suppose that the algorithm terminates in t phases. By Proposition 8.5.7 we have $|\tilde{A}_{t-1}| = 1$. By Lemma 8.5.3 and Lemma 8.5.8 it follows that for $p \geq 2$,

$$|\tilde{A}_p| \leq |\tilde{A}_{p-2}| - |\tilde{A}_{p-1}|$$

and hence

$$|\tilde{A}_{p-2}| \geq |\tilde{A}_{p-1}| + |\tilde{A}_p|.$$

Let F_i denote the i-th Fibonacci number. We claim that for all $0 \leq i \leq t$,

$$F_i \leq |\tilde{A}_{t-i}|.$$

The claim is true for $i = 0$ and $i = 1$. Proceeding inductively, suppose that the claim is true up to i and that $i + 1 \leq t$. Then

$$|\tilde{A}_{t-(i+1)}| \geq |\tilde{A}_{t-i}| + |\tilde{A}_{t-(i-1)}| \geq F_i + F_{i-1} = F_{i+1}$$

and the induction step is complete. It follows that

$$F_t \leq |\tilde{A}_0| = N$$

and by known estimates for F_t we conclude that

$$t \leq \frac{\log N}{\log \psi} + O(1)$$

where $\psi = \frac{1}{2}(1 + \sqrt{5}) = 1.61803$. Thus the algorithm terminates within about $\frac{\log N}{\log \psi} = 1.44 \log N$ phases.

Finally observe that in each phase p, the phase-p messages either collide or are intercepted (stopped) at some processor that is active in the current phase and necessarily sent its phase-p message out away in the other direction. Thus at most one phase-p message will travel over each link in the ring and, consequently, each phase requires at most N messages. It follows that the algorithm uses at most $1.44N \log N + O(N)$ messages.

$\square$

8.6 Leader Election on Complete Networks

In this section we study leader election in a complete network of processors, in which each pair of processors is connected by a communication link. In contrast with the ring network, in which the diameter is large, but each processor knows exactly its neighborhood, here we have the other extreme, where the diameter is the smallest possible, i.e., equal to 1, but each processor has no knowledge of its environment.

We present an $O(N \log N)$ algorithm for the leader election problem on a complete network with N nodes, and show a matching $\Omega(N \log N)$ lower bound.

8.6.1 $O(N \log N)$ Algorithm

In the algorithm, nodes attempt to capture other nodes, thus increasing their domains. A node that captures all other nodes will be the leader. A captured node keeps its domain, without trying to augment it. A node can be part of more than one domain, but it will remember its current domain, by having a designated edge pointing to its master.

Let $size_P$ and id_P denote the size of the domain and the identity of processor P. The capturing process proceeds as follows. As long as a processor is a candidate, it sends a message, trying to capture another node. Suppose node A sends a message to node B. Node B transfers the message to its master C (might be B itself). Now, C compares $(size_A, id_A)$ with $(size_C, id_C)$. If $(size_A, id_A) > (size_C, id_C)$, that is, either $size_A > size_c$ or $size_A = size_C$ and $id_A > id_C$, it stops being a candidate, informs B of this, B joins A's domain, and then A continues its capturing process. If $(size_A, id_A) < (size_C, id_C)$, then node C informs B, but node B does not inform A about anything. In order to keep the number of messages low, node B forwards only one message at a time to its master; only after hearing it, will it transfer other messages.

We assume that each processor arbitrarily labels its adjacent edges with numbers $1, 2, \ldots, N-1$, and that edge 0 is a self-loop. The algorithm in processor P uses the following data structures:

$state$ – $\in \{candidate, non\text{-}candidate, leader\}$.
$unused\text{-}edges$ – the set of unused edges of $\{1, 2, \ldots, N-1\}$.
$size_P$ – denotes the number of processors in P's domain.
$edge\text{-}to\text{-}master$ – $\in \{1, 2, \ldots, N\}$, denotes the edge connecting P to its
    current master; including 0 which connects P to itself.
$waiting$ – A Boolean variable, indicating whether a node waits for a response
    from its master, or it is free to forward messages to it.
$queue$ – contains messages that are waiting to be forwarded to the master.
pop(m), **push**(m) – procedures for maintaining this $queue$.

The following messages are used:
(win, e) – a message sent along edge e by a processor who is joining P's domain.
$(try1, size, id, e)$ – a message sent along edge e by a processor, carrying its size
    and identity, in an attempt to capture a new processor.
$(try2, size, id, e, e')$ – a message forwarded along edge e' by a processor to its
    master.
$(lose, e)$ – a message sent along edge e from the master, when it won a
    capturing attempt.
$(back, e, e')$ – a message sent back along edge e' from a master, when it lost
    a capturing attempt.

Any processor that receives a message for the first time starts to execute the code. This includes a non-empty set of processors that start the algorithm by receiving a $(win, 0)$ message. The edge on which a message is received was written as the last coordinate of the message; thus, $(try1, size, id, e)$ is sent on edge e.

ALGORITHM COMP-NET
    $state := candidate;$
    $edge\text{-}to\text{-}master := 0;$
    $size := 0;$
    $waiting := false;$

```
unused-edges := {1, ..., N − 1};
repeat
  receive(m);
  if m = (win) then
    size := size + 1;
    if size = N then
      state := leader
    else
      send(try1, size, id_P, size);
    end if
  else if m = (try1, size, id, e) then
    if not waiting then
      send(try2, size, id, e, edge-to-master);
      waiting := true
    else
      push(m)
    end if
  else if m = (try2, size, id, e, e') then
    if (size, id) < (size_P, id_P) then
      send(lose, e')
    else
      state := non-candidate; send(back, e, e');
    end if
  else if m = (lose, e) then
    waiting := false
    if queue ≠ ∅ then
      pop(try1, size, id, e);
      waiting := true;
      send(try2, size, id, e, edge-to-master);
    end if
  else if m = (back, e, e') then
    edge-to-master := e;
    waiting := false;
    send(win, e);
    if queue ≠ ∅ then
      pop(try1, size, id, e);
      waiting := true;
      send(try2, size, id, e, edge-to-master)
    end if
  end if
until state := leader;
```

Lemma 8.6.1. *Each execution of the algorithm COMP-NET terminates.*

Proof. The relation between the messages is the following:

$$win \rightarrow try1, \; try1 \rightarrow try2, \; try2 \rightarrow lose, \; lose \rightarrow try2,$$

$$try2 \rightarrow back, \; back \rightarrow try2, \text{ and } back \rightarrow win,$$

where $m_1 \rightarrow m_2$ means that in response to receiving a message of type m_1 a processor can send a message of type m_2, and that no other messages can be sent except one of those above.

In every execution of the algorithm, *win* messages are received at most N times by each processor. Hence they are also sent only a finite number of times. Hence we have also a finite number of *try1* and *back* messages. It can also be shown that there are also a finite number of *lose* messages, and of *try2* messages.

$\square$

Lemma 8.6.2. *In each execution of the algorithm COMP-NET, at least one processor is elected as a leader.*

Proof. Assume that upon termination (which occurs, according to Lemma 8.6.1) we have no leader (in other words, that the algorithm reached a deadlock situation, in which no more messages are sent, but no leader exists in the network).

Let P be the processor with the largest domain s upon termination; if there is more than one processor with a domain of that size, we take as P the one with the largest identity. When P received its last *win* message, it first increased its size to s. Since we assumed $s \neq N$, this means that P sent a *try1* message. This *try1* message is eventually forwarded as a *try2* message to processor Q; if Q discovers that its $(size, id)$ is smaller than P's, it will send a *lose* message, thus P will receive a *win* message, and its size will increase, contradicting the assumption. Therefore, we conclude that Q's $(size, id)$ is larger than P's, a contradiction.

$\square$

To show that no more than one leader exists we show the following lemma, which will also prove useful in the complexity analysis of the algorithm. We number the atomic steps in the execution by $1, 2, \ldots$, and say that the step i occurred at time i. Let $domain_t(P)$ denote the set of vertices that have already sent a *win* message to P by time $\leq t$.

Lemma 8.6.3. *If $|domain_t(P)| = |domain_{t'}(P')|$, then*

$$domain_t(P) \cap domain_{t'}(P') \neq \emptyset.$$

Proof. Assume that $Q \in domain_t(P) \cap domain_{t'}(P')$, and without loss of generality that Q is the first node in the domain of P that joins the domain of

P', and that this is the smallest time t' when this happens. Since the domain of P does not change after t', and since the domain of P' became larger than that of P at the moment t' when Q joined the domain of P', then it is not possible that $|domain_t(P)| = |domain_{t'}(P')|$, a contradiction.

□

The following deals with the message complexity of the algorithm:

Theorem 8.6.4. *The message complexity of the COMP-NET algorithm is* $O(N \log N)$.

Proof. Let the sizes of the domains of the processors, after the algorithm has terminated, be $s_1 = N \geq s_2 \geq \ldots \geq s_N$. By Lemma 8.6.3, $s_2 \leq \frac{N}{2}$ (since otherwise the two domains, of sizes s_1 and s_2, will not be disjoint). Similarly, $s_k \leq \frac{N}{k}$ for every k. Since the number of messages sent by each processor is bounded by four times its domain, we get that the total number of messages sent is bounded by

$$4 \cdot (s_1 + s_2 + \cdots + s_N) \leq 4 \cdot N(1 + \frac{1}{2} + \frac{1}{3} + \cdots + \frac{1}{N}) = O(N \log N),$$

and this completes the proof.

□

8.6.2 $\Omega(N \log N)$ Lower Bound

Let A be a distributed algorithm acting on a graph $G = (V, E)$. An execution of A consists of *atomic steps*, each entity receiving a message, doing some local computation and then sending messages to neighbors. With each execution we can associate a sequence

$$send = \langle send_1, send_2, \ldots, send_k \rangle$$

that includes all the sent messages, in their order of occurrence (if there are no such events then *send* is the empty sequence). In the case that two or more messages are sent at the same time, order them randomly (thus, in such a case many *send* sequences may correspond to the same execution). Each event $send_i$ we identify with the triple $(v(send_i), e(send_i), m_i)$, containing the message sent m_i, the node sending the message, $v(send_i)$, and the edge used by it $e(send_i)$. We assume that $send_1$ occurred at time 0, and $send_i$ at time $\tau_i \geq 0$.

Let $send(t)$ be the prefix of length t of the sequence $send$, namely

$$send(t) = \langle send_1, send_2, \ldots, send_t \rangle$$

($send(0)$ is the empty sequence). If $t < t'$ then we say that $send(t')$ is an *extension* of $send(t)$, and we note that $send(t) < send(t')$. $send$ is called a *completion* of $send(t)$. Note that a completion of a sequence is not necessarily unique.

Let $new = new(send)$ be the subsequence $\langle new_1, new_2, \ldots, new_r \rangle$ of the sequence $send$ that consists of all the events in $send$ that use previously unused edges. (An edge is $used$ if a message has already been sent along it from either side.) This means that the message $send_i = (v(send_i), e(send_i), m_i)$ belongs to new if and only if $e(send_i) \neq e(send_j)$ for all $j < i$. $new(t)$ denotes the prefix of size t of the sequence new.

Define the graph $G(new(t)) = (V, E(new(t)))$, where $E(new(t))$ is the set of edges used in $new(t)$, and call it the graph $induced$ by the sequence $new(t)$. If for every execution of the algorithm A the corresponding graph $G(new)$ is connected then we term this algorithm $global$. Note that all the graphs $G(new)$ above have a fixed set V of vertices (some of which may be isolated).

For each algorithm A and graph G we define the $exhaustive$ set of A $with$ $respect$ to G, denoted by $EX(A, G)$ (or $EX(A)$ when G is clear from the context), as the set of all the sequences $\sigma = new(t)$ corresponding to possible executions of A.

From the model used in this subsection the following facts, defined below as axioms, hold for every algorithm A and every graph G. Note that these axioms reflect only some properties of distributed algorithms which are needed here.

Axiom 1: The empty sequence is in $EX(A, G)$.

Axiom 2: If two sequences σ_1 and σ_2, which do not interfere with each other, are in $EX(A, G)$, then so also is their concatenation $\sigma_1 \circ \sigma_2$. (σ_1 and σ_2 do not interfere with each other if no two edges e_1 and e_2 that occur in σ_1 and σ_2, respectively, have a common end-point; this means that the corresponding partial executions of A do not affect each other and hence any of their synchronous mergers corresponds to a legal execution of A.)

Axiom 3: If σ is a sequence in $EX(A, G)$ with a last element (v, e, m), and if e' is an unused edge adjacent to v, then the sequence obtained from σ by replacing e by e' is also in $EX(A, G)$. (This reflects the fact that a node cannot distinguish between its unused edges.)

Note that these three facts do not imply that $EX(A, G)$ contains any non-empty sequence. However, if the algorithm A is global then the following fact holds as well:

Axiom 4: If σ is in $EX(A, G)$ and C is a proper subset of V containing all the non-isolated nodes in $G(\sigma)$, then there is an extension σ' of σ in which the first message (v, e) in σ' but not in σ satisfies $v \in C$. (This reflects the facts that some unused edge will eventually carry a message and that isolated nodes in $G(\sigma)$ may remain asleep until some message from an already awakened node reaches them.)

Definition 8.6.5. *The edge complexity $e(A)$ of an algorithm A acting on a graph G is the maximal length of a sequence new over all executions of A. The message complexity $m(A)$ of an algorithm A acting on a graph G is the maximal length of a sequence sent over all executions of A.*

Clearly $m(A) \geq e(A)$.

The following lemma is needed in the sequel:

Lemma 8.6.6. *Let A be a global algorithm acting on a complete graph $G = (V, E)$, and let $U \subset V$. Then there exists a sequence of messages $\sigma \in EX(A, G)$ such that $G(\sigma)$ has one connected component whose set of vertices is U and the vertices in $V - U$ are isolated.*

Proof. A desired sequence σ can be constructed in the following way. Start with the empty sequence (using Axiom 1). Then add a message along a new edge that starts in a vertex in U (Axiom 4) and that does not leave U (Axiom 3 and the completeness of G). This is repeated until a graph having the desired properties is constructed.

$\square$

We now prove the lower bound for global algorithms.

Theorem 8.6.7. *Let A be a global algorithm acting on a complete graph G with N nodes. Then the edge complexity $e(A)$ of A is $\Omega(N \log N)$.*

Proof. For a subset U of V we define $e(I)$ to be the maximal length of a sequence σ in $EX(A, G)$ which induces a graph that has a connected component whose set of vertices is U and isolated vertices otherwise (such a sequence exists). Define $e(k)$, $1 \leq k \leq N$, by

$$e(k) = \min\{e(U) \mid U \subseteq V, \ |U| = k\}.$$

Note that $e(N)$ is a lower bound on the edge complexity of the algorithm A.

The theorem will follow from the inequality

$$e(2k + 1) \geq 2e(k) + k + 1$$

for $k < \frac{N}{2}$.

Let U be a disjoint union of U_1, U_2 and v, such that $|U_1| = |U_2| = k$, and $e(U) = e(2k + 1)$. We note that $C = U_1 \cup U_2$. Let σ_1 and σ_2 be sequences in $EX(A, G)$ of lengths $e(U_1), e(U_2)$ inducing subgraphs G_1, G_2 that have one connected component with vertex sets U_1, U_2 (and all other vertices are isolated), respectively. These two sequences do not interfere with each other, and therefore (by Axiom 2) their concatenation $\sigma = \sigma_1 \circ \sigma_2$ is also in $EX(A, G)$. The proper subgraph C of $G(\sigma)$ satisfies the assumption of Axiom 4. Note that each node in C has at least k adjacent unused edges within C. By Axiom 4 there is an extension of σ by a message (v, e), where $v \in C$. By Axiom 3 we may choose the edge 3 in such a way that it connects two vertices in C. This process can be repeated until at least one vertex in C saturates all its edges to other vertices in C. This requires at least k messages along previously unused edges. One more application of Axiom 4 and Axiom 3 results in a message from some node in C to the vertex v. The resulting sequence σ' induces a graph that contains one connected component on the set of vertices U and isolated vertices otherwise. Thus we have

$$e(2k + 1) = e(U) \geq e(U_1) + e(U_2) + k + 1 \geq 2e(k) + k + 1.$$

The above inequality implies that for $N = 2^i - 1$ and the initial condition $e(1) = 0$ we have

$$e(N) \geq \frac{N + 1}{2} \log \left(\frac{N + 1}{2} \right).$$

Since it is obvious that $e(m) \geq e(N)$ for $m > N$, this implies the theorem.

$\square$

From this theorem if follows that:

Theorem 8.6.8. *Let A be a global algorithm acting on a complete graph G with N nodes. Then the message complexity $m(A)$ of A is $\Omega(N \log N)$.*

8.7 Leader Election on Hypercubes

8.7.1 Preliminary Notions

In the following we need these definitions and notions.

For $i \leq n$, by $\mathbf{e}_i$ we denote the binary n-vector with a 1 in the i-th position and 0's elsewhere, and by e_i the corresponding binary string. Without loss of generality we assume that at each vertex local link labels are from $\{e_1, \ldots, e_n\}$, but there is no correlation between local link labels at different vertices.

Intuitively, by a *world according to* x we mean the hypercube as it is seen from the vertex x. More formally, by a world according to x we mean an assignment of n-bit binary representations (in the rest of this chapter referenced as *addresses*) to all vertices of Q_n and an assignment of link labels from $\{e_1, e_2, \ldots, e_n\}$ to all links, such that x is addressed by 0^n, links adjacent to x are labeled by initial local labels, and assignments of vertex addresses and link labels are globally consistent in the following sense: Addresses of two neighboring vertices differ in exactly one bit position, determined by the label of the link that connects them.

A fixed world according to x can be described by a pair (σ_x, β_x), where $\sigma_x(v)$ denotes a permutation that maps local link labels at v to link labels in the world according to x, and β_x is a naming function which assigns all vertex addresses according to x: $\beta_x(y)$ denotes the vertex with address y in the world according to x. So by $\beta_x^{-1}(i)$ for $i \in$ ID we denote an address of the vertex with *id* i in the world according to x. The local label of a link $h = (u, v)$ in the vertex u we denote by h_u. Consequently, $\sigma_x(u)[h_u]$ means the label of the link h in the world according to x.

By $Send(h, message)$ we mean a procedure sending a message *message* along the link h.

8.7.2 Leader Election

We present a leader election algorithm on hypercubes, as a combination of two algorithms – *Warrior* and *King*. These two algorithms are combined following the well-known design principle of accelerated cascading.

At the beginning *Warriors* are born at each vertex that starts a computation of the leader election algorithm (that can be potentially all vertices in Q_n). Initially, the domain of the warrior consists of a singleton vertex at which the warrior was born. Each warrior tries to enlarge its domain to became a sphere of radius 5 (i.e. a set of all vertices of distance 5 from the center of the sphere). When the warrior succeeds, it is promoted to a *King* and then it executes the algorithm for kings.

The goal of the *warrior phase* is to eliminate warriors in an efficient way such that only a small number of them become kings. In order to keep the communication complexity of this phase low, warriors build their domains compactly. The domain of the warrior is always a sphere (possibly without a complete outermost layer) with the center at the vertex where the warrior was born.

In the *king phase* every kingdom tries to expand until it rules the whole universe Q_n. The main difference with respect to the warrior phase is that when two kingdoms meet, the whole defeated kingdom is appended to the winning one. The point is that although such a joining procedure is rather costly, the warrior phase assures that kingdoms are large enough (so there are not so many of them) to keep the total cost of joining procedures low.

8.7.3 Warrior Phase

Warriors have the name *Warrior(size,v)*, where the attribute *size* is the size (i.e., the number of vertices) of the domain of this warrior and the attribute v is the identity of its birthplace. By $\mathcal{D}_v$ we denote the domain of a warrior born at v. In a duel between two warriors, the one with the larger domain is stronger. If the domains are of equal size, the birthplace identities break the tie. Moreover kings are stronger than warriors. This leads to the fact that the resulting leader is not necessarily the one with the maximal initial *id*, but the one whose domain grows fastest.

We present an informal description of the algorithm for the warrior born at v:

1. Mark v by your name.
2. Go to the last explored vertex and using sequential BFS find a vertex x not belonging to $\mathcal{D}_v$. (When expanding a vertex, examine unexplored links one-by-one, not in parallel.)
3. If x is marked by a stronger warrior then die; otherwise follow the links leading to the birthplace w of the former owner of x.

4. If w is marked by a stronger warrior then die; otherwise mark w by your name. (If the warrior from w is still alive, it will die after returning to w.)
5. Return to the vertex x, mark it by your name and label the link leading from x to $\mathcal{D}_v$. This ensures the existence of a tree rooted at v spanning $\mathcal{D}_v$.
6. Return to v. If v is marked by a stronger warrior then die; otherwise increase your *size* attribute by one.
7. If your domain is not a sphere of radius 5 then *go to 2*.
8. Start the algorithm for kings.

Lemma 8.7.1. *Using the warrior algorithm, the cost of building a domain of size k and radius d is $O(dk)$ messages.*

Proof. Each added vertex will be charged the communication cost of the iteration in which it was added to the domain, except the cost of step 2. Since all other steps are clearly in $O(d)$, all we need is to show how to distribute the potentially high cost of step 2 among the vertices such that every vertex is charged $O(d)$.

The cost of step 2 can be divided into two parts:

Exploring links at border vertices: The cost of every trial to find a new neighbor will not be charged to the border vertex, but to the neighbor that was reached by this trial. Because the radius of the domain is d (and we use BFS), no vertex is charged here more than d.

Traversing to another border vertex (when the previous border vertex already explored all its neighbors): Traversal will be charged to the vertex that was fully explored. Because every vertex is fully explored just once, and the path from one such vertex to another one is bounded by $2d$, every vertex is charged here at most additional $O(d)$. □

Lemma 8.7.2. *Let each warrior promoted to king have a domain with size bounded by K and radius bounded by d. Assume d is $O(1)$. Then the total communication complexity of the warrior phase is $O(N \log \log N)$.*

Proof. The proof is based on the previous lemma. Let s_i be the number of domains that reached the size i. It holds that $s_i \leq N/i$. The overall communication complexity of the warrior phase can be bounded by $\sum_{i=1}^{K} c \cdot i \cdot (s_i - s_{i+1})$, where c is the upper bound constant obtained in Lemma 8.7.1 and s_{K+1} is defined as 0. Then $\sum_{i=1}^{K} c \cdot i \cdot (s_i - s_{i+1}) = c \sum_{i=1}^{K} s_i \leq c \sum_{i=1}^{K} N/i = c \cdot N \sum_{i=1}^{K} 1/i \in O(N \log K)$. Because $K < n^d$ and $d \in O(1)$ we get $O(N \log K) = O(N \log n) = O(N \log \log N)$. □

8.7.4 King Phase

The goal of the king phase is to repeatedly merge kingdoms, until there remains only one kingdom, whose capital becomes the leader. The computation is finished by broadcasting the identity of the leader.

Efficient Routing in Unoriented Hypercubes

In the king phase we need to efficiently send messages far apart in the unoriented hypercube. In this subsection we show how to do this.

Lemma 8.7.3. *Consider an unoriented Q_n with an initiator x. Let y_x be an address of a vertex y in Q_n in a world according to x. Then there exists an algorithm which computes an optimal routing path from the initiator x to the destination y using $O(nd^2)$ messages, where d is the distance between x and y in Q_n.*

Proof. We can restrict this to the case $y_x = 1^d 0^{n-d}$. If y_x is not in this form, we can permute link labels of x (in a world according to x) to achieve the appropriate address form. The main idea of the proof is to explore only the following small part of the hypercube: We will successively find vertices with addresses from $\{0^i 1^j 0^{n-i-j} \mid i = 1, \ldots, d, j = 0, \ldots, d - j\}$. The vertex v at the distance i from the initiator x can compute its address from the known addresses of its two neighbors which are at the distance $i-1$ from the initiator x. The address of v is obtained as the binary **or** of the addresses of both of these neighbors.

Algorithm $Route(y_x)$;
Input: (Invoked at the initiator x with an address 0^n.) An address
$\qquad y_x = 1^d 0^{n-d}$ of a destination vertex y.
Output: An implicit routing path from x to y.
Method:
   At the initiator x:
      {*Send link's labels (neighbor's addresses) to the first d neighbors.*}
        $Name_x := 0^n$;
        for $i := 1$ **to** d **pardo** $Send(e_i, (start, e_i, d))$;
   At a vertex v:
      (1) **Upon receiving a message (start, e_i, d) via the link h:**
         {*v is a neighbor of the initiator. Inform all neighbors of v*
         about its address.}
         $Name_v := e_i$;
         $Label_{v,h} := e_i$;
         for $j := 1$ **to** n **pardo**
           if $i \neq j$ **then** $Send(e_j, (Name_v, d))$;
      (2) **Upon receiving a message (x, d):**
         {*For vertices that did not yet compute their names.*}

Remember this message.

Proceed if you had received two messages of the form $0^i1^j0^{n-i-j}$ (received on h_1)

and $0^{i+1}1^j0^{n-i-j-1}$ (received on h_2) where $0 \le i \le d-j$.

$\text{Name}_v := 0^i1^{j+1}0^{n-i-j-1}$;

$\text{Label}_{v,h_1} := e_{i+j+1}$;

$\text{Label}_{v,h_2} := e_{i+1}$;

if $j+1 = d$ then stop

else

    for $k := 1$ to n pardo Send(e_k, (Name_v, d));

(3) **Upon receiving a message (x, d) via the link h:**

{*For vertices that had already computed their names.*}

$\text{Label}_{v,h} := x \textbf{ xor } \text{Name}_v$;

Correctness: Let

$$V_d = \{0^i1^j0^{n-i-j} \mid i = 1, \ldots, d, j = 0, \ldots, d-i\} \cup \{0^n\}$$

and

$$E_d = \{(x, y) \mid x, y \in V_d\}.$$

To prove the correctness of the *Route*() it is sufficient to show

(1) $\forall v \in V_d : \text{Name}_v = \beta_x^{-1}(v)$,

(2) $\forall h \in E_d, h = (u, v) : \text{Label}_{u,h_u} = \sigma_x(u)[h_u]$.

We prove (1) by contradiction. Let w be the first vertex for which it holds that $\text{Name}_w \ne \beta_x^{-1}(w)$ in *Route*(). Its name is computed as a Boolean **or** of names of some neighboring vertices u and v. By assumption, u and v had correctly computed their names. Let

$$\text{Name}_u = 0^i1^j0^{n-i-j}, \text{ and } \text{Name}_v = 0^{i+1}1^j0^{n-i-j-1}.$$

There are just two vertices that can receive direct messages from both u and v. These vertices have addresses

$$\text{Name}_u \textbf{ and } \text{Name}_v = 0^{i+1}1^{j-1}0^{n-i-j}$$

and

$$\text{Name}_u \textbf{ or } \text{Name}_v = 0^i1^{j+1}0^{n-i-j-1}.$$

As w had wrongly computed its name, it should have the address

$$w' = \text{Name}_u \textbf{ and } \text{Name}_v = 0^{i+1}1^{j-1}0^{n-i-j}.$$

But according to the algorithm, u will compute its name after receiving messages from two vertices, one of which has the name w'. (The same holds for v.) This means that the vertex w' had already computed its name and cannot be

w. Thus, for w there remains the only possibility, to have the address $Name_u$ or $Name_v$. But in this case w correctly computes its name, a contradiction.

Further, (2) follows directly from (1) and the description of the algorithm.

Communication complexity: Every vertex from V_d sends n messages, initiator x sends d messages and other vertices send no messages. Thus, the total communication complexity is $d + n|V_d| \in O(nd^2)$, because $|V_d| \in O(d^2)$.

□

Lemma 8.7.3 allows us to call subroutine $FarSend(destination, message)$. This subroutine first calls algorithm $Route(destination)$ of Lemma 8.7.3 and then sends *message* via the optimal route to its *destination*. Note, that if we need to send a message via this optimal path more than once, it is sufficient to use only the second part of $FarSend()$ which we refer to as $FarJump()$. We just need to keep in mind that the cost of $FarSend()$ is not d, but $O(nd^2)$ messages.

To avoid the confusion caused by multiple and concurrent calls of the subroutine $FarSend()$, we mark internal messages of each such a call by the name of the sending vertex v and the counter c_v which is incremented in each call of $FarSend()$ in v. (Computed link labels are also marked in this way, the subsequent $FarJump()$'s will not be confused if trails of multiple $Route()$'s intersect.)

We need also procedure $FarOrient(x)$, which will (when invoked at v) label links of x consistently with link labels at v (in the world according to v). $FarOrient()$ can be implemented as follows: Send the address not only to x but also to all its neighbors. Then x asks all its neighbors about their addresses and from the received replies and its own address it computes the labels of the incident links. Such an implementation of $FarOrient()$ requires $n + 1$ calls of $FarSend()$ and an additional $2n$ messages for local lookup to compute the labels. Thus $FarOrient()$ can be implemented using $O((nd)^2)$ messages.

The Structure of a Kingdom

During the whole king phase the algorithm constructs and maintains kingdoms in such a way that each kingdom $\mathcal{K}$ is a union of spheres $S_i^{\mathcal{K}}$, each of radius at most 5. These spheres may not have complete outermost layers (e.g., a sphere of radius 5 contains all vertices at a distance up to 4 from its center and some of the vertices at the distance 5). All these spheres are disjoint, so the kingdoms are disjoint too.

The structure of a kingdom $\mathcal{K}$ is stored in the following way:

1. Let c_i be the center of the sphere $S_i^{\mathcal{K}}$ and let c_0 denote the capital of $\mathcal{K}$. *Map*, stored in c_0, is a set $\{(\beta_{c_0}^{-1}(c_i), |S_i^{\mathcal{K}}|) \mid i \geq 1\}$.
2. There is a tree $\mathcal{T}_i^{\mathcal{K}}$ spanning $S_i^{\mathcal{K}}$ and rooted towards c_i, which is maintained at vertices of $S_i^{\mathcal{K}}$ by variables *Parent*.

3. There is a spanning tree $TC^{\mathcal{K}}$ over all c_i, rooted towards c_0. The depth of $TC^{\mathcal{K}}$ is bounded by n. (Links of $TC^{\mathcal{K}}$ correspond to paths between centers, they are created using $FarSend()$ and can be optimally traversed using $FarJump()$.) The implementation of $TC^{\mathcal{K}}$ is the following: If c_i is a son of c_j in $TC^{\mathcal{K}}$, then a variable σ at c_i has value $\sigma_{c_j}(c_i)$ and variable $NextJump$ at c_i has value $\beta_{c_j}^{-1}(c_i)$.

4. Trees $T_i^{\mathcal{K}}$ and $TC^{\mathcal{K}}$ form a directed graph, denoted as $ST^{\mathcal{K}}$, that spans the whole kingdom $\mathcal{K}$. (The depth of $ST^{\mathcal{K}}$ is bounded by $O(n^2)$ since each link of $TC^{\mathcal{K}}$ corresponds to the path that is not longer than n, and $T_i^{\mathcal{K}}$, for all i, has a depth of at most 5.)

Brief Description of the King Algorithm

The $King$ algorithm is described in a token-driven framework. There is one token reserved for each kingdom. When two kingdoms meet, after performing the duel only one token survives.

We present the algorithm from the point of view of activities performed by the token, instead of the more common view of computation executed at each vertex. A brief description of the $King$ algorithm consists of four basic activities:

1. *Build your basic domain:* Mark your basic domain (a sphere of radius 5 inherited from the warrior phase) using sequential BFS and build its BFS spanning tree. When finished, or the mark of another king has been found, return to your capital. Add recently built sphere to your *Map*.

2. *Resolve duel:* If there are attacker kings at your capital waiting for duels with you, call $ResolveDuel()$.

3. *Resolve collision:* If you have collided with another kingdom, call $ResolveCollision()$.

4. *Build a colony:* Find a place for a new colony. Choose an address x such that the distance from your kingdom is at least 5. (This can be done by a local computation at your capital using *Map*.) If you cannot find such an address, you are the only surviving king. Broadcast the identity of your capital. Halt. Otherwise establish a colony.

   Set up a center of the colony using $FarOrient(x)$.

   Expand this colony using sequential BFS (and build its spanning tree) until either it becomes a sphere of radius 4 or a mark of another *king* has been found.

   Return to your capital.

   Add recently built sphere to your *Map*.

   Go to 2.

In the procedure $ResolveDuel()$, the weaker kingdom surrenders to the stronger one and its token is discarded. The *Map* and the spanning tree of the winning kingdom are updated – the defeated capital becomes a son of the winner capital in its $\mathcal{TC}^{\mathcal{K}}$. The kingdom is weaker if its size is smaller or the sizes are equal and the identity of its capital is smaller.

In the procedure *ResolveCollision()*, the king returns back to the place of collision, then follows the spanning tree of the attacked kingdom until he comes to its capital. During this travel the attacking king keeps track of his position relative to his capital and of the relationship of the local orientation to the orientation of his capital. At the attacked capital, the attacking king waits until the attacked king returns home (at that moment *ResolveDuel()* is called) or he immediately knows that he is weaker and surrenders (if he is weaker according to local *Map*) – this is needed to avoid cyclical waiting.

Detailed Description of the King Algorithm

For brevity, the centers of colonies and capitals (of active and defeated kingdoms) are called *essential* vertices.

8.7.4.1. Variables used at a vertex of the hypercube Q_n.

Owner – integer expressing the identification number of the owner of this vertex. Used for testing collisions and for BFS. Initially set to *nil*.

Parent – a link of the distributively stored rooted tree of a sphere. Values are from $\{\mathbf{e_1}, \dots, \mathbf{e_n}, nil\}$, initially set to *nil*.

σ – at an essential vertex: a translation (permutation) from the local labeling of this vertex to the labeling used at its parent in $\mathcal{TC}^{\mathcal{K}}$. Initially set to the identity permutation.

NextJump – at an essential vertex: a relative address of the parent (in $\mathcal{TC}^{\mathcal{K}}$) of this vertex, given in the world according to this parent. Initially set to *nil* (no parent). σ together with *NextJump* allow the attacking king to efficiently follow the links of $\mathcal{TC}^{\mathcal{K}}$ and keep track of its position and local orientation.

TowardCollision – used for distributively stored path from the capital towards the place of collision. Values are from $\{\mathbf{e_1}, \dots, \mathbf{e_n}, nil\}$, initially set to *nil*.

Map – at an active capital: an information about the topology of the kingdom in the form of a set of pairs *(center,sphere-size)*. Initially, *Map* is empty.

AttackedBy – at an active capital: a list of arrived kings waiting for the return of the home king to perform a duel. Initially, *AttackedBy* is empty.

And others – used by BFS, *FarSend()*, *FarJump()* and *FarOrient()*.

8.7.4.2. Attributes of a token.

Each active kingdom has its token (called *king*) that travels through the graph and tries to expand its kingdom. The token has the following attributes:

1. *status* – expressing what the king is currently doing.
2. *id* – identity of the capital of the kingdom.

3. *size* – integer, expressing the size of the kingdom, i.e., the sum of sizes of all spheres listed in *Map*.
4. *pos* – relative address determining a position of the *king* according to his capital. Initially set to **0**.
5. σ – permutation determining the relationship between a local orientation and the orientation at the capital. Initially set to the identity permutation.
6. *Map* – a set giving information about the defeated capital to the winner. *Map* is usually stored at the capital and only when the king is defeated does he take his *Map* and bring it to the winner.
7. And some others used during the BFS of the base kingdom and the colonies.

8.7.4.3 Building a base kingdom.

When a *warrior* is promoted to *king*, his base kingdom consists of a single vertex – the capital c_0. The *king* builds his base kingdom using BFS to depth 5. During his traversal the *king* marks new visited vertices by setting their *Owner* variable to *king.id* and constructs the BFS spanning tree by setting vertex's *Parent* to the last traversed link. The number of newly added vertices is stored in *king.added*. After a successful BFS up to depth 5 or a collision the king returns to his capital.

8.7.4.4. Establishing a colony.

Choose an address x such that $\forall S_i = (y_i, size_i) \in Map : d(x, y_i) > 5 + r(size, n)$, where $r(s, n)$ is the radius of a sphere (with a possibly incomplete outermost layer) containing s vertices, in Q_n. If such an address cannot be chosen, you are the only surviving king. (See that there could be a place for a new colony (of radius 4) not computable from the *Map* due to the incomplete outermost layers of the spheres, but certainly there is not enough place for a new kingdom (of radius 5).) In this case, broadcast the identity of the capital (the leader) of the kingdom and halt.

FarOrient() is used to orient the center of the colony consistently with the capital, then the king travels to this center using *FarJump()* and expands the colony. The colony is built using BFS from its center to depth 4 in a similar way as the base kingdom was built from the capital. After a successful BFS or collision the king returns to his capital.

8.7.4.5. Resolving collisions.

It may happen that during the process of building the base kingdom or colony, the *king* finds the mark of another kingdom (by testing *Owner* $\neq$ *king.id*). In such a case, the *king* performs the following actions:

1. Break the current BFS. Return to your capital using the spanning tree of your kingdom and mark on the way the path towards the place of collision.

2. Update the *Map* by adding the pair $(c, king.added)$, where c is the center of the sphere currently being built and *king.added* is the number of vertices successfully added to this sphere before the collision happened.
3. Call *ResolveDuel*() if there are kings waiting for you $(AttackedBy \neq \emptyset)$. (You may die here.)
4. First follow the links marked towards the place of collision and then follow the spanning tree of the attacked kingdom until you come to its capital. Cast *FarOrient*() on vertex one step ahead until you come to the first essential vertex of the attacked kingdom, so you always know where you are going and what is the relationship between the orientation at the current vertex and the orientation at your capital. When you follow the links of the $\mathcal{TC}^\mathcal{K}$ of the attacked kingdom $\mathcal{K}$, calling *FarOrient()* is not necessary, as the variables σ and *NextJump* help keep track of your position and orientation.
5. Call *PreliminaryDuel*() at the capital of the attacked kingdom.

8.7.4.6. Resolving a duel.

ResolveDuel() is performed whenever *king* returns home and there are attackers waiting for a duel $(AttackedBy \neq \emptyset)$. If the strongest attacker is stronger than the local king then:

1. The token of the local king is discarded.
2. The attacker makes the defeated capital the son of his capital (by setting the local variables σ and *NextJump* – he knows the required information because he kept track of it during his attack).
3. The attacker takes the local *Map* and returns home using *FarSend*() – this means that all subsequent traversals of this newly added logical link of its $\mathcal{TC}^\mathcal{K}$ can be done effectively using *FarJump*(). At home he updates his *Map* using the *Map* of the updated kingdom. (Here the information about the position and orientation of the defeated capital is also used.)

If the attacker is weaker, he returns home. At home he checks whether there is a stronger attacker waiting for a duel. If there is one, he forgets about his unsuccessful attack and surrenders to this attacker (see previous case). Otherwise he surrenders to the king he attacked (setting his capital to be the son of the winning capital, and bringing the *Map* of his kingdom to the winner). Note that he can arrive to an already defeated capital, so he should follow the spanning tree of the winning kingdom (and keep track of his position and orientation) in the same manner as when attacking.

8.7.4.7. Preliminary duel.

PreliminaryDuel() is performed whenever the attacking king arrives at the capital of the attacked kingdom. The attacker compares its size with the last known size of the attacked kingdom (stored at the local *Map*).

If the attacker is stronger he waits for the attacked king to return home (by being added to *AttackedBy*). It may happen that during his expedition the king of the attacked kingdom enlarged his kingdom, so the rule "bigger wins" would not be ensured if the attacker just left the mark *Die!* and continued his own computation. If the attacker is weaker, he proceeds as in *ResolveDuel()*.

A full description of the *King* algorithm is given later.

Correctness and Communication Complexity of the King Algorithm

Lemma 8.7.4. *The structure of kingdoms is correctly built and maintained.*

The proof follows from the following arguments (see Section 8.7.4):

1. Each part that is added to the kingdom (i.e., when it is added to its *Map*) is either a sphere or a set of spheres (when appending a defeated kingdom), so the kingdom is always a set of spheres. Check the full algorithm to see that *Map* always contains the correct information.
2. The spanning trees are built during the building of the base kingdom and the colonies.
3. The edges of this spanning tree are created either when the colony is established or the defeated kingdom is connected to the winner. Check the full algorithm to see that the variables σ and *NextJump* are correctly set.
4. It is sufficient to show that the (logical) depth of $\mathcal{TC}^{\mathcal{K}}$ is bounded by n. All inner edges of $\mathcal{TC}^{\mathcal{K}}$ are directed from the defeated capital to the winner capital. Since the larger kingdom always wins, the parent vertex corresponds to the kingdom which is at least twice as big as the defeated one. There are 2^n vertices, so the depth is bounded by n.

Lemma 8.7.5 (Correctness). *The King algorithm always terminates and there is exactly one vertex selected as the leader.*

Proof. First note that cyclic waiting (deadlock) cannot occur. Consider the oriented graph (dependency graph) of the waiting relations. Each edge leads from the bigger vertex (kingdom) to the smaller one, because the only waiting is when a larger kingdom waits for a smaller one (the sizes of the *Maps* are compared) and the attacker always carries the actual size that is stored at its capital (that is the explanation for the back-up return home after a collision). This ensures that cyclic waiting is indeed avoided.

It is easy to see that there is always at least one active king. A king dies only when he performs a duel with another king. In such a case, one king (that with the larger kingdom) always survives.

Next, we need to show that if there are at least two kingdoms, they always collide, resulting in the death of one king. This follows from the fact that kingdoms grow (there is no deadlock) and they are disjoined.

Each active kingdom grows, so there will be eventually a kingdom that cannot find a place for another colony. Since there is no place for another colony, there is no place for another kingdom, so this kingdom is a single surviving kingdom and it can declare its capital to be the leader.

$\square$

Lemma 8.7.6 (Communication Complexity). *If $O(N/n^5)$ warriors become kings, then the total number of messages in the* King *phase is $O(N)$.*

Proof. The total communication can be accounted for by some of these tasks:
(1) Building base kingdoms.
(2) Establishing colonies that successfully grow to full size.
(3) Establishing colonies that do not grow to full size.
(4) Resolving collisions and joining the kingdoms.
(5) Final broadcasting.

1. The cost of building the base kingdom is clearly $O(n^5)$ – the size of an initial kingdom. Since there are $O(N/n^5)$ kingdoms (kingdoms do not overlap), the total cost is $O(N)$.
2. There are at most $O(N/n^4)$ such colonies, because the colonies do not overlap. The cost of building one such colony is $O(n^4)$ for the initial establishment and orientation (complexity of $FarOrient()$), and $O(n^4)$ for the actual growth. So the overall communication complexity of this computation is $O(N)$.
3. The cost of establishment and growth of each such colony is again $O(n^4)$, the difference is in the argument about their number: each colony that does not grow to full size caused a collision and the death of one kingdom. (In fact there may be two colonies responsible for one death, because we did not consider attacking when attacked.) The number of such colonies is thus bounded by the initial number of kingdoms $O(N/n^5)$; thus the overall cost of this computation is $O(N/n)$.
4. The cost of resolving one collision is bounded by $O(n^3)$. This follows from the fact that spanning trees of kingdoms have depths bounded by $O(n^2)$. The single $O(n^3)$ action in resolving a collision is $FarSend()$ between a defeated and a winning capitals. The number of calls of $FarOrient$ at the beginning of an attack is bounded by 10 – it is necessary to call them only between the center of the sphere of collision and the first essential vertex of the attacked kingdom (both centers of spheres of radius of at most 5) and the cost of each such $FarOrient()$ is $O(n^2)$, since $d = 1$.
There are at most $O(N/n^5)$ joins, so the overall cost of this case is $O(N/n^2)$.
5. The broadcasting algorithm requires $O(N)$ messages.

Counting altogether we get that the total communication complexity of the king phase is $O(N)$ messages. $\square$

All messages are of size $O(n)$ bits, except when Map is being sent. Information about each sphere listed in Map is sent over at most $O(n^2)$ links (the depth of $\mathcal{ST}^{\mathcal{K}}$), there are at most $O(N/n^4)$ different spheres (either colonies or base kingdoms), so we can bound the total bit complexity for sending Maps to $O(n^2) \cdot O(n) \cdot O(N/n^4) = O(N/n)$. We get:

Lemma 8.7.7. *The total bit communication complexity of the king phase is $O(N/n)$ bits.*

Taking this together with the results from the warrior phase, we get the final theorem:

Theorem 8.7.8. *There exists a leader election algorithm on an unoriented hypercube that uses*

$$O(N \log \log N)$$

messages in the worst case, with a total bit complexity of

$$O(N \log N \log \log N).$$

King Algorithm

Algorithm $King(king)$; {*Called at the capital of the kingdom.*}
begin
At the capital:
   $x := \mathbf{0}$; {x *is the center of the last processed sphere*}
   Init(king);
   repeat
      {update Map and $size$}
      if king.added > 0 **then** Map := Map$\cup\{(x,$king.added$)\}$;
      king.size := ComputeSize(Map);
      if AttackedBy $\neq \emptyset$ **then** ResolveDuel(king); continue;
      {*start the next iteration of the loop*}
      {*There was a collision, attack!*}
      if king.status $=$ Collision **then**
         king.status := Attack;
         {*Launch Attack() from the center of the collision sphere.*}
         if $x \neq \mathbf{0}$ **then** $FarJump(x,king)$ **else** $Attack(king)$;
         wait for the return of your king;
         {*unsuccessful attack*}
         if king.status $=$ Defeated **then**
            {*Forget about your attack, here is somebody to surrender to*}
            if $\exists$ *attacker* $\in$ AttackedBy : (attacker.size $>$ king.size)
               or (attacker.size $=$ king.size) **and** (attacker.id $>$ king.id)
            then
               attacker.status := Winner;

```
                    attacker.Map := TranslateMap(Map, attacker.σ, attacker.pos);
                    σ := attacker.σ;
                    NextJump := attacker.pos;
                    FarSend(attacker.σ⁻¹(attacker.pos), attacker);
                    take kings from AttackedBy and transfer them
                    following links towards the winner capital; halt.
                    {token of the local king dies here}
                else {surrender to the king you unsuccessfully attacked}
                    σ := king.σ⁻¹;
                    king.status := Surrender;
                    NextJump := king.σ⁻¹(king.pos);
                    {You can't translate Map, because recipient
                      may be meanwhile defeated}
                    king.Map := Map;
                    Follow(king);
                    {Follow spanning tree of the winner kingdom.}
                    if AttackedBy ≠ ∅ then
                        take kings from AttackedBy and transfer them following
                        links towards the winner capital;
                    fi;
            else {king.status = Winner}
                Map := Map∪king.Map;
                king.Map := ∅; {Just to save bit complexity.}
                king.size := ComputeSize(Map);
                if AttackedBy ≠ ∅ then
                    ResolveDuel(king); continue;
                    {start the next iteration of the loop}
                fi;
            fi;
        fi;
    fi;
    choose an address x such that
    ∀Sᵢ = (yᵢ, sizeᵢ) ∈ Map : d(x, yᵢ) > 5 + diam(sizeᵢ, n),
    where diam(s, n) is the diameter of a sphere of size s in Qₙ.
    if there is no such x then {you are the only surviving king;}
        Broadcast(c₀); halt.
    else {establish new colony with center x}
        king.status := Colonize;
        king.added := 0;
        king.pos := x; {center of the currently processed colony}
        king.σ := identity permutation;
        FarOrient(x); FarJump(x,king);
        wait for the return of your king to his capital;
    fi;
until true;
end;
```

Procedure PrepareDuel(attacker)
{*At the active capital upon arrival of the enemy king attacker.*}
begin
    if attacker.status = Attack **then**
        if (attacker.size < ComputeSize(Map))
            or (attacker.size = ComputeSize(Map))
            and (attacker.id < Owner) **then** {attacker weaker}
            attacker.status := Defeated;
            FarSend(attacker.σ^{-1}(attacker.pos), attacker);
            {*attacker returns home to his capital*}
        else Add attacker to the list *AttackedBy*;
    else
        if attacker.status = Surrender **then** {*Attacker dies – no Send()*}
        Map := Map∪
        TranslateMap(attacker.Map, attacker.σ^{-1},
                    attacker.σ^{-1}(attacker.pos));
end;

Procedure ResolveDuel(king); {*When local king returns to his capital.*}
begin
    while AttackedBy $\neq \emptyset$ **do**
        choose attacker from AttackedBy;
        if (attacker.size < king.size) **or**
        (attacker.size = king.size) **and** (attacked.id < king.id)
        then
        attacker.status := Defeated;
        FarSend(attacker.σ^{-1}(attacker.pos), attacker);
        remove attacker from AttackedBy;
    else
        attacker.status := Winner;
        attacker.Map := TranslateMap(Map, attacker.σ, attacker.pos);
        σ := attacker.σ;
        NextJump := attacker.pos;
        FarSend(attacker.σ^{-1}(attacker.pos), attacker);
        take kings from AttackedBy and let them follow the
        links towards the winner capital;
        halt. {*token of the local king dies here*}
      fi;
end;

Procedure Attack(king);
{*From the point of view of the attacking king, starting at the center of the*
collision sphere, until he arrives at the enemy capital.}
begin
    while TowardCollision is set **do** {*within my colony*}
        king.σ := king.$\sigma \circ \sigma$;
        king.pos := king.pos + king.σ(TowardCollision);
        FarOrient(TowardCollision,king);
        clear *TowardCollision* after departure of the king;
    od;
    {*coming to the first enemy's essential vertex*}
    while not at essential vertex **do**
        king.σ := king.$\sigma \circ \sigma$;
        king.pos :=king.pos + king.σ(Parent);
        FarOrient(Parent, king);
    od;
    Follow(king);
end;

Procedure Follow(king); {*Following the links of* TC^K. }
begin
    while not at active capital **do**
        king.σ = king.$\sigma \circ \sigma$;
        king.pos = king.pos + king.σ(NextJump);
        FarJump(σ^{-1}(NextJump), king);
    od;
end;

Procedure Init(king);
Initial marking of the kingdom; BFS traversal from the capital to
depth 5 with exit when a mark of another kingdom is found
at the capital of the king:
begin
    king.status := Init;
    king.id := c_0;
    king.added := 1;
    {*breadth-first search from the capital on layers up to depth 5*}
    for king.maxlen := 1 **to** 5 **do**
        king.len := 1;
        while (Unexplored$\neq \emptyset$) **and** (king.status $\neq$ Collision) **do**
            choose $l \in$ Unexplored;
            Unexplored := Unexplored$-\{l\}$;
            Send(king, l);
            Wait for the return of *king* from l with king.id = *Owner*;

```
            if king.status = Collision then
                TowardCollision := l;
                exit.
            fi;
        od;
        Unexplored := links to all neighbors;
    od;
    Map := {(0, king.added)};
    king.size := king.added;
end;
```

At a vertex v upon receiving king with status Init via a link h:

```
begin
    {a mark of another kingdom found}
    if (Owner is set) and (Owner ≠ king.id) then
        king.status = Collision;
        Send(king, h);
    else
        if king.len = king.maxlen then
        {new vertex added at outermost layer}
            Owner := king.id;
            Parent := h;
            king.added := king.added+1;
            Send(king, h);
        else {inner vertex at the distance len from the capital}
            king.len := king.len+1;
            while (Unexplored ≠ ∅) and (king.status ≠ Collision) do
                choose l ∈Unexplored;
                Unexplored := Unexplored−{l};
                Send(king,l);
                Wait for the return of king from l with king.id = Owner;
                if king.status = Collision then
                    TowardCollision := l;
                    Send(king, h);
                    exit.
                fi;
            od;
            king.len := king.len − 1;
            Unexplored := all neighbors;
            Send(king, h);
    fi;
end;
```

Procedure Colonize(x);
{BFS traversal from x to depth 4 with exit when a mark of
another kingdom is found}
At the center of the colony upon receiving a king with status Colonize:
begin
    { *The center of a colony lies inside another kingdom, nothing is added
    to the kingdom, so we can directly proceed with Attack()*}
    if Owner is set **then**
        king.status := Attack;
        Attack(king);
    else {*breadth-first search from the center on layers up to depth 4*}
        king.added := 1;
        for king.maxlen := 1 **to** 4 **do**
            king.len = 1;
            while (Unexplored$\neq \emptyset$) **and** (king.status $\neq$ Collision) **do**
                choose $l \in$ Unexplored;
                Unexplored := Unexplored$-\{l\}$;
                Send(king, l);
                Wait for the return of *king* from l with king.id = *Owner*;
                if king.status = Collision **then**
                    TowardCollision := l;
                    FarJump(king.σ^{-1}(king.pos), king);
                    {*send king back home for backup*}
                    exit.
                fi;
            od;
            Unexplored := all neighbors;
        od;
        FarJump(king.σ^{-1}(king.pos), king);
        {*return to your capital*}
    fi;
end;
At a vertex v upon receiving a king *with status Colonize via a link h:*
    execute the same code as in the case of receiving a *king* with status *Init* in
    the procedure *Init*();

Procedure ComputeSize(*Map*) computes the sum of the sizes of all the
spheres in *Map*.
Procedure TranslateMap(Map, σ, address) – the aim is to translate the
coordinates of the centers in such a way that *Map* can be directly added to
the map of the winner, whose relative address (in his orientation given by σ)
is *address*.

8.8 Leader Election on Synchronous Rings

We present the Chang-Roberts election algorithm for synchronous rings with unknown size.

In the Chang-Roberts election algorithm each processor sends a message over the ring containing its own identity. A processor, say with identity i, that receives a message $\langle j \rangle$ acts as follows:

If $j > i$, the message is purged.

If $j < i$, the message is forwarded.

If $j = i$, the processor obtained the *elected* status and sends a special message around the ring to abort all activity in other processes and force them to obtain the *defeated* status.

Proof. To establish the correctness of the Chang-Roberts algorithm, let m be the smallest identity of any processor. A processor different from m does not obtain the *elected* status because its message does not pass processor m. Processor m obtains the *elected* status because all processors forward its message. This algorithm uses $\Omega(N^2)$ messages in the worst case, but only $O(N \log N)$ messages in the average case. There is a variant, in which processor i also purges a message $\langle j \rangle$ if i has earlier received a message $\langle j' \rangle$ with $j' < j$.
□

9

Fault-Tolerant Broadcast in Distributed Networks

*Forget about yourself
and the world will remember you.*

Jack London

9.1 Consensus Problem

Many algorithms that are needed in the implementation of distributed networks use broadcasting as a basic mechanism for sending an item of information from a source processor to a number of others. This is done, for example, in order to ensure synchronization in a distributed network or consistency in a distributed system, to reach a consensus in a distributed system or to manage a transaction that involves a distributed database. The problem was discussed in the previous chapter when the distributed network was reliable. But in the face of unreliabilities the problem is more complex and forms one of the central issues in distributed computing; it is known under different denotations – as the problem of interactive consistency, as the Byzantine agreement, the problem of the Byzantine Generals, the consensus problem, or the unanimity problem. We shall refer to it as the consensus problem.

To give a precise formulation of the *Consensus Problem*, consider a network of N processors that can communicate by exchanging messages along bidirectional links. The aim of consensus is to ensure the reliable broadcast of messages, meaning that when a source processor disseminates any item of information, all other reliable processors receive that item unchanged, or in other words have the same "perception" of that item. This can be explicitly expressed in the form of a pair of constraints that must be satisfied in every solution of the consensus problem: that when any processor despatches any item of information to the others

C1: all reliable processors receive the same item; and
C2: if the source processor is reliable, then the received item is identical to the issued item.

Clearly, if the source processor is reliable, then constraint (C1) follows from (C2) and it is easy to implement a protocol that will satisfy both constraints, but a processor that receives a broadcast item does not a priori know whether or not the source processor is reliable. The difficulty of the problem lies in the types of faults that can occur and the absence of the knowledge of whether the processors concerned have failed or not.

Consider networks to be completely connected. Among the various types of processor failure, the first to be considered is crash failure (with the processor neither sending further messages nor responding to enquiries from outside). With this type of failure any assumption concerning continued progress by the processors is violated and, provided there is a known upper bound on the transmit delay of messages between any pair of processors, can be detected by requiring all messages to be acknowledged and timing the interval between issue and receipt of acknowledgement. The second type of out-of-course halting of processors is intermittent, with "pauses" during which they are out of contact with their environment. This presents a more difficult problem and for an algorithm that must be resilient to this type of failure there must be some synchronization between its various components, to ensure that the chronological order of the messages is known correctly. There is also the possibility of what may be called misbehavior by a faulty processor to be countered; such a processor – not known to be faulty – may send different values to different recipients, and perhaps no value to some. Given all these possibilities, no assumptions can be made about a processor that achieves consensus among the processors – that is, with respect to the constraints (C1), (C2) – in spite of this.

For resilience to malfunctioning of the source processor the constraint (C2) requires a protocol that enables the receivers to agree among themselves on the identity of the value received, and for this they must exchange the values they have received. However, as the processors cannot be assumed to be completely reliable, malfunctionings may occur during these exchanges and values may be transmitted that are different from those received. Thus nothing can be concluded from the results of the first round of exchanges: if a reliable processor receives in this exchange a value that differs from what it received in the original broadcast it does not know whether it is the source of the fault or whether the processor concerned in the exchange is faulty; and equally if it receives two identical values – there is nothing to show that a faulty source processor has not sent out two different values and a faulty intermediate processor has not changed the value that it received. A single round is therefore insufficient and further stages must follow, synchronized so as to counter possible failures, whether complete or intermittent, of some of the processors.

The number of stages of exchanges necessary to achieve a consensus with regard to (C1) and (C2) is a measure of the time complexity of the solution; another important measure is the cost, in terms of the number of messages that have to be exchanged.

Let t be the maximum number of simultaneously failed processors – either final or intermittent – that the algorithm is required to be resilient to. This is a fundamental characteristic of the algorithm and [DS83] have shown that every solution will need at least $t + 1$ stages of message exchange to arrive at a consensus, which is therefore the measure of the minimum time complexity. Thus the distinction between different solutions rests on the different total numbers of messages they require.

As already shown, one of the difficulties of the problem lies in possible malfunctioning of a processor so that it may change the value of items sent to it. The question arises of whether this type of behavior can be eliminated or, equivalently, this can be made detectable by the receiving processor; and the answer is that this can be done if each process can "sign", in an unforgeable manner, every message that it issues. The signature is added to the message, and it and the message are encoded in such a way that the receiver can check the origin and the authenticity and no processor can generate the signature of any other; this enables the receiver to detect any modification of the content. Thus a processor can detect when a value received has been corrupted by a faulty intermediate processor that has relayed it, and so ignore this. The conclusion is that the effects of the malfunctioning of a processor can be countered by the use of encoding and signatures, and therefore that the only type of failure that needs to be considered is out-of-course halting, either permanent or intermittent, of processors. In an actual system where there is no intentional malfunctioning of processors – no sabotage, in fact – a simple error-correcting code will suffice in place of the signature mechanism.

The general situation is that the difficulties of the consensus problem arise from the types of failure to which the solution must be resilient and the absence of a priori knowledge concerning these. We shall look at several solutions in this chapter: we consider an algorithm in which the idea of message exchange stages is made precise, then we consider methods using signed messages, restricting our study to two out of the many solutions that have been proposed.

9.1.1 Fault-Tolerant Broadcast with Unsigned Messages

We assume that there are N processors among which at most t can be malfunctioning at any time of computation, and that they operate in a reliable communication network, i.e., that any correctly operating processor can always send a message to any other and the messages are received unchanged in the order in which they were submitted. To ensure such a reliability in the communication, if there could be up to k simultaneous link failures there must be at least $k + 1$ independent routes between every pair of processors – a redundancy (k-connectivity) that is the price to be paid to guarantee the resilience against such a simultaneous failure. The links are assumed to be bidirectional, and the network can be represented by a complete graph of N vertices.

Further, we assume that messages are not signed, so that a malfunctioning processor may modify their contents. The identity of the source of a message is known to the receiver – this is easily ensured when the graph of the network is complete. To counter the effects of the complete stopping at any processor it must be possible to detect the absence of a message that should have been received, which again is easily achieved.

Impossibility Criterion

There is no solution to the consensus problem unless the maximum number t of processors that can be malfunctioning at any one time is strictly less than one-third of the total number N in the network; that is, if messages are unsigned then $N \geq 3t + 1$. Now, we shall illustrate the impossibility argument by considering the case of three processors, one of which is faulty.

Let P_0, P_1, P_2 be the three processors with P_0 being the source and P_2 the faulty processor. In the first stage, P_0 sends a value a both to P_1 and P_2 and in the second stage P_1 and P_2 exchange the values they have received; but P_2, being faulty, changes the value to b and sends this to P_1. Thus at the end of stage 2, P_1 has received the values a from P_0 and b from P_2. The second value is what P_2 is presumed to have received from P_0 in stage 1 and might modify if it were faulty; if, being faulty, it had not performed the exchange operation required of it in stage 2, then P_1 would have needed to consider some default value, say b. Having received two different values, P_1 knows that there are faulty processors in the system but does not know how many there are, or their identities; and if it is to respect the consensus constraints (C1), (C2) it must, if it can, choose the value a.

Consider now the same system but with P_0 the faulty processor, P_1 and P_2 working reliably. The two stages are now as follows: in the 1st stage P_0 sends a to P_1 and b to P_2, and in the 2nd stage P_1 sends a to P_2 and P_2 sends b to P_1. P_1 has the same perception of its environment in this case as before and, therefore, having no means of distinguishing between the two, must come to the same conclusion, which is to accept the value a as in the first case. In other words, whatever its environment, if P_1 receives a from the source P_0 it must accept this as correct. By the same argument, P_2 will accept as correct the value b it has received from the faulty P_0. Thus the two reliable processors P_1, P_2 have not reached a consensus and the constraint (C1) has been violated; so there is no solution when $N = 3$ and $t = 1$.

It is important to realize that in this case nothing is gained by making further exchanges – these can only repeat the results already obtained and so give no new information.

Underlying Principles of the Solution

Apart from showing that no solution is possible in the case of three processors of which one is faulty, the above example brings out the mechanism used in all solutions – a sequence of stages of synchronized value exchanges; and we have stated that if up to t processors can be faulty then at least $t + 1$ such stages are necessary. The first stage consists of the source processor P_0 sending out the value to be transmitted to each of the other, receiver processors. In the subsequent stages these values exchange among themselves, as follows. Every processor attaches its identifier to every message that it transmits, so a further identifier is added at each stage and the message becomes of the form

$$\langle v; P_0, P_{i_1}, \ldots, P_{i_k} \rangle$$

where v is the message value.

When the processor P_i receives such a message, it interprets it as follows: at stage j the processor P_{i_j} has received the message

$$\langle v; P_0, P_{i_1}, P_{i_2}, \ldots, P_{i_{j-1}} \rangle,$$

has added its own identifier P_{i_j}, and sent the message on to the other processors including $P_{i_{j+1}}$. The message now says "P_{i_k} says that $P_{i_{k-1}}$ has told it that $P_{i_{k-2}}$ has told it that ... that P_{i_1} has told it that P_0 has sent it the value v".

All the reliable processors operate the exchange of messages in this way at each stage; those that are operating unreliably may deliver messages incorrectly, alter them, not deliver them at all or behave sometimes correctly and sometimes incorrectly.

It is essential that all the processors agree at the start on a default value for any action, say v_{def}, to be taken whenever a value is expected but fails to appear; v_{def} must belong to the set of values that can be transmitted.

The Algorithm

We need a *majority* function which, given a set of values as its argument, returns the value that occurs most frequently – the majority vote; if no such value exists, the function returns v_{def}.

Algorithm $UM_n(t)$

1: Source P_0 sends its value to each of the other $N-1$ processors $P_1, \ldots, P_{N-1}$;

2: Each receiver saves the value received from P_0 or records the default value v_{def} if it has not received anything;

if $t > 0$ **then**

   for every receiver processor P_i **do**

      3.1: Let v_i be the value saved by P_i in stage 2; P_i acts as the source and sends v_i by $UM_{N-1}(t-1)$ to all $N-2$ processors $P_1, \ldots, P_{i-1}, P_{i+1}, \ldots, P_{N-1}$;

      3.2: Let v_j, $j \neq i$, be the value received by P_i from P_j at the end of stage 3.2, or v_{def} if nothing is received; P_i records the value $majority(v_1, v_2, \ldots, v_{N-1})$;

   end for

end if

Correctness Proof

Proving that $UM(t)$ works as intended consists in showing that it ensures the broadcasting of a value satisfying the constraints (C1), (C2). In view of the criterion concerning unsigned messages, if a maximum of t processors can be faulty at any one time there must be at least $3t+1$ processors altogether in the system to which an algorithm relates; and we shall show that with these assumptions $UM(t)$ does in fact solve the problem. The proof, like the formulation of the algorithm given above, is based on recursion; we need first a lemma that concerns (C2) alone, in which the source processor is assumed to be fault free, from which a theorem concerning both constraints follows.

Lemma 9.1.1. *For any integers t and p, the algorithm $UM_N(t)$ satisfies the constraint (C2) provided that the total number N of processors exceeds $2p+t$ and the maximum number of simultaneously faulty processors is p.*

Proof. The proof is by recursion on t, and there is the underlying assumption that the communication links are reliable.

The lemma is clearly true for $t = 0$, for $UM_N(0)$ means that all the processors are reliable and therefore (C1) holds trivially. Suppose the lemma is true for $t - 1$. Then in the stage 1 the source P_0, which can be assumed to be reliable since we are concerned here only with (C2), sends a value v to the $N - 1$ other processors. From $N > 2p + t$ we have

$$N - 1 > 2p + t - 1$$

and we can apply the recurrence: each of the $N - 1$ algorithms $UM_{N-1}(t-1)$ invoked in this stage satisfies (C2), so at the end of the stage each will have received the value v from each reliable processor P_j, and $v_j = v$ (step 3.2 in the program). Since there are at most p unreliable processors and $N - 1 > 2p + t - 1 > 2p$, a majority of these $N - 1$ processors are reliable ($p < \frac{1}{2}N$) and therefore each of the reliable processors will receive v as the majority value at the end of the stage. Thus the result holds for t, and therefore also generally. □

Theorem 9.1.2. *For a positive integer t the algorithm $UM_N(t)$ satisfies the constraints (C1) and (C2) provided that the total number of processors N exceeds $3t$ and the maximum number of simultaneously faulty processors is t.*

Proof. The proof is again by recursion on t. If $t = 0$ the theorem holds trivially. Suppose it holds for $t - 1$. Two cases are distinguished, whether the source P_0 is reliable or not.

Case 1: P_0 *is reliable:* If $t = p$ the previous lemma gives the result that with $N > 3p$, $UM_N(t)$ satisfies (C2); and the guarantee of (C1) follows since P_0 is reliable. So the theorem holds for t, and therefore generally.

Case 2: P_0 *is faulty:* There are up to t faulty processors, including also P_0; so among the receivers there may be up to $t - 1$ faulty processors. From the assumption $N > 3t$ it follows that the total number of processors apart from P_0 exceeds $3t - 1$, which is greater than $3(t - 1)$. The theorem can therefore be applied to each of the $N - 1$ applications of $UM_{N-1}(t - 1)$ of step 3.1 in the program and satisfies (C1) and (C2); and at step 3.2 all the reliable processors will have received the same values v_j, for all j, and will therefore have obtained the same value from the application of the majority function. This means that (C1) is satisfied, and again the theorem is proved.

□

Complexity Analysis

For the number of stages, we have seen that if there is a maximum of t faulty processors then the number of stages needed is $t + 1$; as mentioned previously this is the minimum number of stages needed by any algorithm that solves the consensus problem.

For the number of messages, $UM(t)$ applied to a system of N processors, the first stage causes the submission of $N - 1$ messages, each of which initiates an execution of $UM(t - 1)$; each of these then causes the sending of $N - 2$ messages, each of which initiates an execution of $UM(t - 2) \ldots$ until $UM(0)$ is reached. There can thus be up to

$$(N - 1)(N - 2) \ldots (N - (t + 1))$$

messages in all, which gives $O(N^{t+1})$ messages.

9.1.2 Fault-Tolerant Broadcast with Signed Messages

The principle of the basic solution is simple. The source processor P_0 sends a signed message containing the value v that it intended to be broadcast to all other $N - 1$ processors. If P_0 is faulty, it may send arbitrary values in different messages. Then each receiving processor adds its own signature and passes the augmented message to all those processors that have not already signed it, and so on. When at the $(t+1)$-st stage a processor receives no further messages it applies the function *choice* to the set of values it has received so as to obtain the consensus value. As before, the k-th stage of this procedure is defined as the stage in which the message in circulation carries k signatures.

Algorithm $SM(t)$;

  1: *Start of stage 1:*
  Source processor P_0 signs the message containing the value and sends it to the other $N - 1$ receiver processors.
  2: *At stage k, $1 \le k \le t + 1$:*
  while P_i, $1 \le i \le N - 1$, receives a message m containing value v and bearing k signatures **do**
    $V_i := V_i \cup \{v\}$

if $k < t + 1$ **then**
    sign message m;
    transmit m to all processors that have not yet signed m;
end if
end while
3: *At end of stage $t + 1$:*
For i, $1 \leq i \leq N - 1$: P_i records the value $choice(V_i)$.

Correctness Proof

We have to show that for all values of t the algorithm satisfies the consensus conditions (C1), (C2). We consider only the case $N \geq t + 2$, where N is the total number of processors. We give the informal proof.

Consider first (C1). If this holds then by implication all the reliable processors assemble the same set of values V_i and therefore obtain the same value as a result of applying the function *choice*. If P_i is a reliable processor which receives a value at stage k, where $k \leq t$, it will record this in its set V_i, and the same value will have been received by k processors during the preceding stages, each of which will have recorded this value in its own set V. P_i being reliable, it will sign the message containing this value and send it to the processors that have not yet received it; so at stage $k + 1$ all these processors will have received the value. Consider now a value recorded in V_i at stage $t + 1$. The message carrying this will have $t + 1$ signatures, and there are at most t faulty processors; therefore a reliable processor has received the value at some previous stage k with $k < t + 1$; and this processor will have sent the value to all those processors that had not yet received it. Thus all the reliable processors have the same set of values at the end of the procedure, so (C1) holds.

Proving that (C2) holds is a matter of studying the case in which the source processor is reliable. If it is, it sends a single value v; the receiving processors can neither change this before relaying it to other processors nor construct new values to replace it – any alterations would be detected (and no changes are introduced by the communication links, as these are assumed to be fault-free). Therefore all the sets V_i contain only the single value v, which proves the theorem.

Complexity Analysis

In terms of number of stages, this is $t+1$, which, as we have already mentioned, is the smallest possible number.

In terms of the number of messages, in the first stage each of $N - 1$ processors relays a message to each of $N - 2$, and so on, so the total number for the whole procedure is

$$(N - 1)(N - 2) \ldots (N - (t + 1)).$$

In terms of size of messages, each carries a maximum of $t + 1$ signatures in addition to the value broadcast, so the size is $O(t)$.

Thus the complexity is the same as that of $UM(t)$ for unsigned messages, and has therefore not been reduced by the signature; the advantage their use has brought is replacement of the constraint $N > 3t$ by the much weaker $N \geq t + 2$.

The number of messages required by $UM(t)$ can be the major disadvantage for practical applications, so there is an incentive to find an algorithm with a lower order of complexity in this respect. By restricting the circumstances in which a processor has to retransmit a message it has received, one can obtain an algorithm with a lower polynomial order of complexity in terms of the number of messages needed.

This reduction is achieved as follows. First, a processor retransmits a message it has received if and only if it has not previously transmitted the value contained in that message; then if the source processor P_0 is fault-free, each of the other processors will receive the same value and will relay it only once to each of the remaining $N - 2$. Thus the total number of messages is

$$(N - 1) + (N - 1)(N - 2)$$

which is $O(N^2)$ rather than the previous $O(N^{t+1})$.

Next, if P_0 is faulty and sends different values to certain processors – possibly a different value to each of the $N - 1$ – the receivers will then retransmit as before. If in this exchange a reliable processor receives two different values it can conclude that P_0 is faulty (because signed messages cannot be altered) and it will therefore choose a default value, which will be the same for all processors. Thus the total number will be the same as before.

This algorithm is in fact based on the same principle of value exchanges as $SM(t)$, and is an improvement over that algorithm.

Algorithm $DS(t)$;

  1: *Start of stage* 1:
Source processor P_0 signs the message containing the value and sends it to the $N - 1$ other processors (P_0 assumed to be functioning correctly)

  2: *Stage k,* $1 \leq k \leq t + 1$:
while P_i, $1 \leq i \leq N - 1$, receives a message m containing value v **and** bearing k signatures **do**
    $V_i := V_I \cup \{v\}$;
    if v is one of the two first distinct values received **and** has not already been transmitted by P_i **and** $(k < t + 1)$ **then**
      sign message m;
      transmit m to all messages that have not yet signed it;
    end if
  end while
  3: *At end of stage* $t + 1$:
For i, $1 \leq i \leq N - 1$:

if $|V_i| = 1$ **then**
  P_i records the value in V_i
else
  P_i records v_{def}
end if

The number of stages is again $t + 1$. Each processor sends at most 2 messages to each of the others, so the total number of messages involved is $2N^2$. The value recorded by a processor is either the value v sent out by the source processor P_0 if this processor is fault free (or seen to be so by the fault-free receiving processor), or the agreed default value v_{def} when P_0 is seen to be faulty.

The correctness proof follows the same lines as that given for $SM(t)$. If P_0 is fault free only the source value v is recorded in the sets V_i, while if it is not then all the reliable processors still arrive at the same decision; the essence is that it can be shown that the first two different values received by a reliable processor are received by all such processors, so that if one takes the value v as the consensus value then all the others take the same value.

9.2 Broadcasting in Synchronous Networks with Dynamic Faults

We consider the problem of dissemination of information in a network with faulty links. The network is synchronized, in the sense that all the processing units in nodes operate according to a global clock. The operations at nodes are divided into consecutive steps and are performed concurrently in consecutive time units determined by the common clock. Some of the links may be faulty in the sense that if one node attempts to broadcast over a faulty link then the message does not go through, but the processing units at nodes are in no way affected. If a node knows the information, then it may broadcast it in one step simultaneously to *all* its adjacent nodes connected by operational links. The global broadcasting protocol is simple: Once a vertex receives the message, it keeps attempting to broadcast it further to all its adjacent vertices, until all the nodes in the network obtain the information. The status of the link may change in the course of a communication process: It may alternate between "faulty" and "operational", but is fixed in one step. The number of links faulty at every given step is at most k. The *edge-connectivity* of a graph G is the minimum number l such that removing some l edges disconnects G. Broadcasting in the presence of k dynamic faults can be accomplished if and only if the edge-connectivity of the underlying network exceeds k. Thus, assume that all considered graphs have this property. The *distance* between two vertices x and y of graph G, denoted $dist(x, y)$, is the length of a shortest path in G between the vertices. The *diameter* $D(G)$ of a graph G is the maximum distance between a pair of vertices in G. If all links are always operational, then our broadcasting protocol disseminates the information in time equal to the diameter $D(G)$ of a graph G.

Now define precisely the time bound. It is a worst-case measure of the speed of proliferation of information in a graph. The broadcasting process may be considered as a game played by one player on a graph. The starting position of the game is a graph G with a distinguished *source* vertex v. Initially, the information is in v. In each step, the player is allowed to block k edges. If two vertices v_1 and v_2 are connected by an edge that is not blocked in a given step, and one of the vertices knows the information, then after this step, both vertices know it. For each strategy of the player, there is a number of steps after which all the vertices have learned the information. Our measure of broadcasting time $T(G, v, k)$, for a given source v and k dynamic faults, is the maximum of these numbers taken over all strategies of the player. The time measure $B(G, k)$, depending only on G and k, is the maximum of $T(G, v, k)$ over all source vertices v of G. This number is called the *broadcasting number* of graph G in the presence of k dynamic faults.

There is a straightforward upper bound of the time of broadcasting in a graph with N vertices: It is $N - 1$, because at each step there is at least one new vertex that learns the information. This bound can be improved to $O(N/k)$, where k is the number of dynamic faults.

In this section we show that broadcasting must always be completed in time $O(d^{k+1})$, for a fixed number of faults k, and time $O(k^{d/2-1})$, for a fixed diameter d. We also show that these orders of magnitude cannot be improved in general.

A communication network may be considered resilient to faults if their presence does not overly increase broadcasting time, i.e., the broadcasting number of the network is close to its diameter. This notion can be formalized as follows: Let $\mathcal{G} = \{G_N \mid N \geq 1\}$ be a family of graphs with edge-connectivity m_N. Then $\mathcal{G}$ is said to be *fault-resilient for broadcasting* if there is a constant $c > 0$ such that

$$B(G_N, m_N - 1) \leq c \cdot D(G_N)$$

for all $N \geq 1$. Below we show that multidimensional tori, hypercubes and star networks are fault-resilient for broadcasting.

Upper Bound for a Fixed Diameter

We start with some notions and a technical lemma. Let a labeled graph G have vertices named $v_1, v_2, \ldots, v_N$. Graph G is said to have the *canopy over v_i equal to c*, for an integer c, if there are c edges in G of the form (v_k, v_j), where $k < i < j$. If graph G is visualized as having all the vertices placed on a line in order of increasing indices and the edges are depicted as arcs drawn above the line, then the canopy over a vertex is the number of arcs drawn above that vertex. The *canopy of a graph* is the maximum of all canopies over vertices for all possible labelings of vertices.

Lemma 9.2.1. *A graph with diameter at most d and canopy at most c has at most $(2c)^{d/2}(1+o(1))$ vertices for an even d and at most $\sqrt{2} \cdot (2c)^{d/2}(1+o(1))$ vertices for an odd d, where the bounds are functions of c for a fixed d.*

We now show a general upper bound on the broadcasting number of a graph, for a fixed diameter d. Since a graph with diameter 1 is a clique, its broadcasting number is $O(1)$. The following theorem deals with graphs of larger diameter.

Theorem 9.2.2. *The broadcasting number of a graph G with fixed diameter $d > 1$, in the presence of k dynamic faults, is $O(k^{d/2-1})$.*

Proof. Fix any strategy of the player who blocks k edges at every step. With respect to this strategy, define layers $L_1, L_2, L_3, \ldots$, where L_j is the set of vertices which get information after exactly j steps. Coalesce the layers into blocks of 150 consecutive layers, call them *B-layers*, and denote them $B_1, B_2, \ldots$. Following the lemma

Lemma 9.2.3. *Assume the occurrence of k dynamic faults. Then, during $5 \cdot i$ consecutive steps, at least*

$$i \cdot (k + 6)/15 + 4 \cdot i$$

new vertices get informed.

for $i = 30$ implies that in 150 consecutive steps at least $2k+3$ new vertices get informed. Hence, each B-layer has at least $2k+3$ elements. Consider the graph whose vertices are B-layers, and two B-layers are adjacent if some of either elements are adjacent in the underlying graph. Since in step $i + 1$ all edges joining vertices from B_x with vertices from B_y, for $x \leq i$, $i + 2 \leq y$, must be blocked, the canopy over any B-layer (in the graph of B-layers) cannot exceed k. Next, coalesce the B-layers into groups of three consecutive ones and call them 3B-layers. As among the vertices in a B-layer B_i there is at least one adjacent only to vertices in B_{i-1}, B_i, and B_{i+1}, because otherwise either the canopy over B_{i-1} or over B_{i+1} (in the graph of B-layers) would be greater than k; then each 3B-layer has a vertex adjacent only to other vertices in this 3B-layer. Call this vertex *local*. Consider the graph of 3B-layers (defined similarly as the graph of B-layers) with vertices labeled in increasing order. The canopy of the graph of 3B-layers is at most k, similarly as before. Its diameter is at most $d - 2$: Since the distance between local vertices in two 3B-layers is at most d, the distance between those 3B-layers (in the graph of 3B-layers) is at most $d - 2$. It follows from Lemma 9.2.1 that the number of 3B-layers is at most

$$\sqrt{2}(2k)^{d/2-1}(1 + o(1))$$

for odd d and at most

$$(2k)^{d/2-1}(1 + o(1))$$

for even d. There are $3 \cdot 150$ layers in each $3B$-layer and the number of steps after which all vertices of the underlying graph have learned the information is equal to the number of layers. Hence, the broadcasting number of the graph G is at most

$$450 \cdot \sqrt{2}(2k)^{d/2-1}(1 + o(1))$$

for odd d and at most

$$450 \cdot (2k)^{d/2-1}(1 + o(1))$$

for even d. This completes the proof of the theorem.

$\square$

Upper Bound for a Fixed Number of Faults

Here we show an upper bound on the broadcasting number that is a function of the diameter of the graph, for a fixed number of dynamic faults.

Lemma 9.2.4. *If G is 2-edge-connected, then $D(G - \{e\}) \leq 2 \cdot D(G)$, for any edge e of G.*

Let $V_k(d)$ be the maximum broadcasting number for graphs of diameter d in the presence of k dynamic faults. Fix k and d and consider any graph G of diameter d, where v_0 is the source of broadcasting. Let $P(v_0, v)$ be a path of length at most d joining v_0 with a given vertex v. Suppose that a fixed w on $P(v_0, v)$ received the information: Let u be the next vertex on $P(v_0, v)$ and let $e = (u, v)$. Consider the next $V_{k-1}(2d)$ steps. If e is not blocked in one of them, the information reaches u. If e is blocked all the time, only $k - 1$ edges in the graph $G - \{e\}$ can be blocked. By Lemma 9.2.4, the inequality $D(G - \{e\}) \leq 2d$ holds; hence, after at most $V_{k-1}(2d)$ steps, the information reaches vertex u. This can be translated into the following recurrence:

$$V_0(d) \leq d,$$

$$V_k(d) \leq d \cdot V_{k-1}(2d).$$

By solving this recurrence, one obtains $V_k(d) \leq 2^{k(k+1)/2} \cdot d^{k+1}$. This proves the following:

Theorem 9.2.5. *The broadcasting number of a graph G with diameter d in the presence of a fixed number k of dynamic faults is $O(d^{k+1})$.*

Tightness of Upper Bounds*

Theorem 9.2.6. 1. *Fix d. There exists a family $\{G_k \mid k \geq 1\}$ of graphs of diameter d such that the broadcasting number of G_k in the presence of k dynamic faults is $\Theta(k^{d/2-1})$.*
2. *Fix k. There exists a family $\{G_d \mid d \geq 1\}$ of graphs of diameter d such that the broadcasting number of G_d in the presence of k dynamic faults is $\Theta(d^{k+1})$.*

9.2.1 Broadcast in Hypercubes with Dynamic Faults

In this subsection we consider the *shouting* communication mode in which any node can inform all its neighbors in one time unit. Under this assumption it is immediate to see that broadcasting can be accomplished in a number of time units equal to the diameter of the network, and this bound is clearly optimal. Now, assume that at any time instant a number of message transmissions (calls) less than the edge-connectivity of the network can fail. The problem is to find an upper bound on the number of time units necessary to complete broadcasting under this additional assumption.

Let $\{0,1\}^d$ be the set of all binary vectors of length d. The d-dimensional hypercube is the graph $H_d = (V, E)$ with $V = \{0,1\}^d$ and $E = \{(x, y) \mid x, y \in V, dist_H(x, y) = 1\}$, where $dist_H(x, y)$ is the Hamming distance between x and y, that is, the number of components on which x and y differ. For any $x \in V$ and integer i, $1 \le i \le d$, let us denote by $x(i)$ the vertex of H_d which differs from x only on the i-th coordinate.

Let $T(d)$ be the minimum number possible of time units necessary to broadcast in H_d under the assumption of the shouting model and that at each time instant at most $d - 1$ calls may fail.

Lemma 9.2.7. $d + 2 \le T(d)$.

Proof. Let us consider two nodes of H_d, say x and y, such that $dist_H(x, y) = d$, and consider two neighbors $x(i)$ and $y(j)$ of i and j, respectively. Let us assume that $x(i)$ is the originator of the broadcast. It is easy to see that if the set of failures is concentrated around $x(i)$, leaving fault-free only the transition from $x(i)$ to x, and this holds during all time steps but the last one, where the set of failures is concentrated around $y(j)$, leaving fault-free only the transmission from y to $y(j)$, then at least $d + 2$ time units are necessary.

$\square$

Let H be a subgraph of a graph G. The vertex boundary of H is the set of vertices not in H joined to some vertex in H. The vertex isoperimetric problem can be stated as follows: For a fixed m, of all subgraphs of G of cardinality m, which one has the smallest vertex boundary?

We now give the solution to the vertex isoperimetric problem for the hypercube.

Let A be a subset of vertices of H_d. Let $\mathcal{S}(A)$ denote the set of all vertices in distance at most 1 from A in H_d, i.e., $\mathcal{S}(A)$ is the union of the vertex boundary of A with A.

Lemma 9.2.8. *Let $1 \le |A| \le 2^d - 1$. The number $|A|$ has a unique representation in the form*

$$|A| = \sum_{i=r+1}^{d} \binom{d}{i} + \sum_{i=s}^{r} \binom{d_i}{i}$$

where $1 \le s \le d_s < d_{s+1} < \cdots < d_r < d$. Then

$$|S(A)| \geq \sum_{i=r}^{d} \binom{d}{i} + \sum_{i=s}^{r} \binom{d_i}{i-1}.$$

The inequality above is the best possible and can be achieved taking $|A|$ vertices in the simple order, i.e., $x = (x_{d-1}, \ldots, x_1, x_0)$ precedes $y = (y_{d-1}, \ldots, y_1, y_0)$ if either

$$\sum_{i=0}^{d-1} x_i < \sum_{i=0}^{d-1} y_i$$

or

$$\sum_{i=0}^{d-1} x_i = \sum_{i=0}^{d-1} y_i$$

and y precedes x lexicographically. A special case

$$|A| = \sum_{i=0}^{r} \binom{d}{i}$$

is more constructive. Then

$$|S(A)| \geq \sum_{i=0}^{r+1} \binom{d}{i}.$$

This means that the set with the minimal vertex boundary is the set of all vertices at most r from $(0, 0, \ldots, 0)$.

Optimal Upper Bound

We use Lemma 9.2.8 to determine the number of time steps which are sufficient to complete the broadcasting in the presence of dynamic faults.

Theorem 9.2.9. *The minimum broadcasting time $T(d)$ in the d-dimensional hypercube with $d - 1$ dynamic link faults satisfies*

$$T(d) \leq \begin{cases} 1 & \text{if } d = 1 \\ 3 & \text{if } d = 2 \\ d + 2 & \text{if } d > 2 \end{cases}$$

Proof. The broadcasting scheme is simple. In each time step each node sends the message to all its neighbors. The analysis follows. Let $d \geq 10$. By A_k denote suitable sets of nodes which know the message after the k-th time step. Note that A_k need not be necessarily the set of all vertices which are aware of the information after the k-th step. Observe that the number of nodes that know the message after the k-th step is at least $|S(A_{k-1})| - d + 1$. Clearly, there exist sets A_0, A_1 and A_2 such that $|A_0| = 1$, $|A_1| = 2$ and $|A_2| = d + 1$. Consider the set $|A_2|$. Its cardinality is

$$|A_2| = \binom{d}{d} + \binom{d}{d-1}.$$

By Lemma 9.2.8

$$|S(A_2)| \geq \binom{d}{d} + \binom{d}{d-1} + \binom{d}{d+2}.$$

Because of the $d - 1$ faulty links, we have

$$|S(A_2)| - d + 1$$

$$\geq \binom{d}{d} + \binom{d}{d-1} + \binom{d}{d-2} - (d-1)$$

$$\geq \binom{d}{d} + \binom{d}{d-1} + \binom{d-1}{d-2} + \binom{d-2}{d-3}.$$

Define A_3 to be a subset of nodes that know the message after the third step and that satisfies

$$|A_3| = \binom{d}{d} + \binom{d}{d-1} + \binom{d-1}{d-2} + \binom{d-2}{d-3}.$$

We claim that for $3 \leq k \leq \lfloor (d+3)/2 \rfloor$, there exists a set A_k such that

$$|A_k| = \sum_{i=d-k+2}^{d} \binom{d}{i} + \binom{d-1}{d-k+2} + \binom{d-k+2}{d-k+1}.$$

By Lemma 9.2.8, setting $r = d - k + 2$, $s = d - k + 1$ and $d_r = d - k + 2$ we get

$$|S(A_{k-1})| \geq \sum_{i=d-k+2}^{d} \binom{d}{i} + \binom{d-1}{d-k+1} + \binom{d-k+2}{d-k}.$$

Because of the $d - 1$ faulty links, we have for $k - 1 < \lfloor (d+3)/2 \rfloor$

$$|S(A_{k-1})| - d + 1$$

$$\geq \sum_{i=d-k+2}^{d} \binom{d}{i} + \binom{d-1}{d-k+1} + \binom{d-k+2}{d-k} - (d-1)$$

$$\geq \sum_{i=d-k+2}^{d} \binom{d}{i} + \binom{d-1}{d-k+1} + \binom{d-k+1}{d-k}.$$

Note that the last inequality holds for $d \geq 10$ only.

Define A_k to be a subset of nodes which know the message after the k-th step and satisfies

$$|A_k| = \sum_{i=d-k+2}^{d} \binom{d}{i} + \binom{d-1}{d-k+1} + \binom{d-k+1}{d-k}.$$

which proves the claim.

Now we use a dual argument. By B_k we will denote suitable sets of nodes, which do not know the message after the k-th step. Again note that B_k need not be the set of all vertices that are not aware of the information after the k-th step. Assume that after the $(d+2)$-th step there exists at least one node which does not know the message. Observe that the number of nodes that do not know the message after the $(k-1)$-th step is at least $|S(B_k)| - d + 1$. Clearly there exist suitable sets B_{d+2}, B_{d+1} and B_d such that $|B_{d+2}| = 1$, $|B_{d+1}| = 2$ and $|B_d| = d + 1$. This results in a similar backward recurrent computation as for $|A_k|$, starting with the set B_d which satisfies

$$|B_d| = \binom{d}{d} + \binom{d}{d-1} + \binom{d-1}{d-2} + \binom{d-2}{d-3}.$$

Thus for $k \geq \lfloor (d+3)/2 \rfloor$ we find a set B_k with

$$|B_d| = \sum_{i=k}^{d} \binom{d}{i} + \binom{d-1}{k-1} + \binom{k-1}{k-2}.$$

Finally, set $k = \lfloor (d+3)/2 \rfloor$. Then

$$2^d \geq |A_k| + |B_k| \geq 2^d + \binom{d-k+1}{d-k} + \binom{k-1}{k-2},$$

a contradiction.

Cases $d \leq 9$ follow directly.

$\square$

9.2.2 Broadcast in Tori with Dynamic Faults

In this section we consider broadcasting in the d-dimensional k-ary even torus in the shouting communication mode, with the assumption that at most $2d-1$ dynamic faults are allowed in each time step. We prove that the broadcasting can be done in an optimal time *diameter* plus 2.

Let $C_{d,k}$ be a network of processors connected as the d-dimensional k-ary torus defined as the Cartesian product of d cycles C_k, for k even. The network $C_{d,k}$ has k^d nodes, is regular of degree $2d$, with edge connectivity $2d - 1$. Its diameter is $dk/2$ and it is both edge and vertex symmetric. The computation is synchronous, with bidirectional links and the all-port (shouting) mode. In each broadcast step at most $2d - 1$ links are crash faulty, i.e., the message transmitted along the faulty link is not delivered. The faults are dynamic in the sense that the set of faulty links can change during the execution. Initially, a source node of $C_{d,k}$ knows a message which has to be transmitted to all other nodes in the network. The problem is to determine the minimal time to perform the broadcasting in the torus $C_{d,k}$, provided that the faults are distributed in the worst possible manner.

We use the following notation. The vertices of C_k are $\{-k/2, -k/2 + 1, \ldots, k/2-1\}$. Thus two vertices $u, v \in C_k$ are adjacent if $|u-v| \pmod{k} = 1$. The vertices of $C_{d,k}$ are d-tuples $(x_1, x_2, \ldots, x_d)$, where $-k/2 \leq x_i \leq k/2 - 1$, for $i = 1, 2, \ldots, d$. Define $\mathbf{0} = (0, 0, \ldots, 0)$.

Let $dist(u, v)$ be the distance between u and v in $C_{d,k}$. Denote

$$S(r) = \{v \in C_{d,k} \mid dist(v, \mathbf{0}) = r\}$$

and

$$V(r) = \{v \in C_{d,k} \mid dist(v, \mathbf{0}) \leq r\} = \cup_{i=0}^{r} S(r).$$

Clearly, $|S(r)|$ is the number of integer solutions of the equation

$$|x_1| + |x_2| + \cdots + |x_d| = r$$

where $-k/2 \leq x_r \leq k/2 - 1$. Since k is even we have

$$|S(r)| = |S(\frac{dk}{2} - r)|$$

which implies

$$|V(r)| + |V(\frac{dk}{2} - r - 1)| = d^k.$$

Moreover, $|S(0)| = 1$, $|S(1)| = 2d$ and $|S(2)| = 2d^2$, for $k \geq 6$.

Let A be a subset of vertices of $C_{d,k}$. Let $\partial(A)$ denote the set of all vertices in distance at most 1 from A in $C_{d,k}$, i.e., $\partial(A)$ is the union of the vertex boundary of A with A. The following isoperimetric inequality[1] is the central fact in estimating the optimal upper bound on the broadcasting time in tori with dynamic faults.

Lemma 9.2.10. Let $C_{d,k}$ be the d-dimensional k-ary torus with k even, and let A be a nonempty set of vertices of $C_{d,k}$. If $|A| = |V(r)| + \alpha|S(r+1)|$, for some r and $0 \leq \alpha < 1$, then

$$|\partial(A)| \geq |V(r+1)| + \alpha|S(r+2)|.$$

---

[1] whose proof is omitted here.

Optimal Upper Bound on the Broadcasting Time

The proof is based on Lemma 9.2.10. First we need the following technical lemma:

Lemma 9.2.11. *Denote*

$$X = 1 - \frac{3}{d} - \sum_{r=3}^{\lfloor \frac{dk}{4} \rfloor} \frac{4d}{|S(r)|}.$$

The value X is nonnegative for $k \geq 6$ and $d \geq k + 4$ or $k \geq d \geq 10$.

The main result follows.

Theorem 9.2.12. *Assume that $k \geq 6$ and $d \geq k + 4$ or $k \geq d \geq 10$. The minimum broadcasting time T in the d-dimensional k-ary torus, for k even, with $2d - 1$ dynamic link faults satisfies*

$$T \leq \frac{dk}{2} + 2.$$

The bound is the best possible.

Proof. The broadcasting scheme is simple. Initially, the source node **0** contains the message to be disseminated. In each step each node sends the message to all its neighbors.

The analysis is as follows. By A_m denote a suitable set of nodes which know the message after the m-th step. Observe that the number of nodes that know the message after the m-th step is at least $|\partial(A_{m-1})| - (2d - 1)$. Clearly, there exist sets A_0, A_1 and A_2 such that $|A_0| = 1$, $|A_1| = 2$ and $|A_2| = 2d + 1 = |V(1)|$.

According to Lemma 9.2.10

$$|\partial(A_2)| \geq |V(2)| = 1 + 2d + 2d^2.$$

Because of the $2d - 1$ faulty links, we have

$$|\partial(A_2)| - (2d - 1) \geq 2d^2 + 2 = |V(1)| + \alpha_3 |S(2)|,$$

where $\alpha_3 = 1 - 1/d$. Define A_3 to be a subset of nodes that know the message after the third step and satisfies

$$|A_3| \geq |V(1)| + \alpha_3 |S(2)|.$$

Similarly,

$$|\partial(A_3)| \geq |V(2)| + \alpha_3 |S(3)|.$$

Because of the $2d - 1$ faulty links, we have

$$|\partial(A_3)| - (2d - 1) \geq |V(2)| + \alpha_3 |S(3)| - (2d - 1) \geq |V(2)| + \alpha_4 |S(3)|,$$

where $\alpha_4 = \alpha_3 - \frac{2d}{|S(3)|}$. Define A_4 to be a subset of nodes that know the message after the fourth step and satisfies

$$|A_4| \geq |V(2)| + \alpha_4 |S(3)|.$$

We prove by induction that for $4 \leq m \leq \frac{dk}{2} - 1$

$$|A_m| \geq |V(m - 2)| + \alpha_m |S(m - 1)|,$$

where

$$\alpha_m = 1 - \frac{1}{d} - \frac{2d}{|S(3)|} - \frac{2d}{|S(4)|} - \cdots - \frac{2d}{|S(m - 1)|}.$$

Assume the claim holds for some $3 \leq m - 1 \leq \frac{dk}{2} - 2$. Lemma 9.2.10 implies

$$|\partial(A_{m-1})| - (2d - 1) \geq |V(m - 2)| + \alpha_m |S(m - 1)|$$

where $\alpha_m = \alpha_{m-1} - \frac{2d}{|S(m-1)|}$. Define A_m to be a subset of nodes that know the message after the m-th step and satisfies

$$|A_m| \geq |V(m - 2)| + \alpha_m |S(m - 1)|.$$

Clearly,

$$\alpha_m = 1 - \frac{1}{d} - \sum_{i=3}^{m-1} \frac{2d}{|S(i)|}.$$

Now we use a dual argument. By B_m denote suitable subsets of nodes, which do not know the message after the m-th step. Assume that after the $(dk/2 + 2)$-th step there exists at least one node which does not know the message. Observe that the number of nodes that do not know the message after the $(m - 1)$-th step is at least $|\partial(B_m)| - (2d - 1)$. Clearly there exist sets $B_{\frac{dk}{2}+2}$, $B_{\frac{dk}{2}+1}$ and $B_{\frac{dk}{2}}$ with cardinality $1, 2, 2d + 1$, respectively. Similarly as for A_3 determine $|B_{\frac{dk}{2}-1}| = 2d^2 + 2$.

Now compute

$$|A_{\frac{dk}{2}-1}| + |B_{\frac{dk}{2}-1}|$$

$$\geq k^d + 1 + 2d^2 \left(1 - \frac{3}{d} - \sum_{r=3}^{\lfloor \frac{dk}{4} \rfloor} \frac{4d}{|S(r)|} \right).$$

By Lemma 9.2.11 the expression in the brackets is nonnegative, which implies a contradiction.

Finally, we show that there is a distribution of faults in each step which forces the broadcasting time $dk/2 + 2$. Consider vertices u, v such that $dist(u, v) = dk/2$. Let u', v' be neighbors of u, v, respectively. Initially, let u' know the message. In the first step we place faults on all edges adjacent to u', except for $\{u', u\}$. From now on we place faults on all edges adjacent to v', except for $\{v', v\}$. After $dk/2 + 1$ steps, the message reaches the vertex v and one additional step is necessary to complete the broadcasting.

□

9.2.3 Broadcast in Star Graphs with Dynamic Faults

In this section we will show that broadcasting in the d-dimensional star networks in the all-port (shouting) communication mode, and under the hypothesis that at most $d-2$ message transmissions can fail during each time instant, can be accomplished in at most the diameter plus 11 time units. This result is within 9 time units of the known lower bound.

Let Σ_d be the set of all permutations of symbols in $\{1, 2, \ldots, d\}$. Given $\mathbf{u} = u_1 u_2 \ldots u_d \in \Sigma_d$ and $i \in \{2, \ldots, d\}$, denote by $u\langle i \rangle = u_i \ldots u_1 \ldots u_d$ the element of Σ_d obtained from u by permuting the first symbol u_1 with the i-th symbol u_i. The d-star graph S_d has vertex set $V(S_d) = \Sigma_d$ and edge set $E(S_d) = \{(u, u\langle i \rangle) \mid u \in \Sigma_d, \, 2 \leq i \leq d\}$. S_d has $d!$ vertices, is a Cayley graph, is regular of degree $d - 1$, is both edge and vertex symmetric, has diameter $diam(S_d)$ equal to $\lfloor 3(d-1)/2 \rfloor$ and has edge and vertex connectivity equal to $d - 1$. We shall make use of the following result:

Lemma 9.2.13. *For any two vertices $u, v \in V(S_d)$, there exist $d - 1$ edge-disjoint paths between u and v of total length at most $(d - 3)\lfloor \frac{3}{2}d \rfloor + 2d$.*

Let $T(d)$ be the minimum possible number of time units necessary to broadcast in S_d under the assumption of the shouting model and that at each time instant at most $d - 2$ calls may fail. $T(d)$ is obviously lower bounded by the diameter of S_d. The following theorem is the main result of this section:

Theorem 9.2.14. $diam(S_d) + 2 \leq T(d) \leq diam(S_d) + 11$.

Proof. Given the d-star S_d and an integer i, $1 \leq i \leq d$, let us denote by $\mathbf{i}_d$ the $(d-1)$-substar of S_d induced by all $(d-1)!$ vertices of S_d having the symbol i in the d-th position. The $d!$ nodes of the d-star graph S_d can be partitioned among the d different $(d-1)$-substars $\mathbf{1}_d, \mathbf{2}_d, \ldots, \mathbf{d}_d$. Let $\mathbf{Id} = 12 \ldots d$ be the identity permutation. We assume that $\mathbf{Id}$ is the originator of the broadcast. We shall prove the theorem through a sequence of steps.

Claim. At time step 4 at least $2 + \lceil (d-2)/2 \rceil$ $(d-1)$-substars have an informed vertex.

Proof. At time unit 1 at least a neighbor of $\mathbf{Id}$ receives the message from $\mathbf{Id}$. Without loss of generality, let us assume that $\mathbf{Id}\langle d \rangle$ is such a vertex. For any $(d-1)$-substar $\mathbf{i}_d$, $2 \leq i \leq d - 1$, there exist two disjoint paths of length 2

$$\mathbf{Id} \rightarrow \mathbf{Id}\langle i \rangle \rightarrow \mathbf{Id}\langle i \rangle\langle d \rangle$$

$$\mathbf{Id}\langle d \rangle \rightarrow \mathbf{Id}\langle d \rangle\langle i \rangle \rightarrow \mathbf{Id}\langle d \rangle\langle i \rangle\langle d \rangle \tag{9.1}$$

whose terminal vertices belong to $\mathbf{i}_d$. Therefore, in order to prevent a generic $(d-1)$-substar $\mathbf{i}_d$ having an informed vertex at time 4, it is necessary that in at least a time step between 2 and 4 both paths in (9.1) be affected by faults. Since the number of faults in any time instance is $d - 2$ we have that at time unit 4 at least $\lceil (d-2)/2 \rceil (d-1)$-substars receive the information through the paths in (9.1). Considering also the $(d-1)$-substars to which the informed vertices **Id** and **Id**$\langle d \rangle$ belong to, the claim is proved.

$\square$

Claim. At time unit $\lfloor 3(d-1)/2 \rfloor + 5$ at least a whole $(d-1)$-substar of S_d is informed.

Proof. Consider the $2 + \lceil (d-2)/2 \rceil$ $(d-1)$-substars that we know have an informed vertex at time 4. We want to prove that at least one of them is totally informed at time unit $\lfloor 3(d-1)/2 + 5 \rfloor$. Suppose not, that is, it is possible to find $2(2 + \lceil (d-2)/2 \rceil)$ vertices x_i and y_i, $1 = 1, \ldots, 2 + \lceil (d-2)/2 \rceil$ such that all the x_i's are informed at time 4, all the y_i's are uninformed at time $\lfloor 3(d-1)/2 + 5 \rfloor$, and each pair $\{x_i, y_i\}$ belongs to the same $(d-1)$-substar. We know that between each x_i and y_i there are $(d-2)$ edge-disjoint paths, of total length at most

$$(d-4)\lfloor \frac{3}{2}(d-1) \rfloor + 2(d-1)$$

and after

$$4 + \frac{(\lceil \frac{1}{2}(d-2) \rceil + 2)[(d-4)\lfloor \frac{3}{2}(d-1) \rfloor + 2(d-1)]}{(\lceil \frac{1}{2}(d-2) \rceil + 2)(d-2) - (d-2)}$$

$$- \frac{(\lceil \frac{1}{2}(d-2) \rceil + 2)(d-2) - 1}{(\lceil \frac{1}{2}(d-2) \rceil + 2)(d-2) - (d-2)}$$

$$\leq \lfloor \frac{3}{2}(d-1) \rfloor + 5$$

time units at least one of the y_i will be informed. The obtained contradiction proves the claim.

$\square$

Claim. At time unit $\lfloor 3(d-1)/2 \rfloor + 7$ at most $(d-1)(d-2)$ vertices of S_d are uninformed.

Proof. Without loss of generality, we assume that $\mathbf{1}_d$ is the $(d-1)$-substar informed at time $\lfloor 3(d-1)/2 \rfloor + 5$. Given $i, j \in \{1, \ldots, d\}$, $i \neq j$, let us denote by $i * j$ a generic permutation of the symbols $1, 2, \ldots, d$ that has symbol i in the first position and symbol j in the last position. Clearly, for any pair i and j there are $(d-2)!$ such kinds of permutations. To prove the claim, we first partition the vertex set of S_d into three disjoint sets: L_1, L_2, and L_3. Set L_1 contains all vertices of the $(d-1)$-substar $\mathbf{1}_d$. Set L_2 contains all permutations of kind $1 * 2, 1 * 3, \ldots, 1 * d$. Set L_3 contains all permutations of kind $(i * i)\langle j \rangle$, for $2 \leq i \leq d$ and $2 \leq j \leq d - 1$. It is clear that L_1, L_2 and L_3 are pairwise disjoint. Moreover, $|L_1| = (d-1)!$, $|L_2| = (d-1)(d-2)!$, and $|L_3| = (d-1)(d-2)(d-2)!$; therefore, $|L_1| + |L_2| + |L_3| = d!$ and thus they constitute a partition of the vertices of S_d. Let us now consider the edges among the sets L_1, L_2, L_3. We first notice that each vertex $1 * i \in L_2$ is adjacent to the vertex $i * 1 \in L_1 = \mathbf{1}_d$, for any i, $2 \leq i \leq d$. Moreover, each vertex $(1 * i)\langle j \rangle \in L_3$ is adjacent to vertex $(1 * i) \in L_2$, but no vertex in L_3 is adjacent to any node in L_1. Recalling that at time unit $T = \lfloor 3(d-1)/2 \rfloor + 5$ all vertices in $l_1 = \mathbf{1}_d$ are informed and all nodes in L_2 are adjacent to vertices in L_1, we get that at time unit $T + 1$ at most $d - 2$ vertices in L_2 are uninformed. Since any vertex in L_2 is adjacent to $(d-2)$ vertices in L_3, we obtain that at time unit $T + 2$ the total number of uninformed vertices in S_d is at most $(d-2)^2 + (d-2) = (d-1)(d-2)$.

□

Claim. At time unit $\lfloor 3(d-1)/2 \rfloor + 11$ all vertices of S_d are informed.

Proof. Let us consider time unit $T + 2$ and an uninformed vertex v of S_d. Vertex v has $d + (d-1)(d-2)$ vertices at distance at most 2 (including also v). Since we know that at time unit $T + 2$ at most $(d-1)(d-2)$ vertices of S_d are uninformed, we can conclude that there are at least d informed vertices at distance 2 from v; therefore, at time $T + 3$ at least a neighbor of v is informed. Since any uninformed vertex v has at least an informed neighboring vertex, we can say that at time $T + 4$ a total of at most $d - 2$ vertices in S_d still will be uninformed. We now prove that 2 additional time units are sufficient to inform all vertices in S_d. We prove this claim by contradiction. Suppose that at time $T + 6$ there exists an uninformed vertex x. Since the number of faulty calls is $(d-2)$, we know that there exists at time $T + 5$ a neighbor y of x that is also uninformed. The set of neighbors of x is disjoint from the set of neighbors of y (since S_d is bipartite). Therefore, the union of these two sets contains $2(d-1)$ elements. Since we have already proved that at time $T + 4$ at most $(d-2)$ vertices were uninformed, we also have that at time $T + 4$ at least $2(d-1) - (d-2) = d$ neighbors either of x or y are informed. This contradicts the fact that at time $T + 5$ both x and y are uninformed.

□

To conclude the proof of the theorem, we just notice that in the cases $d = 4$ and d=6 it is easy to prove directly that $T(4) \leq diam(S_4) + 11 = 6 + 11$ and $T(6) \leq diam(S_6) + 11 = 9 + 11$.

The proof of the lower bound is similar to that for hypercubes.

□

References

[AK89]   S.B. Akers, B. Krishnamurthy: A group-theoretic model for symmetric interconnection networks. *IEEE Transactions on Computers* 38 (1989), No. 4, 555–566.

[ABR90]   F. Annexstein, M. Baumslag, A.L. Rosenberg: Group action graphs and parallel architectures. *SIAM Journal on Computing* 19 (1990), No. 3, 544–569.

[ACGK+99]   G. Ausiello, P. Crescenzi, G. Gambosi, V. Kann, A. Marchetti-Spaccamela, M. Protasi: *Complexity and Approximation – Combinatorial Optimization Problems and Their Approximability Properties.* Springer, 1999.

[AKS83]   M. Ajtai, J. Komlós, E. Szemerédi: An $O(n \log n)$ sorting network. *Proc. 15th ACM Symposium on Theory of Computing*, pp. 1–9, 1983.

[ALMN91]   W.A. Aiello, F.T. Leighton, B.M. Maggs, M. Newman: Fast algorithms for bit-serial routing on a hypercube. *Mathematical Systems Theory* 24 (1991), No. 4, 253–271.

[BAPR+91]   Y. Ben-Asher, D. Peleg, R. Ramaswami, A. Schuster: The power of reconfiguration. *Journal of Parallel and Distributed Computing* 13 (1991), No. 2, 139–153.

[BDP97]   P. Berman, K. Diks, A. Pelc: Reliable broadcasting in logarithmic time with Byzantine link failures. *Journal of Algorithms* 22 (1997), 199–211.

[Be64]   V. Beneš: Permutation groups, complexes, and rearrangeable multistage connecting networks. *Bell System Technical Journal*, Vol. 43 (1964), pp. 1619–1640.

[Be65]   V. Beneš: *Mathematical Theory of Connecting Networks and Telephone Traffic.* Academic Press, New York, NY, 1965.

[BFP96]   P. Berthomé, A. Ferreira, S. Pérennès: Optimal information dissemination in star and pancake networks. Extended abstract in: *Proc. 5th IEEE Symp. on Parallel and Distributed Processing (SPDP '93)*, 1993, 720–724. Full version in: *IEEE Transactions on Parallel and Distributed Systems*, Vol. 7, No. 12 (1996), pp. 1292–1300.

[BF97]   D. Barth, P. Fraigniaud: Approximation algorithms for structured communication problems. In: *Proc 9th Annual Symposium on Parallel Algorithms and Architectures (SPAA '97)*, ACM, pp. 180–188, 1997.

[BF00]   D. Barth, P. Fraigniaud: Scheduling jobs in O(congestion+dilation) with applications to multi-point communication problems. Technical Report LRI-1239, Univ. Paris-Sud, Orsay, France, 2000.

[BGNS00]   A. Bar-Noy, S. Guha, J. Naor, B. Schieber: Multicasting in heterogeneous networks. *SIAM Journal on Computing*, Vol. 30 (2000), No. 2, 347–358.

[BHLP92]   J.-C. Bermond, P. Hell, A.L. Liestman, J.G. Peters: Broadcasting in bounded degree graphs. *SIAM Journal on Discrete Mathematics* 5 (1992), No. 1, 10–24.

[BHMS90]   A. Bagchi, S.L. Hakimi, J. Mitchem, E. Schmeichel: Parallel algorithms for gossiping by mail. *Information Processing Letters* 34(4) (1990), 197–202.

[Bo79]   B. Bollobás: *Graph Theory*. Springer, New York, 1979.

[BP85]   C.A. Brown, P.W. Purdom: *The Analysis of Algorithms*. Holt, Rinehart and Winston, New York, 1985.

[BP88]   J.-C. Bermond, C. Peyrat: Broadcasting in deBruijn networks. In: *Proc. 19th Southeastern Conference on Combinatorics, Graph Theory and Computing*, Congressus Numerantium 66 (1988), 283–292.

[BP89]   J.C. Bermond, C. Peyrat: deBruijn and Kautz networks: A competitor for the hypercube? In: F. Andre, J.P. Verjus, editors, *Hypercube and Distributed Computers*, North-Holland, Amsterdam, 1989, 279–294.

[BS72]   B. Baker, R. Shostak: Gossips and telephones. *Discrete Mathematics* 2 (1972), No. 1, 191–193.

[BS00]   R. Beier, J.F. Sibeyn: A powerful heuristic for telephone gossiping. In: *Proc. 7th Colloquium on Structural Information and Communication Complexity*, pp. 17–35, Carleton Scientific, 2000.

[CC84]   C. Choffrut, K. Culik II: On real-time cellular automata and trellis automata. *Acta Informatica* 21 (1984), No. 4, 393–407.

[CDP94]   B.S. Chlebus, K. Diks, A. Pelc: Sparse networks supporting efficient reliable broadcasting. *Nordic Journal of Computing* 1 (1994), 332–345.

[CF00]   J. Cohen, P. Fraigniaud: Broadcasting and multicasting in trees. Technical Report LRI-1265, Université Paris-Sud, France, 2000.

[CFM99]   J. Cohen, P. Fraigniaud, M. Mitjana: Scheduling calls for multicasting in tree-networks. Short paper at the *10th ACM-SIAM Symp. on Discrete Algorithms* (SODA '99), 1999.

[CFM02]   J. Cohen, P. Fraigniaud, M. Mitjana: Polynomial-time algorithms for minimum-time broadcast in trees. *Theory of Computing Systems* 35(2002), No. 6, 641–665.

[CG98]   F. Comellas, G. Giménez: Genetic programming to design communication algorithms for parallel architectures. *Parallel Processing Letters*, vol. 8 (No. 4) (1998) pp. 549–560.

[CGS83]   K. Culik II, J. Gruska, A. Salomaa: Systolic automata for VLSI on balanced trees. *Acta Informatica* 18 (1983), No. 4, 335–344.

[CGS84]   K. Culik II, J. Gruska, A. Salomaa: Systolic trellis automata. Part I. *International Journal of Computer Mathematics* 15 (1984), Nos. 3-4, 195–212.

[CGV89]   R.M. Capocelli, L. Gargano, U. Vaccaro: Time bounds for broadcasting in bounded degree graphs. In: *Proc. 15th Int. Workshop on Graph-Theoretic Concepts in Computer Science (WG '89)*, Lecture Notes in Computer Science 411, Springer 1989, 19–33.

[CM98]   K. Ciebiera, A. Malinowski: Efficient broadcasting with linearly bounded faults. *Discrete Applied Mathematics* 89 (1998), 99–105.

[CLR90]   T.H. Cormen, C.E. Leiserson, R.L. Rivest: *Introduction to Algorithms*, McGraw-Hill, 1990.

[CP93]   R. Cypher, C.G. Plaxton: Deterministic sorting in nearly logarithmic time on the hypercube and related computers. *Journal of Computer and System Sciences*, 47 (1993), 501–548.

[CS78]   M. Cutler, Y. Shiloach: Permutation layout. *Networks*, Vol. 8 (1978) 253–278.

[CSW84]   K. Culik II, A. Salomaa, D. Wood: Systolic tree acceptors. *R.A.I.R.O. Theoretical Informatics* 18 (1984), 53–69.

[dB46]   N.G. deBruijn: A combinatorial problem. *Koninklijke Nederlandsche Akademie van Wetenschappen Proc.* 49 (1946), 758–764.

[DDKP+98]   K. Diks, S. Dobrev, E. Kranakis, A. Pelc, P. Ružička: Broadcasting in unlabeled hypercubes with linear number of messages. *Information Processing Letters*, Vol. 66 (1998), 181–186.

[DDSV93]   K. Diks, H.N. Djidjev, O. Sýkora, I. Vrťo: Edge separators of planar and outerplanar graphs with applications. *Journal of Algorithms* 14 (1993), 258–279.

[DFKP+02]   S. Dobrev, P. Flocchini, R. Královič, G. Prencipe, P. Ružička, N. Santoro: Black hole search by mobile agents in hypercubes and related networks. In: *Proc. 6th International Conference on Principles of Distributed Systems* (OPODIS), pp. 169–180, 2002.

[DP92]   K. Diks, A. Pelc: Almost safe gossiping in bounded degree networks. *SIAM J. Disc. Math.* 5 (1992), 338–344.

[DP00]   K. Diks, A. Pelc: Optimal adaptive broadcasting with a bounded fraction of faulty nodes, *Algorithmica* 28 (2000), pp. 37–50.

[DLOY98]   A. Dessmark, A. Lingas, H. Olsson, H. Yamamoto: Optimal broadcasting in almost trees and partial k-trees. In: *Proc. 15th Symposium on Theoretical Aspects of Computer Science* (STACS '98), *Lecture Notes in Computer Science* 1373, Springer, pp. 432–443, 1998.

[DP95]   K. Diks, A. Pelc: Broadcasting with universal lists. Extended abstract in: *Proc. 28th Hawaii Int. Conf. on System Sciences (HICSS '95)*, 1995, Vol. 2, 564–573. Full version in: *Networks* 27 (1996), 183–196.

[dR94]   J. de Rumeur: *Communications dans les réseaux de processeurs.* Collection Etudes et Recherches en Informatique, Masson, Paris, 1994.

[DR97]   S. Dobrev, P. Ružička: Linear broadcasting and N log log N election in unoriented hypercubes. In: *Proc. 4th SIROCCO, Series in Informatics* 1, Carleton Scientific, Ottawa, 1997, pp. 53–68.

[DR98a]   S. Dobrev, P. Ružička: Broadcasting on anonymous unoriented tori. In: *Proc. 24th WG, Lecture Notes in Computer Science* 1517, Springer, Berlin, 1998, pp. 50–62.

[DR98b]   S. Dobrev, P. Ružička: On the communication complexity of strong time-optimal distributed algorithms. *Nordic Journal of Computing*, Vol. 5 (1998), pp. 87–104.

[DR98c]   S. Dobrev, P. Ružička: Yet another modular technique for efficient leader election. In: *Proc. 25th SOFSEM, Lecture Notes in Computer Science* 1521, Springer, Berlin, 1998, pp. 312–321.

[DRT98]   S. Dobrev, P. Ružička, G. Tel: Time and bit optimal broadcasting on anonymous unoriented hypercubes. In: *Proc. 5th SIROCCO, Series in Informatics*, Carleton Scientific, Ottawa, 1998, pp. 173–187.

[DS87]   W. Dally, C. Seitz: Deadlock-free message routing in multiprocessor interconnection networks. *IEEE Transactions on Computers* 36 (1987), No. 5, 547–553.

[DS93]   S. Djelloul, D. Sotteau: *Personal communication*, 1993.

[EK02]   M. Elkin, G. Kortsarz: A combinatorial logarithmic approximation algorithm for the directed telephone broadcast problem. In: *Proc. ACM STOC* 2002, pp. 438–447.

[EK03a]   M. Elkin, G. Kortsarz: A sublogarithmic approximation algorithm for the undirected telephone broadcast problem: a path out of a jungle. In: *Proc. SODA* 2003, pp. 76–85.

[EK03b]   M. Elkin, G. Kortsarz: An approximation algorithm for the directed telephone multicast problem. In: *Proc. 30th International Colloquium on Automata, Languages and Programming (ICALP)* 2003, *Lecture Notes in Computer Science* 2719, Springer 2003, pp. 212-223.

[EK03c]   M. Elkin, G. Kortsarz: Logarithmic inapproximability of the radio broadcast problem. *Journal of Algorithmics* 52, No. 1 (2004), 8–25.

[EM89]   S. Even, B. Monien: On the number of rounds necessary to disseminate information. In: *Proc. 1st ACM Symp. on Parallel Algorithms and Architectures (SPAA '89)*, 1989, pp. 318–327.

[ES79]   R.C. Entringer, P.J. Slater: Gossips and telegraphs. *Journal of the Franklin Institute* 307 (1979), 353–360.

[Ev79]   S. Even: *Graph Algorithms*. Pitman, London, 1979.

[Fa80]   A.M. Farley: Minimum-time line broadcast networks. *Networks* 10 (1980), 59–70.

[FHMP79]   A.M. Farley, S. Hedetniemi, S. Mitchell, A. Proskurowski: Minimum broadcast graphs. *Discrete Mathematics* 25 (1979), No. 2, 189–193.

[FHMM+94]   R. Feldmann, J. Hromkovič, S. Madhavapeddy, B. Monien, P. Mysliwietz: Optimal algorithms for dissemination of information in generalized communication modes. Extended abstract in: *Proc. 4th Int. Conf. on Parallel Architectures and Languages Europe (PARLE '92)*, *Lecture Notes in Computer Science* 605, Springer 1989, pp. 115–130. Full version in: *Discrete Applied Mathematics* 53 (1994), Nos. 1–3, 55–78.

[FKRR+00]   P. Flocchini, R. Královič, A. Roncato, P. Ružička, N. Santoro: On time versus space for monotone dynamic monopolies in regular topologies. In: *Proc. 7th SIROCCO, Series in Informatics* 7, Carleton Scientific, Ottawa, 2000, pp. 111–125.

[FKRR+03]   P. Flocchini, R. Královič, A. Roncato, P. Ružička, N. Santoro: On time versus size for monotone dynamic monopolies in regular topologies. *Journal of Discrete Algorithms*, 1(2) (2003), 129–150.

[FL94]   P. Fraigniaud, E. Lazard: Methods and problems of communication in usual networks. *Discrete Applied Mathematics* 53 (1994), Nos. 1–3, 79–133.

[FLS94]   P. Fraigniaud, A.L. Liestman, D. Sotteau: Open problems. *Parallel Processing Letters* 3 (1994), No. 4, 507–524.

[FLM+92]   R. Funke, R. Lüling, B. Monien, F. Lücking, H. Blanke-Bohne: An op-
            timized reconfigurable architecture for Transputer networks. In: *Proc.
            25th Hawaii Int. Conf. on System Sciences (HICSS '92)*, 1992, Vol. 1,
            pp. 237–245.

[FM92]   R. Feldmann, P. Mysliwietz: The shuffle exchange network has a
            hamiltonian path. Extended abstract in: *Proc. 17th Math. Founda-
            tions of Computer Science* (MFCS'92), *Lecture Notes in Computer
            Science* 629, pp. 246–254. Full version in: *Mathematical Systems The-
            ory* 29 (1996), 471–486.

[FP80]   A.M. Farley, A. Proskurowski: Gossiping in grid graphs. *Journal of
            Combinatorics, Information and System Sciences* 5 (1980), No. 2, 161–
            172.

[FP99a]   M. Flammini, S. Pérennès: Lower bounds on systolic gossiping. *Infor-
            mation and Computation*, to appear.

[FP99b]   M. Flammini, S. Pérennès: Lower bounds on the broadcasting and
            gossiping time of restricted protocols. Technical Report 3612, INRIA
            Sophia, 1999, later: *SIAM Journal of Mathematics* 17 (2004), No. 4,
            521–540.

[FP01]   M. Flammini, S. Pérennès: On the optimality of general lower bounds
            for broadcasting and gossiping. *SIAM Journal of Discrete Mathemat-
            ics* 14(2):267–282, 2001.

[FPRU90]   U. Feige, D. Peleg, P. Raghavan, E. Upfal: Randomized broadcast in
            networks. In: *International Symposium SIGAL'90. Lecture Notes in
            Computer Science* 450, pp. 128–137, 1990.

[Fr94]   P. Fraigniaud: *Vers un principe de localité pour les communications
            dans les réseaux d'interconnection.* Habilitation Thesis, École Normale
            Supérieure de Lyon, 1994.

[Fr00]   P. Fraigniaud: Approximation algorithms for collective communica-
            tions with limited link and node-contention. Technical Report LRI-
            1264, Université Paris-Sud, France, 2000.

[Fr01a]   P. Fraigniaud: Approximation algorithms for minimum-time broad-
            cast under the vertex-disjoint paths mode. In: *Proc. 9th Annual Euro-
            pean Symposium on Algorithms (ESA '01), Lecture Notes in Computer
            Science* 2161, pp. 440–451, 2001.

[Fr01b]   P. Fraigniaud: Minimum-time broadcast under edge-disjoint paths
            modes. In: *Proc. 2nd International Conference on Fun with Algorithms
            (FUN '01)*, Carleton Scientific, pp. 133–148, 2001.

[FU92]   R. Feldmann, W. Unger: The cube-connected-cycles network is a sub-
            graph of the butterfly network. *Parallel Processing Letters* 2 (1992),
            No. 1, 13–19.

[FV96]   P. Fraigniaud, S. Vial: Approximation algorithms for information dis-
            semination problems. In: *Proc. IEEE Second International Conference
            on Algorithms and Architectures for Parallel Processing (ICAAPP-
            96)*, IEEE Singapore, pp. 155–162, 1996.

[FV97a]   P. Fraigniaud, S. Vial: Heuristic algorithms for personalized commu-
            nication problems in point-to-point networks. In: *Proc. SIROCCO '97
            (4th International Colloquium on Structural Information and Com-
            munication Complexity)*, pp. 240–252, 1997.

[FV97b]   P. Fraigniaud, S. Vial: Approximation algorithms for broadcasting and gossiping. *Journal of Parallel and Distributed Computing* 43 (1997), 47–55.

[FV99]   P. Fraigniaud, S. Vial: Comparison of heuristics for one-to-all and all-to-all communications in partial meshes. *Parallel Processing Letters* 9(1), 1999, 9–20.

[GP96]   L. Gasieniec, A. Pelc: Adaptive broadcasting with faulty nodes. *Parallel Computing* 22 (1996), 903–912.

[GP98]   L. Gasieniec, A. Pelc: Broadcasting with linearly bounded transmission faults. *Discrete Applied Mathematics* 83 (1998), 121–133.

[GG81]   O. Gabber, Z. Galil: Explicit construction of linear-sized superconcentrators. *Journal of Computer and System Sciences* 22 (1981), 407–420.

[GJ79]   M.R. Garey, D.S. Johnson: *Computers and Intractability – A Guide to the Theory of NP-Completeness.* W.H. Freeman, San Francisco, 1979.

[Go94]   C. Gowrisankaran: Broadcasting in recursively decomposable Cayley graphs. *Discrete Applied Mathematics* 53 (1994), Nos. 1–3, 171–182.

[GHR90]   D.S. Greenberg, L.S. Heath, A.L. Rosenberg: Optimal embeddings of butterfly-like graphs in the hypercube. *Mathematical Systems Theory* 23 (1990), 61–77.

[GRW83]   J. Gruska, P. Ružička, J. Wiedermann: Systolické systémy. In: *Proc. SOFSEM*, 1983, pp. 267–308

[Ha72]   F. Harary: *Graph Theory.* Addison-Wesley, Reading, MA, 1972.

[HHL88]   S.M. Hedetniemi, S.T. Hedetniemi, A.L. Liestman: A survey of gossiping and broadcasting in communication networks. *Networks* 18 (1988), 319–349.

[HJM90]   J. Hromkovič, C.D. Jeschke, B. Monien: Optimal algorithms for dissemination of information in some interconnection networks. Extended abstract in: *Proc. 15th Int. Symp. on Mathematical Foundations of Computer Science (MFCS '90), Lecture Notes in Computer Science* 452, Springer 1990, 337–346. Full version in: *Algorithmica* 10 (1993), 24–40.

[HJM93]   J. Hromkovič, C.-D. Jeschke, B. Monien: Note on optimal gossiping in some weak-connected graphs. *Theoretical Computer Science* 127 (1994), No. 2, 395–402.

[HKL95]   J. Hromkovič, P. Kanarek, K. Loryś: Unpublished manuscript.

[HKMP95] J. Hromkovič, R. Klasing, B. Monien, R. Peine: Dissemination of information in interconnection networks (broadcasting and gossiping). In: *Combinatorial Network Theory* (D.-Z. Du and F. Hsu, eds.), Kluwer Academic Publishers, 1995, 125–212.

[HKMU91] R. Heckmann, R. Klasing, B. Monien, W. Unger: Optimal embedding of complete binary trees into lines and grids. Extended abstract in: Proc. 17th Int. Workshop on Graph-Theoretic Concepts in Computer Science (*WG '91*), *Lecture Notes in Computer Science* 570, Springer 1991, 25–35. Full version in: *Journal of Parallel and Distributed Computing* 49 (1998), No. 1, 40–56.

[HKPU+94] J. Hromkovič, R. Klasing, D. Pardubská, W. Unger, H. Wagener: The complexity of systolic dissemination of information in interconnection networks. Extended abstract in: *Proc. 1st Canada-France Conference* on Parallel Computing (*CFCP '94*), *Lecture Notes in Computer Science* 805, Springer 1994, 235–249. Full version in: *R.A.I.R.O. Theoretical Informatics and Applications* 28 (1994), Nos. 3–4, 303–342.

[HKPU+95]   J. Hromkovič, R. Klasing, D. Pardubská, W. Unger, J. Waczulik, H. Wagener: Optimal systolic gossiping in cycles and grids. In: *Proc. 10th Int. Conference on Fundamentals of Computation Theory* (*FCT '95*), *Lecture Notes in Computer Science*, Springer 1995, 273–282.

[HKS93]   J. Hromkovič, R. Klasing, E.A. Stöhr: Dissemination of information in vertex-disjoint paths mode. Extended abstract in: *Proc. 19th Int. Workshop on Graph-Theoretic Concepts in Computer Science* (*WG '93*), *Lecture Notes in Computer Science* 790, Springer 1993, 288–300. Full version in: *Computers and Artificial Intelligence* 15 (1996), No. 4, 295–318.

[HKSW93]   J. Hromkovič, R. Klasing, E.A. Stöhr, H. Wagener: Gossiping in vertex-disjoint paths mode in d-dimensional grids and planar graphs. Extended abstract in: *Proc. 1st European Symposium on Algorithms* (*ESA '93*), *Lecture Notes in Computer Science* 726, Springer 1993, 200–211. Full version in: *Information and Computation* 123 (1995), No. 1, 17–28.

[HKUW94]   J. Hromkovič, R. Klasing, W. Unger, H. Wagener: Optimal algorithms for broadcast and gossip in the edge-disjoint path modes. Extended abstract in: *Proc. 4th Scandinavian Workshop on Algorithm Theory* (*SWAT '94*), *Lecture Notes in Computer Science* 824, Springer 1994, 219–230. Full version in: *Information and Computation* 133 (1997), No. 1, 1–33.

[HLKK+95]   J. Hromkovič, K. Loryś, P. Kanarek, R. Klasing, W. Unger, H. Wagener: On the sizes of permutation networks and consequences for efficient simulation of hypercube algorithms on bounded-degree networks. In: *Proc. 12th Symposium on Theoretical Aspects of Computer Science* (*STACS '95*), *Lecture Notes in Computer Science* 900, Springer 1995, 255–266.

[HMS72]   A. Hajnal, E.C. Milner, E. Szemeredi: A cure for the telephone disease. *Canadian Mathematical Bulletin* 15 (1972), 447–450.

[Ho97]   D.S. Hochbaum (ed.): *Approximation Algorithms for NP-Hard Problems*. PWS Publishing, Boston, MA, 1997.

[HOS92]   M.C. Heydemann, J. Opatrny, D. Sotteau: Broadcasting and spanning trees in deBruijn and Kautz networks. *Discrete Applied Mathematics* 37/38 (1992), 297–317.

[Hr03]   J. Hromkovič: *Algorithmics for Hard Problems*, 2nd edition. Springer, 2003.

[HS74]   F. Harary, A.J. Schwenk: The communication problem on graphs and digraphs. *J. Franklin Institute* 297 (1974), 491–495.

[IK84]   O.H. Ibarra, S.M. Kim: Characterizations and computational complexity of systolic trellis automata. *Theoretical Computer Science* 29 (1984), 123–153.

[IKM85]   O.H. Ibarra, S.M. Kim, S. Moran: Sequential machine characterizations of trellis and cellular automata and applications. *SIAM Journal on Computing* 14 (1985), 426–447.

348   References

[IPK85]   O.H. Ibarra, M.A. Palis, S.M. Kim: Fast parallel language recognition
           by cellular automata. *Theoretical Computer Science* 41 (1985), 231–
           246.
[JM95]   K. Jansen, H. Müller: The minimum broadcast time problem for sev-
           eral processor networks. *Theoretical Computer Science* 147 (1995),
           69–85.
[JRS98]   A. Jakoby, R. Reischuk, C. Schindelhauer: The complexity of broad-
           casting in planar and decomposable graphs. *Discrete Applied Mathe-
           matics* 83 (1998), 179–206.
[KCV86]   D.W. Krumme, G. Cybenko, K.N. Venkataraman: Simultaneous
           broadcasting in multiprocessor networks. In: *Proc. Int. Conf. on Par-
           allel Processing*, 1986, pp. 555–558.
[KCV92]   D.W. Krumme, G. Cybenko, K.N. Venkataraman: Gossiping in min-
           imal time. *SIAM Journal on Computing* 21 (1992), No. 1, 111–139.
[KK79]   P. Kermani, L. Kleinrock: Virtual cut-through: A new computer com-
           munication switching technique. *Computer Networks* 3 (1979), No. 4,
           267–286.
[KKL88]   J. Kahn, G. Kalai, N. Linial: The influence of variables on boolean
           functions (extended abstract). In: *Proc. 29th IEEE Symp. on Foun-
           dations of Computer Science (FOCS '88)*, 1988, 68–80.
[KKR03]   R. Královič, R. Královič, P. Ružička: Broadcasting with many faulty
           links. *Proceedings in Informatics* 17, *Proc. 10th International Col-
           loquium on Structural Information & Communication Complexity
           (SIROCCO)*, pp. 213–224 Carleton Scientific, 2003.
[Kl94]   R. Klasing: The relationship between gossiping in vertex-disjoint paths
           mode and bisection width. Extended abstract in: *Proc. 19th Int. Symp.
           on Mathematical Foundations of Computer Science (MFCS '94)*, Lec-
           ture Notes in Computer Science 841, Springer 1994, pp. 473–483. Full
           version in: *Discrete Applied Mathematics* 83 (1998), Nos. 1–3, 229–246.
[KL92]   M. Klawe, F.T. Leighton: A tight lower bound on the size of planar
           permutation networks. *SIAM J. Disc. Math.* 5(4) (1992), pp. 558–563.
[KLM90]   R. Klasing, R. Lüling, B. Monien: Compressing cube-connected-cycles
           and butterfly networks. Extended abstract in: *Proc. 2nd IEEE Sym-
           posium on Parallel and Distributed Processing (SPDP '90)*, 858–865.
           Full version in: *Networks* 32 (1998), 47–65.
[KLOW94]   M. Kutyłowski, K. Loryś, B. Oesterdiekhoff, R. Wanka: Fast and feasi-
           ble periodic sorting networks of constant depth. In: *Proc. 35th Annual
           Symposium on Foundations of Computer Science (FOCS '94)*, ACM
           1994, pp. 369–380.
[KMPS92]   R. Klasing, B. Monien, R. Peine, E.A. Stöhr: Broadcasting in butterfly
           and deBruijn networks. Extended abstract in: *Proc. 9th Symposium on
           Theoretical Aspects of Computer Science (STACS '92)*, Lecture Notes
           in Computer Science 577, Springer 1992, pp. 351–362. Full version in:
           Discrete Applied Mathematics 53 (1994), Nos. 1–3, 183–197.
[Kn68]   D.E. Knuth: *The Art of Computer Programming, Vol. 1: Fundamental
           Algorithms*. Addison-Wesley, Reading, Massachusetts, 1968.
[Kn75]   W. Knödel: New gossips and telephones. *Discrete Mathematics* 13
           (1975), No. 1, 95.
[KP94]   G. Kortsarz, D. Peleg: Traffic-light scheduling on the grid. *Discrete
           Applied Mathematics* 53 (1994), Nos. 1–3, 211–234.

[KP95]   G. Kortsarz, D. Peleg: Approximation algorithms for minimum time broadcast. *SIAM Journal on Discrete Mathematics* 8 (1995), 401–427.

[KR]   R. Královič, P. Ružička: Minimum feedback vertex sets in shuffle-based interconnection networks, *Proceedings in Informatics* 13, *Proc. 9th International Colloquium on Structural Information & Communication Complexity (SIROCCO)*, pp. 237–246, Carleton Scientific.

[Kr92]   D.W. Krumme: Fast gossiping for the hypercube. *SIAM Journal on Computing* 21 (1992), No. 2, 365–380.

[KR99]   R. Královič, P. Ružička: Rank of graphs: The size of acyclic orientation cover for deadlock-free packet routing. In: *6th SIROCCO, Series in Informatics* 5, Carleton Scientific, Ottawa, 1999, pp. 181–193.

[KR01]   R. Královič, P. Ružička: On immunity and catastrophic indices of graphs, *Proceedings in Informatics* 11, *Proc. 8th International Colloquium on Structural Information & Communication Complexity (SIROCCO)*, pp. 231–242, Carleton Scientific, 2001.

[KR03]   R. Královič, P. Ružička: Minimum feedback vertex sets in shuffle-based interconnection networks. *Information Processing Letters* 86 (2003), 191–196.

[KRR99]   R. Královič, B. Rovan, P. Ružička: Interval routing on layered cross product of trees and cycles. In: *Proc. 5th Euro-Par, Lecture Notes in Computer Science* 1685, Springer, Berlin, 1999, pp. 1231–1240.

[KRRS98]   R. Královič, B. Rovan, P. Ružička, D. Štefankovič: Efficient deadlock-free multi-dimensional interval routing in interconnection networks. In: *Proc. 12th DISC, Lecture Notes in Computer Science* 1499, Springer, Berlin, 1998, pp. 273–287.

[KRRS02]   R. Královič, B. Rovan, P. Ružička, D. Štefankovič: Efficient deadlock-free multi-dimensional interval routing in interconnection networks. *Computing and Informatics* 21, 265–287 (2002).

[KRS97]   R. Královič, P. Ružička, D. Štefankovič: The complexity of shortest-path and dilation bounded interval routing schemes. In: *3rd Euro-Par, Lecture Notes in Computer Science* 1300, Springer, Berlin, 1997, pp. 258–265.

[KRS00]   R. Královič, P. Ružička, D. Štefankovič: The complexity of shortest path and dilation bounded interval routing. *Theoretical Computer Science*, Vol. 234, Nos. 1–2 (2000), 85–107 (Preliminary version appeared in *Resource Material in Compact Routing: Research Meeting on Structural Information and Communication Complexity*, Siena, June 1997).

[KS83]   C.P. Kruskal, M. Snir: The performance of multistage interconnection networks for multiprocessors. *IEEE Transactions on Computers* 32 (1983), 1091–1098.

[Ku79]   H.T. Kung: Let's design algorithms for VLSI systems. In: *Proc. Caltech Conference of VLSI* (C.L. Seifz, Ed.), Pasadena, California, 1979, pp. 65–90.

[LADF+92]   C.E. Leiserson, Z.S. Abuhamdeh, D.C. Douglas, C.F. Feynman, M.N. Ganmuki, J.V. Hill, W.D. Hillis, B.C. Kuszmaul, M.A. St. Pierre, D.S. Wells, M.C. Wong, S.-W. Yang, R. Zak: The network architecture of the Connection Machine CM-5. In: *Proc. 4th ACM Symp. on Parallel Algorithms and Architectures (SPAA '92)*, 1992, pp. 272–285.

[Le92]   F.T. Leighton: *Introduction to Parallel Algorithms and Architectures: Array, Trees, Hypercubes.* Morgan Kaufmann Publishers, 1992.

[LHL94]   R. Labahn, S.T. Hedetniemi, R. Laskar: Periodic gossiping on trees. *Discrete Applied Mathematics* 53 (1994), Nos. 1–3, 235–246.

[Li85]   A.L. Liestman: Fault-tolerant broadcast graphs. *Networks* 15 (1985), 159–171.

[LP88]   A.L. Liestman, J.G. Peters: Broadcast networks of bounded degree. *SIAM Journal on Discrete Mathematics* 1 (1988), No. 4, 531–540.

[LR93a]   A.L. Liestman, D. Richards: Network communication in edge-colored graphs: Gossiping. *IEEE Transactions on Parallel and Distributed Systems* 4 (1993), No. 4, 438–445.

[LR93b]   A.L. Liestman, D. Richards: Perpetual gossiping. Extended abstract in: *Proc. 4th Int. Symp. on Algorithms and Computation (ISAAC '93), Lecture Notes in Computer Science* 762, Springer 1993, 259–266. Full version in: *Parallel Processing Letters* 3 (1993), No. 4, 347–355.

[LR93c]   G. Lerman, L. Rudolph: *Parallel Processors: Studies in the Design of Parallel Machines.* Plenum, New York, 1993.

[LR97]   R. Labahn, A. Raspaud: Periodic gossiping in back-to-back trees. *Discrete Applied Mathematics* 75 (1997), 157–168.

[LT79]   R.J. Lipton, R.E. Tarjan: A separator theorem for planar graphs. *SIAM Journal on Applied Mathematics* 36 (1979), No. 2, 177–189.

[LW90]   R. Labahn, I. Warnke: Quick gossiping by multi-telegraphs. In: R. Bodendiek, R. Henn, eds., *Topics in Combinatorics and Graph Theory,* Physica-Verlag, Heidelberg, 1990, pp. 451–458.

[LW94]   R. Labahn, I. Warnke: Quick gossiping by telegraphs. *Discrete Mathematics* 126 (1994), 421–424.

[MDFK+95]   B. Monien, R. Diekmann, R. Feldmann, R. Klasing, R. Lüling, K. Menzel, T. Römke, U.-P. Schroeder: Efficient use of parallel & distributed systems: From theory to practice. In: J. van Leeuwen, ed., *Computer Science Today, Lecture Notes in Computer Science* 1000, Springer 1995, 62–77.

[MFKL93]   B. Monien, R. Feldmann, R. Klasing, R. Lüling: Parallel architectures: Design and efficient use. In: *Proc. 10th Symposium on Theoretical Aspects of Computer Science (STACS '93), Lecture Notes in Computer Science* 665, Springer 1993, pp. 247–269.

[Mi93]   M. Middendorf: Minimum broadcast time is NP-complete for 3-regular planar graphs and deadline 2. *Information Processing Letters* 46 (1993) 281–287.

[MLL92]   B. Monien, R. Lüling, F. Langhammer: A realizable efficient parallel architecture. In: *Proc. 1st Heinz Nixdorf Symposium: Parallel Architectures and their Efficient Use, Lecture Notes in Computer Science* 678, Springer 1992, pp. 93–109.

[MR01]   M. Makúch, P. Ružička: On the complexity of path layouts in bounded degree ATM networks a case study for butterfly networks. *Proceedings in Informatics* 11, *Proc. 8th International Colloquium on Structural Information & Communication Complexity (SIROCCO),* pp. 243–258, Carleton Scientific, 2001

[MRSR+95]   M.V. Marathe, R. Ravi, R. Sundaram, S.S. Ravi, D.J. Rosenkrantz, H.B. Hunt III: Bicriteria network design problems. In: *Proc. 22nd Int.*

	Colloquium on Automata, Languages and Programming, Lecture Notes in Computer Science 944, Springer, 1995, pp. 487–498.
[MS90]	B. Monien, I.H. Sudborough: Embedding one interconnection network in another. *Computing Supplementum* 7 (1990), 257–282.
[MS92]	V.E. Mendia, D. Sarkar: Optimal broadcasting on the star graph. *IEEE Transactions on Parallel and Distributed Systems* 3 (1992), No. 4, 389–396.
[NSV02]	G. Narayanaswamy, A. Shende, P. Vipanarayanan. 4-Systolic broadcasting in a wrapped butterfly network. In: *Proc. 4th Int. Workshop on Distributed Computing (IWDC 2002), Lecture Notes in Computer Science* 2571, Springer 2002, pp. 98–107.
[Pe68]	M.G. Pease: An adaptation of the fast Fourier transform for parallel processing. *Journal of the ACM* 15 (1968), 252–264.
[Pe95]	S. Pérennès: *The relation between gossiping in the disjoint-paths modes and the forwarding index.* Technical Report, Université de Nice-Sophia Antipolis, France, 1995.
[Pe96]	S. Pérennès: Lower bound on broadcasting time of deBruijn type networks. In: *Proceedings of the Euro-Par'96 Parallel Processing, Lecture Notes in Computer Science* 1123, Springer, 1996, pp. 325–332.
[Pe98]	S. Pérennès: Broadcasting and gossiping on deBruijn, shuffle-exchange and similar networks. *Discrete Applied Mathematics*, Vol. 83 (1998), Nos. 1–3, pp. 247–262.
[Pi77]	N. Pippenger: Superconcentrators. *SIAM Journal of Computing*, No. 6 (1977), 298–304.
[Pr81]	A. Proskurowski: Minimum broadcast trees, *IEEE Transactions on Computers* C-30, 5(1981), 363–366;
[PR89]	I. Prívara, P. Ružička: Variácie na tému unifikácia. In: *Proc SOFSEM*, 1989, pp. 175–204 (in Slovak).
[PR91a]	I. Prívara, P. Ružička: Variácie na tému unifikácia I. *Informačné systémy* 19 (1991), 2 (in Slovak).
[PR91b]	I. Prívara, P. Ružička: Variácie na tému unifikácie II. *Informačné systémy* 19, 3 (1991), 151–173 (in Slovak).
[PR93]	I. Prívara, P. Ružička: On efficiency of equation solving with constraints. *Workshop ALTEC*, Smolenice, 1993 (abstract).
[PR97]	I. Prívara, P. Ružička (eds.): Mathematical Foundations of Computer Science 1997. *Proceedings*, Springer, Berlin. 1997, 517 p.
[PRS93]	I. Prívara, P. Ružička, J. Šturc: Constraint Equational Programming. In: *Proceedings 20th SOFSEM*, 1993, pp. 223–248.
[PRR94]	I. Prívara, B. Rovan, P. Ružička (eds.): *Mathematical Foundations of Computer Science 1994, Lecture Notes in Computer Science* 841, Springer, Berlin, 1994, 628 p.
[PS89]	D. Peleg, A.A. Schäffer: Time bounds on fault-tolerant broadcasting. *Networks* 19 (1989), 803–822.
[PV76]	N. Pippenger, L. Valiant: Shifting graphs and their applications. *Journal of the ACM* 23, No. 3 (1976), 423–432.
[PV81]	F.P. Preparata, J.E. Vuillemin: The cube-connected-cycles: a versatile network for parallel computation. *Communications of the ACM*, Vol. 24 (1981), pp. 300–309.

352   References

[Ra94]   R. Ravi: Rapid rumor ramification: approximating the minimum broadcast time. In: *Proc. 35th Ann. IEEE Symp. on Foundations of Comput. Sci.*, IEEE Computer Society, 1994, pp. 202–213.

[RL88]   D. Richards, A.L. Liestman: Generalizations of broadcasting and gossiping. *Networks* 18 (1988), 125–138.

[RP89a]   P. Ružička, I. Prívara: An almost linear Robinson unification algorithm. *Acta Informatica*, Vol. 27, No. 1 (1989), 61–71.

[RP89b]   P. Ružička: Peblovanie. *Proc. SOFSEM*, 1989, pp. 205–224 (in Slovak).

[RP93]   Ružička, P., Prívara, I.: On tree pattern unification problems. In: *9th FCT, Lecture Notes in Computer Science* 710, Springer, Berlin, 1993, pp. 418–429

[RS74]   P. Ružička, J. Šturc: Dve triedy LR analyzovateľných nejednoznačných gramatík. *Proc. Integrované informačné systémy, VVS OSN*, Bratislava, 1974, pp. 81–98 (in Slovak).

[RS75]   P. Ružička, J. Šturc: Syntaktická a sémantická analýza. Research report "Metasystémové prostriedky jazykových systémov III. Teoretické problémy kompilátorov", VVS Bratislava, 1975, pp. 19–46 (in Slovak).

[RS00]   P. Ružička, D. Štefankovič: On the complexity of multi-dimensional interval routing schemes. *Theoretical Computer Science* 245 (2000) 255–280.

[Ru73a]   P. Ružička: Niektoré hľadiská efektívnosti posunovo redukčných analyzačných schém. Research report "Metasystémové prostriedky jazykových systémov II", VVS Bratislava, 1973, pp. 9–32 (in Slovak).

[Ru73b]   P. Ružička: Poznámka k veľkosti DeRemerových analyzátorov. Research report "Metasystémové prostriedky jazykových systémov II", VVS Bratislava, 1973, pp. 47–54 (in Slovak).

[Ru74a]   P. Ružička: LR metódy syntaktickej analýzy. In: *Proc. SOFSEM*, 1974, pp. 99–126 (in Slovak).

[Ru74b]   P. Ružička: Kanonické precedenčné schémy so zovšeobecnenou redukciou. *Informačné systémy* 2 (1974) 149–160 (in Slovak).

[Ru75a]   P. Ružička: On the Size of DeRemer's Analyzers. *Kybernetika*, Vol. 11, No. 3 (1975), 207–217.

[Ru75b]   P. Ružička: Local Disambiguating Transformations. In: *Proc. 4th MFCS, Lecture Notes in Computer Science* 32, Springer, Berlin, 1975, pp. 399–405.

[Ru75c]   P. Ružička: Size Complexity of Context-Free Languages. *Elektronische Informationsverarbeitung und Kybernetik – EIK*, Vol. 11, Nos. 4–6, 1975, pp. 296–299 (in German).

[Ru77]   P. Ružička: Exkurzie do porovnávacích problémov. Research report "Efektívnosť algoritmov softwarových systémov II", VVS Bratislava, 1977, pp. 9–32 (in Slovak).

[Ru78a]   P. Ružička: On Grammatical Characterization of Floyd's Operator Precedence Analyzers. In: *Proc. ASA*, Smolenice, 1978.

[Ru78b]   P. Ružička: Optimalizácia deterministických syntaktických analyzátorov. Research report "Efektívnosť algoritmov programovacích systémov", VVS Bratislava, 1978, pp. 68–86 (in Slovak).

[Ru78c]   P. Ružička: Syntaktická analýza a popisná zložitosť. Research report "Popisná zložitosť algoritmov, automatov a jazykov", VVS Bratislava, 1978, pp. 96–98 (in Slovak).

[Ru79a] P. Ružička: O časovej–pamäťovej a časovej–aritmetickej súvislosti problémov vytvárania čiastočných usporiadaní s centrálnym prvkom. Research report "Efektívnosť algoritmov a dátové štruktúry I", VVS Bratislava, 1979, pp. 28–42 (in Slovak).

[Ru79b] P. Ružička: O súvislosti času a pamäti niektorých problémov usporiadania. Proc. SOFSEM, 1979, pp. 390–392 (in Slovak).

[Ru79c] P. Ružička: Space-time trade-offs in producing certain partial orders. Annales Universitatis Scientiarum Budapestinensis De Rolando Eötvös Nominatae 7, Vol. II, 1979, pp. 97–114.

[Ru79d] P. Ružička: Validity test for Floyd's operator precedence parsing algorithms. In: Proc. 8th MFCS, Lecture Notes in Computer Science 74, Springer, Berlin, 1979, pp. 415–424.

[Ru80] P. Ružička: Time and space bounds in producing certain partial orders. In: 9th MFCS, Lecture Notes in Computer Science 88, Springer, Berlin, 1980, pp. 415–424.

[Ru81a] P. Ružička: Validity test for Floyd's operator-precedence parsing algorithm is polynomial in time. Kybernetika, Vol. 17, No. 5 (1981), 368–379.

[Ru81b] P. Ružička: Bounds for On-Line Selection. Kybernetika, Vol. 17, No. 2 (1981), 147–157.

[Ru81c] P. Ružička: Čas a pamäť vo výpočtovej zložitosti. Informačné systémy 9 (1981), 139–156 (in Slovak).

[Ru81d] P. Ružička: Pamäťová zložitosť hry s čiernymi a bielymi kameňmi na stromoch. In: Proc. SOFSEM, 1981, pp. 404–409 (in Slovak).

[Ru81e] P. Ružička: Peblovanie. Research report "Typy dát, programy a algoritmické problémy: Popisná a výpočtová zložitosť", VVS Bratislava, 1981, pp. 52–60 (in Slovak).

[Ru81f] P. Ružička: Vzaimosvjaz meždu vremenjem i pamjaťju dlja problemy vyborki. In: Proc. "Matematičeskije problemy baz dannych", CEMI AN ZSSR Moskva, 1981, pp. 96–114 (in Russian).

[Ru82a] P. Ružička: Algoritmy syntaktickej analýzy. MFF UK Bratislava, 1982, 118 p. (in Slovak).

[Ru82b] P. Ružička: Efektívny algoritmus paralelného výberu. In: Proc. SOFSEM, 1982, pp. 395–397 (in Slovak).

[Ru82c] P. Ružička: Ocenki dlja bystroj vyborki. Programnyje sistemy i voprosy efektivnogo ispolzovania EVM. Sbornik naučnych trudov (A.P. Eršov, ed.), Novosibirsk, 1982, pp. 31–44 (in Russian).

[Ru82d] P. Ružička: Progresívne pokrývanie grafov. Research report "Teória a metodológia programových a dátabázových systémov", VVS Bratislava, 1982, pp. 107–125 (in Slovak).

[Ru83a] P. Ružička: Rozpoznávanie bezkontextových jazykov na systolických obvodoch. In: Proc. MOP, 1983, pp. 169–182 (in Slovak).

[Ru83b] P. Ružička: Systolické algoritmy pre nenumerické problémy. Research report "Distribuované a paralelné systémy" (J. Gruska, ed.), VVS Bratislava, 1983, pp. 72–100 (in Slovak).

[Ru83c] P. Ružička: Výpočtový model pre VLSI. Research report "Distribuované a paralelné systémy" (J. Gruska, ed.), VVS Bratislava, 1983, pp. 37–46 (in Slovak).

354　　　References

[Ru84a]　　P. Ružička: Algoritmus na rozpoznávanie vzoriek v reťazcoch pracujúci v reálnom čase. In: *Proc. SOFSEM*, 1984, pp. 347–350 (in Slovak).

[Ru84b]　　P. Ružička: Vyhľadávanie vzoriek v reťazcoch. In: *Proc. MOP*, 1984, pp. 235–273 (in Slovak).

[Ru85a]　　P. Ružička: Two variants of the black-and-white pebble game. *Computers and Artificial Intelligence*, Vol. 4, No. 3 (1985), pp. 211–221.

[Ru85b]　　P. Ružička: Aké ťažké je generovať cieľový program? In: *Proc. MOP*, 1985, pp. 159–199 (in Slovak).

[Ru86]　　P. Ružička: Distribuované algoritmy na rozsiahlych komunikačných sieťach. In: *Proc. MOP*, 1986, pp. 101–148 (in Slovak).

[Ru87a]　　P. Ružička: Distribuované výpočty. In: *Proc. SOFSEM*, 1987, pp. 215–246 (in Slovak).

[Ru87b]　　P. Ružička: Úlohy o distribuovanej dohode v prítomnosti chýb. Research report "Teoretické aspekty výpočtových systémov nových generácií II", VUSEI-AR Bratislava, 1987, pp. 13–40 (in Slovak).

[Ru88a]　　P. Ružička, I. Prívara: An almost linear Robinson unification algorithm. In: *13th MFCS, Lecture Notes in Computer Science* 324, Springer, Berlin, 1988, pp. 501–511 .

[Ru88b]　　P. Ružička: On efficiency of interval routing algorithms. In: *13th MFCS, Lecture Notes in Computer Science* 324, Springer, Berlin, 1988, pp. 492–500.

[Ru89]　　P. Ružička: Efektívne unifikačné algoritmy. Research report "Efektívna realizácia logických programov", VUSEI-AR Bratislava, 1989, pp. 51–65 (in Slovak).

[Ru90a]　　P. Ružička: Peblovanie – technika analýzy efektívnosti výpočtov. *Informačné systémy* 18 (1990), 287–313 (in Slovak).

[Ru90b]　　P. Ružička: Úloha o semiunifikácii. In: *Proc. SOFSEM*, 1990, pp. 69–78 (in Slovak).

[Ru91a]　　P. Ružička: An efficient decision algorithm for the uniform semi-unification problem. In: *Proc. 16th MFCS, Lecture Notes in Computer Science* 520, Springer, Berlin, 1991, pp. 415–425.

[Ru91b]　　P. Ružička: A note on efficiency of interval routing algorithms. *The Computer Journal*, Vol. 34, No. 5 (1991), 475–476.

[Ru91c]　　P. Ružička: Rozhodnuteľné prípady semi-unifikácie. Research report "Teoretické problémy informatiky", VUSEI-AR Bratislava, 1991, pp. 42–48 (in Slovak).

[Ru92a]　　P. Ružička: An efficient solution of the left-linear semi-unification problem. In: *Proc. 19th SOFSEM*, 1992.

[Ru92b]　　P. Ružička: *Počítače. Nový rozum do vrecka*, Mladé letá, 1992 (in Slovak).

[Ru93]　　P. Ružička: Time, space and rounds for dynamic pebbling. *Workshop ALTEC*, Bordeaux, 1993 (abstract).

[Ru96]　　P. Ružička: Efficient tree-pattern unification algorithm. In: *Proc. 23rd SOFSEM, Lecture Notes in Computer Science* 1175, Springer, Berlin, 1996, pp. 417–424.

[Ru98]　　P. Ružička: Efficient communication schemes. In: *Proc. 25th SOFSEM, Lecture Notes in Computer Science* 1521, Springer, Berlin, 1998, pp. 244–263.

[Ru01]   P. Ružička: On efficiency of path systems induced by routing and communication schemes. *Computing and Informatics*, Vol. 20, No. 2 (2001).

[RW76a]   P. Ružička, J. Wiedermann: How good is the adversary lower bound? In: *Proc. 6th MFCS, Lecture Notes in Computer Science* 53, Springer, Berlin, 1977, pp. 465–475.

[RW76b]   P. Ružička, J. Wiedermann: On the lower bound for minimum comparison selection. In: *Proc. 5th MFCS, Lecture Notes in Computer Science* 45, Springer, Berlin, 1976, pp. 495–502.

[RW77]   P. Ružička, J. Wiedermann: Vyhľadávacie stromy ako štruktúra pre organizáciu súborov. *Informačné systémy* 5 (1977), 435–480 (in Slovak).

[RW78a]   P. Ružička, J. Wiedermann: On the lower bound for minimum comparison selection. *Aplikace matematiky*, Vol. 23, No. 1 (1978), pp. 1–8.

[RW78b]   P. Ružička, J. Wiedermann: Algoritmy neúplného usporiadania I. *Informačné systémy*, Vol. 6, No. 4 (1978), pp. 451–472 (in Slovak).

[RW78c]   P. Ružička, J. Wiedermann: Algorithms of Partial Orderings. In: *Proc. 5th Symposium on Algorithms*, Štrbské pleso, 1978, pp. 380–392.

[RW79]   P. Ružička, J. Wiedermann: On the Conjecture Relating Minimax and Minimean Complexity Norms. *Aplikace matematiky*, Vol. 24, No. 5 (1979), 321–325.

[RW93]   P. Ružička, J. Waczulík: On time-space trade-offs in dynamic graph pebbling. *Lecture Notes in Computer Science* 711, Springer, Berlin, 1993, pp. 671–681.

[RW94]   P. Ružička, J. Waczulík: Dynamic graph pebbling in minimal space. *RAIRO Informatique Théorique*, Vol. 28, No. 6 (1994), 557–565.

[Sc00a]   C. Schindelhauer: Broadcasting time cannot be approximated within a factor of 57/56-epsilon. ICSI Technical Report TR-00-002.

[Sc00b]   C. Schindelhauer: On the inapproximability of broadcasting time. In: *Proc. 3rd International Workshop on Approximation Algorithms for Combinatorial Optimization Problems (APPROX'00)*, 2000, pp. 226–237.

[SCH81]   P.J. Slater, E.J. Cockayne, S.T. Hedetniemi: Information dissemination in trees. *SIAM Journal on Computing* 10(4) (1981), 692–701.

[Si01]   J.F. Sibeyn: Faster gossiping on butterflies. In: *28th International Colloquium on Automata, Languages and Programming, Lecture Notes in Computer Science* 2076, Springer, 2001, 785–796.

[SP89]   M.R. Samatham, D.K. Pradhan: The deBruijn multiprocessor network: A versatile parallel processing and sorting network for VLSI. *IEEE Transactions on Computers* 38 (1989), No. 4, 567–581.

[SS88]   Y. Saad, M.H. Schultz: Topological properties of the hypercube. *IEEE Transactions on Computers* 37 (1988), No. 7, 867–872.

[SS03]   J.F. Sibeyn, M. Šoch : Optimal Gossiping on CCCs of even dimension. *Parallel Processing Letters*, Vol. 13, No. 1 (2003) 35–42.

[St91a]   E.A. Stöhr: Broadcasting in the butterfly network. *Information Processing Letters* 39 (1991), 41–43.

[St91b]   E.A. Stöhr: On the broadcast time of the butterfly network. In: *Proc. 17th Int. Workshop on Graph-Theoretic Concepts in Computer Science (WG '91), Lecture Notes in Computer Science* 570, Springer 1991, 226–229.

[St92] E.A. Stöhr: *An upper bound for broadcasting in the butterfly network.* Technical Report, University of Manchester, United Kingdom, 1992.

[SW84] P. Scheuermann, G. Wu: Heuristic algorithms for broadcasting in point-to-point computer networks. *IEEE Trans. on Computers*, C-33 (1984), 804–811.

[SW93] V.S. Sunderam, P. Winkler: Fast information sharing in a complete network. *Discrete Applied Mathematics* 42 (1993), 75–86.

[Ul84] J.D. Ullman: *Computational Aspects of VLSI* . Computer Science Press, Rockville, MD, 1984, 495 p.

[Va01] V.V. Vazirani: *Approximation Algorithms.* Springer, 2001.

[Wa68] A. Waksman: A permutation network. *Journal of the ACM*, vol. 15, No. 1 (1968), 159–163.

[Wan94] R. Wanka: Paralleles Sortieren auf mehrdimensionalen Gittern. *Ph.D. Dissertation*, Universität Paderborn, 1994 (in German).

[We94] R. Werchner: Personal communication, 1994.

[Wi65] I.H. Wilkinson: *The Algebraic Eigenvalue Problem.* Clarendon Press, Oxford, 1965.

[Wi92] D.B. Wilson: Embedding leveled hypercube algorithms into hypercubes. In: *Proc. 4th ACM Symposium on Parallel Algorithms and Architectures (SPAA '92)*, pp. 264–270.

[XL92] C.-Z. Xu, F.C.M. Lau: Distributed termination detection of loosely synchronized computations. In: *Proc. 4th IEEE Symp. on Parallel and Distributed Processing*, Texas, 1992, 196–203.

[XL94] C.-Z. Xu, F.C.M. Lau: Efficient distributed termination detection for synchronous computations in multicomputers. Technical Report TR-94-04, Department of Computer Science, University of Hong Kong, March 1994.

[Xu01] J. Xu: Topological structure and analysis of interconnection networks. *Network Theory and Applications, Vol. 7*, Kluwer Academic Publishers, October 2001, 352 p.

Index